W9-BGL-510

White-Light
Optical Signal Processing

White-Light Optical Signal Processing

FRANCIS T. S. YU
Electrical Engineering Department
The Pennsylvania State University
University Park

A Wiley-Interscience Publication

John Wiley & Sons

New York / Chichester / Brisbane / Toronto / Singapore

Library of Congress Cataloging in Publication Data:

Yu, Francis T. S., 1934–
White-light optical signal processing.

(Wiley series in pure and applied optics, ISSN 0277-2493)
"A Wiley-Interscience publication."
Includes index.
1. Optical data processing. I. Title. II. Series.

TA1632.Y8 1985 621.36′7 84-29154
ISBN 0-471-80954-3

Printed in the United States of America

10 9 8 7 6 5 4 3 2 1

In memory of my mother,
for her courage and love

Preface

Since the discovery of a strong coherent source, the laser has become a fashionable tool for many scientific uses, particularly for application to coherent optical signal processing. This trend has been largely due to its complex amplitude processing capability. However, coherent optical signal processing systems are plagued with annoying artifact noise, which is regarded as the number one enemy of these systems. Aside from the artifact noise, the coherent sources are generally expensive, and the coherent optical signal processing environment is usually very stringent. For example, it requires a dust-free room and a heavy optical bench.

Recently, we have looked at optical signal processing from a different standpoint, and a question arises: Is it necessarily true that all signal processing operations require a coherent source? The answer to this question is no. There are many optical signal processing operations that can be easily carried out with a reduced coherence source. In other words, if the coherence requirement for a specific processing operation is relaxed, the signal processing can be carried out by a partially coherent or a white-light source. The well-known Van Cittert-Zernike theorem allows us to develop a white-light signal processing system that can operate much like a conventional coherent system in partial coherence mode. In other words, the white-light system can, on one hand, process the signal in complex amplitude like a coherent processor, and at the same time, it suppresses the artifact noise like an incoherent processor.

In short, the major advantages of a white-light signal processing system are: (1) it is capable of suppressing the coherent artifact noise; (2) the white-light source is generally inexpensive; (3) the processing environment is not very stringent; (4) the white-light system is easy and economical to maintain; and (5) it is very suitable for color signal or image processing.

The aim of this book is to provide the readers with a basic background in partially coherent or white-light optical signal processing. It also serves as a reference for optical physicists and engineers who are involved in current research and development of noncoherent and white-light signal processing.

There are seven chapters. The first serves as a basic background for readers who may not be familiar with the partial coherence theory. Chapter 2 discusses the basic concepts of partially coherent signal processing. Chapters 3 to 5 discuss the theoretical aspects of coherence requirement, transfer function, and noise performance of a white-light optical signal processor under partial coherence regime. The last two chapters discuss the techniques and applications of white-light optical signal processing. Since a white-light source emanates all the visible wavelengths, the white-light processor is particularly suitable for color image processing. Thus many of the applications discussed in Chapters 6 and 7, and many of the examples shown, involve the processing of color images.

I believe that white-light optical signal processing will have widespread application. This book sets forth the basic foundation, those principles that have already been established in part. It can guide interested readers toward various imaginative signal processing applications. In view of the great number of contributors to this field, I apologize for possible omissions of appropriate references in various parts of this book.

I am indebted to my students for their encouragement, motivation, and enormous assistance during the preparation of the manuscript. Special mention must be made of M. F. Cao, T. H. Chao, X. X. Chen, M. S. Dymek, D. Fisher, G. Gheen, K. F. Hsu, B. Javidi, T. N. Lin, X. J. Lu, H. M. Mueller, P. H. Ruterbusch, K. S. Shaik, Q. W. Song, J. A. Tome, K. B. Xu, L. N. Zheng, Q. Zhou, and S. L. Zhuang for their excellent work and for proofreading much of the manuscript; I must also acknowledge the work of Miss Julie Corl, who typed the manuscript. Many thanks also due to my wife, Lucy, and my children, Peter, Ann, and Edward, for their encouragement, patience, and love. Without their support, this book could not have been brought to completion.

Finally, the support of the U.S. Air Force Office of Scientific Research in the area of white-light optical signal processing is gratefully acknowledged.

FRANCIS T. S. YU

University Park, Pennsylvania
March 1985

Contents

White-Light
Optical Signal Processing

1

Introduction to Partial Coherence Theory

Most of the optical signal processing techniques reported to date have confined themselves to either extreme coherence or incoherence. And yet we can expect a continuous transition between these two extreme limits. Such a transitional region exists and is known as the field of *partial coherence.*

The earliest investigation of the subject of partial coherence may be that of Verdet in 1869 [1.1]. He studied the region of coherence for light from an extended source. It was, however, in 1890 that Michelson [1.2] first established the relationship between the visibility of interference fringes and the intensity distribution of an extended source. Mention must be made of Berek's significant contribution to the study of partial coherence in 1926 [1.3]. He utilized the concept of correlation in relation to the image formation for a microscope. However, the most important developments of partial coherence theory must be those due to Van Cittert [1.4] in 1934 and Zernike [1.5] in 1938. They determined the *degree of coherence* for light disturbances at any two points on a screen illuminated by an extended light source. The Van Cittert–Zernike method was later simplified and applied to the study of image formation and resolving power by Hopkins [1.6, 1.7] in the 1950s. Nevertheless it was Wolf's mutual coherence function in 1957 [1.8] that made a broader scope of applications possible for coherence theory. This is simply due to the fact that the mutual coherence depicts the correlation properties of the light disturbances at any two points for which the mutual coherence function obeys the general wave equation. These properties of the mutual coherence function are adequate for the analysis of any optical experiment involving the interference and diffraction phenomena of light.

1.1. GENERAL ASPECTS

The widespread application of partially coherent light has made a discussion of the principles of coherence in radiation more significant.

If the radiations from two point sources maintain a fixed phase relation between them, they are said to be *mutually coherent*. An extended source is coherent if all points of the source have fixed phase differences between them. Here we discuss some general aspects of partial coherence theory, as it applies to partially coherent optical signal processing. Thus it is necessary to modify and extend the foregoing definitions.

Classically, in electromagnetic radiation theory, for example, in the development of Maxwell's equations, it is usually assumed that the electric and magnetic fields at any position are measurable at all times. Thus no account need be taken of partial coherence theory. There are problems, however, in which this assumption of known fields cannot be made; in these cases it is often helpful to apply partial coherence theory. For example, if a diffraction pattern that results from several sources of radiation is to be worked out, an exact result cannot be obtained unless the degree of coherence of the separate sources is taken into account. It may be desirable, in such a case, to obtain an average that would represent the statistically most likely result from any such combination of sources. It is therefore more useful to provide a statistical description than to follow the dynamic behavior of a wave field in detail.

Our treatment of partial coherence is based on such averaging. In particular, following Wolf's definition of coherence [1.8, pp. 499–503], we choose the second-order moment as the quantity to be averaged. Thus what we call the *mutual coherence function* is defined as

$$\Gamma_{12}(\tau) \triangleq \langle u_1(t+\tau)u_2^*(t)\rangle, \tag{1.1}$$

where the asterisk denotes the complex conjugate, $u_1(t)$ and $u_2(t)$ are the respective complex wave fields at points P_1 and P_2, and $\Gamma_{12}(\tau)$ is the mutual coherence function between these points for a time delay τ; the symbol $\langle\ \rangle$ denotes a time average. From Eq. (1.1) we can define a *normalized mutual coherence function*:

$$\gamma_{12}(\tau) \triangleq \frac{\Gamma_{12}(\tau)}{[\Gamma_{11}(0)\Gamma_{22}(0)]^{1/2}}; \tag{1.2}$$

$\gamma_{12}(\tau)$ may also be called the *complex degree of coherence*, or the degree of correlation.

We now demonstrate that the normalized mutual coherence function $\gamma_{12}(\tau)$ can be measured using Young's experiment of interference. In Fig. 1.1,

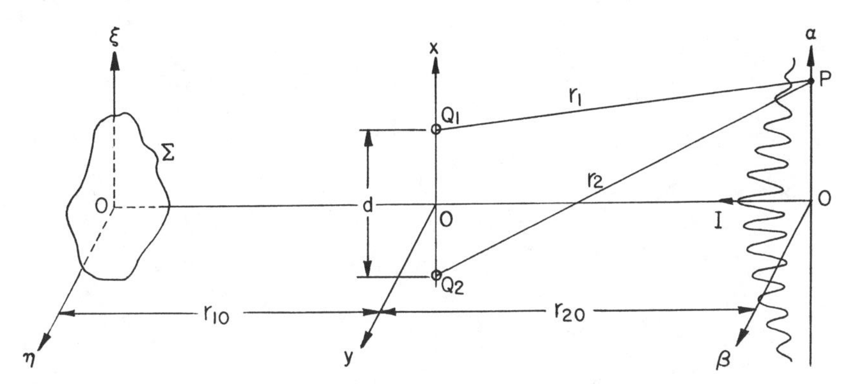

Figure 1.1. Young's experiment. Σ is an extended but nearly monochromatic source.

Σ represents an extended source of light, which is assumed to be incoherent but nearly monochromatic; that is, its spectrum is of finite width, but narrow. The light from this source falls upon a screen at a distance r_{10} from the source, and upon two small apertures (pinholes) in this screen, Q_1 and Q_2, separated by a distance d. On an observing screen at a distance r_{20} from the diffracting screen, an interference pattern is formed by the light passing through Q_1 and Q_2. Now let us suppose that the changing characteristics of the interference fringes are observed as the parameters of Fig. 1.1 are changed. As a measurable quantity, let us adopt Michelson's *visibility* of the fringes, which is defined as

$$V \triangleq \frac{I_{\max} - I_{\min}}{I_{\max} + I_{\min}}, \tag{1.3}$$

where $I_{\max}$ and $I_{\min}$ are the maximum and minimum intensities of the fringes.

For the visibility to be measurable, the conditions of the experiment, such as narrowness of spectrum and closeness of optical path lengths, must be such as to permit $I_{\max}$ and $I_{\min}$ to be clearly defined. Let us assume that these ideal conditions exist.

As we begin our parameter changes, we find, first, that the average visibility of the fringes increases as the size of the source Σ is made smaller. Next, as the distance between the pinholes Q_1 and Q_2 is changed with Σ held constant (and circular in form), the visibility changes in the manner shown in Fig. 1.2. When Q_1 and Q_2 are very close together, the minimum intensity between the fringes falls to zero, and the visibility is unity. As d is increased, the visibility falls rapidly and reaches zero as $I_{\max}$ and $I_{\min}$ become equal. Further increase in d causes a reappearance of fringes, although they are shifted on the screen

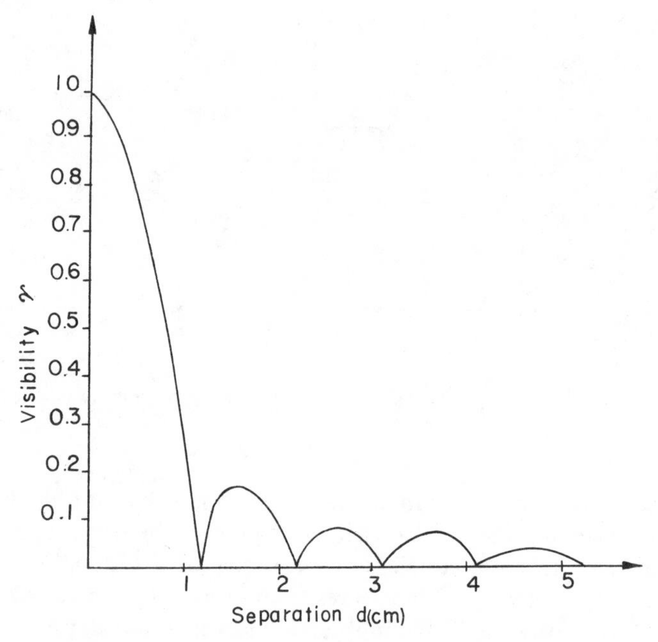

Figure 1.2. Visibility as a function of pinhole separation.

by half a fringe; that is, the previously light areas are now dark, and vice versa. Still further increase in d causes the repeated fluctuations in visibility shown in the figure. A curve similar to that of Fig. 1.2 is obtained if the pinhole spacing is kept constant while the size of Σ is changed. These effects can be predicted from the theorem of Van Cittert and Zernike [1.7, 1.8]. The visibility versus pinhole separation curve is sometimes used as a measure of *spatial coherence*, as discussed in the next few sections. The screen separations r_{10} and r_{20} are both assumed to be large compared with the aperture spacing d and with the dimensions of the source. Beyond this limitation, changes in r_{10} or r_{20} change the scale of effects such as are shown in Fig. 1.2, without changing their general character.

As the point of observation P (see Fig. 1.1) is moved away from the center of the observing screen, the visibility decreases as the path difference $\Delta r = r_2 - r_1$ increases, eventually becoming zero. The effect also depends upon how nearly monochromatic the source is. It is found that the visibility of the fringes is appreciable only for a path difference, that is,

$$\Delta r \simeq \frac{2\pi c}{\Delta\omega}, \tag{1.4}$$

where c is the velocity of light and $\Delta\omega$ is the width of the spectrum of the source of light. Equation (1.4) is often used to define the *coherence length* of the source.

The foregoing example shows that it is not necessary to have completely coherent light to produce an interference pattern, but that under the right conditions such a pattern may be obtained from an incoherent source. This effect may properly be called *partial coherence*, and we need a method for defining and measuring it.

A further development of the preceding equations is helpful. Thus $u_1(t)$ and $u_2(t)$ of Eq. (1.1), the complex wave fields at the points Q_1 and Q_2, are subject to the scalar wave equation in free space:

$$\nabla^2 u = \frac{1}{c^2}\frac{\partial^2 u}{\partial t^2}. \tag{1.5}$$

This is a linear equation, and the wave field at point P on the observing screen is a sum of those from Q_1 and Q_2:

$$u_p(t) = c_1 u_1\left(t - \frac{r_1}{c}\right) + c_2 u_2\left(t - \frac{r_2}{c}\right), \tag{1.6}$$

where c_1 and c_2 are appropriate complex constants. The corresponding irradiance at P may be written

$$I_p = \langle u_p(t) u_p^*(t)\rangle = I_1 + I_2 + 2\mathrm{Re}\left\langle c_1 u_1\left(t - \frac{r_1}{c}\right) c_2^* u_2^*\left(t - \frac{r_2}{c}\right)\right\rangle, \tag{1.7}$$

where I_1 and I_2 are proportional to the squares of the magnitudes of $u_1(t)$ and $u_2(t)$. Let us define the following variables:

$$t_1 = \frac{r_1}{c} \quad \text{and} \quad t_2 = \frac{r_2}{c}. \tag{1.8}$$

Then Eq. (1.7) can be written as

$$I_p = I_1 + I_2 + 2c_1 c_2^* \,\mathrm{Re}\langle u_1(t - t_1) u_2^*(t - t_2)\rangle. \tag{1.9}$$

Thus we see that the averaged quantity in Eq. (1.9) is the cross-correlation of the two complex wave fields.

If we put $t_2 - t_1 = \tau$, Eq. (1.9) can be written as

$$I_p = I_1 + I_2 + 2c_1 c_2^* \,\mathrm{Re}\langle u_1(t + \tau) u_2^*(t)\rangle,$$

and combining this with Eq. (1.1), we obtain

$$I_p = I_1 + I_2 + 2c_1c_2^* \operatorname{Re}[\Gamma_{12}(\tau)]. \tag{1.10}$$

The self-coherence functions (i.e., the autocorrelations) of the radiations from the two pinholes can therefore be defined as

$$\Gamma_{11}(0) = \langle u_1(t)u_1^*(t)\rangle \qquad \text{and} \qquad \Gamma_{22}(0) = \langle u_2(t)u_2^*(t)\rangle. \tag{1.11}$$

If we let the following relations hold:

$$|c_1|^2\Gamma_{11}(0) = I_1 \qquad \text{and} \qquad |c_2|^2\Gamma_{22}(0) = I_2,$$

the intensity at P, Eq. (1.10), can then be written in terms of the degree of complex coherence, Eq. (1.2), that is,

$$I_p = I_1 + I_2 + 2(I_1I_2)^{1/2} \operatorname{Re}[\gamma_{12}(\tau)]. \tag{1.12}$$

Let us write $\gamma_{12}(\tau)$ in the form

$$\gamma_{12}(\tau) = |\gamma_{12}(\tau)|\exp[i\phi_{12}(\tau)], \tag{1.13}$$

and assume also that $I_1 = I_2 = I$, which may be called the *best condition*. Then Eq. (1.12) becomes

$$I_p = 2I[1 + |\gamma_{12}(\tau)|\cos \phi_{12}(\tau)]. \tag{1.14}$$

Thus we see that the maximum value of I_p is $2I[1 + |\gamma_{12}(\tau)|]$ and the minimum value is $2I[1 - |\gamma_{12}(\tau)|]$. By substituting these values into the visibility equation, Eq. (1.3), we find that

$$V = |\gamma_{12}(\tau)|. \tag{1.15}$$

That is, under the best condition, the visibility of the fringes is a measure of the absolute value of the complex degree of coherence.

1.2. MUTUAL COHERENCE FUNCTION

The complex cross-correlation is usually defined as the second-moment time average, that is,

$$\Phi_{12}(\tau) \triangleq \langle f_i^*(t)f_2(t+\tau)\rangle = \lim_{T\to\infty} \frac{1}{2T}\int_{-T}^{T} f_1^*(t)f_2(t+\tau)dt. \tag{1.16}$$

Thus the mutual coherence function is defined as [8]

$$\Gamma_{12}(\tau) \triangleq \langle u_1(t+\tau)u_2^*(t)\rangle = \lim_{T\to\infty} \frac{1}{2T}\int_{-T}^{T} u_1(T, t+\tau)u_2^*(T, t)dt. \quad (1.17)$$

It is now possible, however, to give a more general definition, fundamental to the theory of partial coherence. This can be done by working from two different aspects of the phenomena: first, from a system ensemble viewpoint, and later, from time-averaging.

To obtain an ensemble average, we define the mutual coherence function as

$$\Gamma_{12}(\boldsymbol{\rho}_1, t_1; \boldsymbol{\rho}_2, t_2) = \lim_{N\to\infty} \frac{1}{N}\sum_{n=1}^{N} u_{1n}(\boldsymbol{\rho}_1, t_1)u_{2n}^*(\boldsymbol{\rho}_2, t_2). \quad (1.18)$$

The summation is over an ensemble of N systems, and $\boldsymbol{\rho}_1$ and $\boldsymbol{\rho}_2$ are the respective position vectors.

If the statistics of the ensemble of systems are ergodic *stationary*, then

$$\Gamma_{12}(\boldsymbol{\rho}_1, t_1; \boldsymbol{\rho}_2, t_2) = \Gamma_{12}(\boldsymbol{\rho}_1, \boldsymbol{\rho}_2, \tau), \quad (1.19)$$

where $\tau = t_1 - t_2$. Equation (1.18) can then be rewritten:

$$\Gamma_{12}(\boldsymbol{\rho}_1, \boldsymbol{\rho}_2, \tau) = \lim_{N\to\infty} \frac{1}{N}\sum_{n=1}^{N} u_{1n}(\boldsymbol{\rho}_1, t_2+\tau)u_{2n}^*(\boldsymbol{\rho}_2, t_2). \quad (1.20)$$

It is well known from the theory of random processes [1.9, p. 15] that when the statistics are ergodic stationary the time average and system ensemble are equal. Thus we have

$$\Gamma_{12}(\boldsymbol{\rho}_1, \boldsymbol{\rho}_2, \tau) = \Gamma_{12}(\tau) = \langle u_1(\boldsymbol{\rho}_1, t+\tau)u_2^*(\boldsymbol{\rho}_2, t)\rangle. \quad (1.21)$$

This definition of the mutual coherence function as a time average has been made dependent on having ergodic stationary statistics. It therefore is valid where the source or radiation is periodic. In general, however, the time-average definition may be written as

$$\Gamma_{12}(\tau) \triangleq \langle u(\boldsymbol{\rho}_1, t+\tau)u^*(\boldsymbol{\rho}_2, t)\rangle. \quad (1.22)$$

The *complex degree of coherence* is defined as in Eq. (1.2),

$$\gamma_{12}(\tau) \triangleq \frac{\Gamma_{12}(\tau)}{[\Gamma_{11}(0)\Gamma_{22}(0)]^{1/2}}, \quad (1.23)$$

with the limits $0 \leqq |\gamma_{12}(\tau)| \leqq 1$, where the lower limit represents the complete incoherence and the upper limit represents the complete coherence between the two-point radiations at $\boldsymbol{\rho}_1$ and $\boldsymbol{\rho}_2$. Note that $|\gamma_{12}(\tau)|$ is a function of τ. It is therefore possible that the radiations at two points may be coherent at one value of τ, but incoherent at another value.

The *self-coherence*, or autocorrelation, function is therefore defined by

$$\Gamma_{11}(\tau) \triangleq \langle u_1(\boldsymbol{\rho}_1, t + \tau) u_1^*(\boldsymbol{\rho}_1, t)\rangle. \tag{1.24}$$

We note that the use of the self-coherence function is very important in the analysis of the operation of Michelson's interferometer. Its zero value, $\Gamma_{11}(0)$, is the highest intensity at point $\boldsymbol{\rho}_1$, that is,

$$\Gamma_{11}(0) \geqq \Gamma_{11}(\tau). \tag{1.25}$$

We also note that $\Gamma_{12}(0)$ of the mutual coherence function is called the *mutual intensity function.* Accordingly, $\Gamma_{12}(0)$ is an essential quantity in the examination of the "stellar" form interferometer (e.g., Michelson stellar interferometer). And we use this quantity to evaluate the performance of the white-light optical signal processing.

It will also be found useful to take the Fourier transforms of both the mutual and self-coherence functions. That of the mutual function is called the *mutual power spectrum*, and is given by

$$\Gamma_{12}(\omega) = \begin{cases} \displaystyle\int_{-\infty}^{\infty} \Gamma_{12}(\tau) e^{-i\omega\tau}\, d\tau, & \omega > 0, \\ 0, & \omega < 0 \end{cases} \tag{1.26}$$

That the second of these conditions holds can be seen from the fact that $\Gamma_{12}(\omega)$ is *analytic*. The Fourier transform of a self-coherence function is the power spectrum of that particular radiation:

$$\Gamma_{11}(\omega) = \begin{cases} \displaystyle\int_{-\infty}^{\infty} \Gamma_{11}(\tau) e^{-i\omega\tau}\, d\tau, & \omega > 0, \\ 0, & \omega < 0. \end{cases} \tag{1.27}$$

Of course a similar equation for $\Gamma_{22}(\omega)$ can be written in terms of $\Gamma_{22}(\tau)$.

1.3. SPATIAL AND TEMPORAL COHERENCE

An important point must be made concerning coherence. The phrase *spatial coherence* is applied to those effects that are due to the size, in space, of the source of radiation. If we consider a point source, and look at two points at equal light-path distances from that source, the radiations reaching these points will be exactly the same. The mutual coherence will be equal to the self-coherence at either point. That is, if the points are Q_1 and Q_2,

$$\Gamma_{12}(Q_1, Q_2, \tau) = \langle u(Q_1, t + \tau)u^*(Q_2, t)\rangle = \Gamma_{11}(\tau). \tag{1.28}$$

As the source is made larger, we can no longer claim an equality of mutual coherence and self-coherence. The lack of complete coherence is a *spatial* effect.

Temporal coherence, on the other hand, is an effect due to the finite spectral width of the source. The coherence is complete for strictly monochromatic radiation, but becomes only partial as other wavelengths are added, giving a finite spectral width to the source.

It is never possible to completely separate the two effects (i.e., spatial and temporal coherence), but it is well to name them and point out their significance, as is discussed in the following.

1.3.1. Spatial Coherence

We now utilize Young's experiment to determine the angular size of a spatially incoherent source (e.g., an extended thermal source) as relating to the spatial coherence. Thus we see that the spatial coherence depends on the size of the extended source. Let us now consider the two paths of light, rays from a point of a linearly extended source ΔS, passing through two narrow slits Q_1 and Q_2, as depicted in Fig. 1.3. If we let $r_{10} \gg d$, then the intensity distribution at observation screen P_3, as a consequence of Eq. (1.9), can be written as

$$I_p(\alpha) = I_1 + I_2 + 2\sqrt{I_1 I_2}\cos\left[k\left(\frac{d}{r_{10}}\xi + \frac{d}{r_{20}}\alpha\right)\right], \tag{1.29}$$

where I_1 and I_2 are the corresponding intensity distributions due to the light rays passing through Q_1 and Q_2, respectively.

Equation (1.29) describes a set of parallel fringes along the α-axis due to a single point radiator on extended source ΔS. Thus we see that the other points on the extended source would cause a superposition of the fringes over

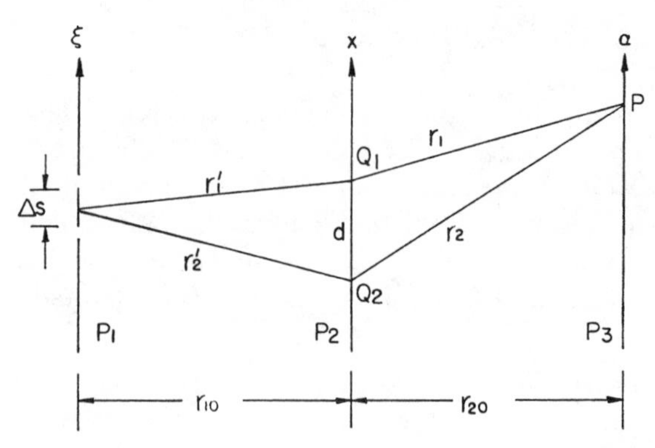

Figure 1.3. Young's experiment for the measurement of spatial coherence.

the observation screen P_3. In a uniformly one-dimensional extended source of size ΔS, the resulting fringe pattern can be written as

$$I'_p(\alpha) = I_1 \Delta S + I_2 \Delta S + 2\sqrt{I_1 I_2} \int_{-\Delta S/2}^{\Delta S/2} \cos\left[k\left(\frac{d}{r_{10}}\xi + \frac{d}{r_{20}}\alpha\right)\right] d\xi$$

$$= \Delta S I_1 + I_2 + 2\sqrt{I_1 I_2}\,\mathrm{sinc}\left[\left(\frac{\pi d}{\lambda r_{10}}\right)\Delta S\right]\cos\left(k\frac{d}{r_{20}}\alpha\right), \qquad (1.30)$$

where $\mathrm{sinc}(\pi\chi) \triangleq \sin \pi\chi/\pi\chi$. With reference to the sinc fact of Eq. (1.30), the fringes vanish at

$$d = \frac{\lambda r_{10}}{\Delta S}. \qquad (1.31)$$

Similarly, we can write Eq. (1.31) in terms of angular size of the source θ_s, that is,

$$d = \frac{\lambda}{\theta_s}, \qquad (1.32)$$

where $\theta_s \triangleq \Delta S/r_{10}$. We note that the sinc factor of Eq. (1.30) is primarily due to the incoherent addition of the point radiators over the extended source ΔS. The result of Eq. (1.30) presents a useful example for the discussion of coherent and incoherent illumination. For completely coherent illumination, it causes interference fringes, while for completely incoherent illumination the interference fringes vanish. Moreover, the expression of Eq. (1.30) shows that the interference pattern (i.e., the cosine factor) is weighted by a broad sinc factor, where the interference fringes go to zero and reappear as a function of d, ΔS, or $1/r_{10}$. If the extended source had been circular, the sinc factor of Eq. (1.30) would have been

$$\frac{J_1(\pi \Delta S d/\lambda r_{10})}{\pi \Delta S d/\lambda r_{10}}, \tag{1.33}$$

where J_1 is the first-order Bessel function. The pinhole separation at which the fringes vanish would then be

$$d = 1.22 \frac{\lambda}{\theta_s}, \tag{1.34}$$

where $\theta_s \triangleq \Delta S/r_{10}$ is the angular size of the circular source.

Since the region of *spatial coherence* vanishes at the point at which the interference fringes vanish, the region of coherence increases as the source size decreases or as the distance of the source increases. In other words, the degree of spatial coherence increases with the distance of wave propagation.

Furthermore, there exists a relationship between the intensity distribution of the source and the degree of spatial coherence, as stated by the Van Cittert–Zernike theorem [1.4, 1.5]. The theorem essentially states that the normalized complex degree of coherence $\gamma_{12}(0)$ between two points at a plane due to an extended incoherent source at another plane is proportional to the normalized inverse Fourier transform of the intensity distribution of the source, that is,

$$r_{12}(0) = \frac{\int_{\Delta S} I(\boldsymbol{\xi}) \exp[ik(\mathbf{x}_1 - \mathbf{x}_2)\boldsymbol{\xi}] d\boldsymbol{\xi}}{\int_{\Delta S} I(\boldsymbol{\xi}) d\boldsymbol{\xi}}, \tag{1.35}$$

where the subscripts 1 and 2 denote the two points at the diffraction screen P_2, $\mathbf{x}_1$ and $\mathbf{x}_2$ are the corresponding position vectors, $\boldsymbol{\xi}$ is the position vector at source plane, $k = 2\pi/\lambda$, $I(\boldsymbol{\xi})$ is the intensity distribution of the source, and the integration is over the source size. Thus it is apparent that, if the intensity distribution of the source is uniform over a circular region, then the mutual coherence function is described by a rotational symmetric Bessel function J_1. On the other hand, if the source intensity is uniform over a rectangular slit, the mutual coherence function is described by a two-dimensional sinc factor.

1.3.2. Temporal Coherence

It is possible for us to split a light wave into two paths, to delay one of these, and then to recombine them to form an interference fringe pattern. In this

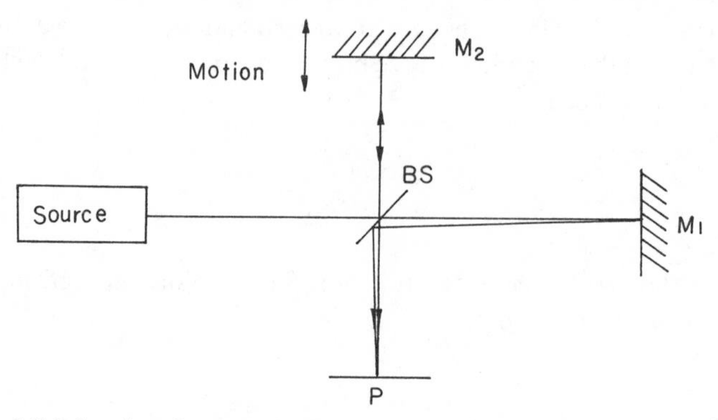

Figure 1.4. Michelson interferometer for the measurement of temporal coherence of a source. *BS*, beam splitter; *M*, mirrors; *P*, observation screen.

way we would be able to measure the *temporal coherence* of the light wave. The degree of temporal coherence is a measure of the cross-correlation of a wave field at one time with respect to another wave field at a later time. Thus the definition of temporal coherence can also be related to *longitudinal coherence*, as opposed to transverse (i.e., spatial) coherence. The maximum optical path length difference between two waves derived from a source is known as the *coherent length* of the source. Since the spatial coherence is determined by the wave field at the transverse direction and the temporal coherence is measured along the longitudinal direction, the spatial and temporal coherences together would describe the degree of coherence of a wave field within a volume of space.

One of the most commonly used techniques to measure the temporal coherence of a light source must be the Michelson interferometer [1.10, pp. 303–308], as shown in Fig. 1.4. With reference to this figure, it is apparent that the beamsplitter *BS* divides the light beam into two paths; one falls to mirror M_1 and the other goes to mirror M_2. By simply varying the path length (i.e., τ) of one of the mirrors, one can observe the variation of the interference fringes at the observation screen *P*. Thus the measure of the visibility of the interference fringes corresponds to a measure of the degree of temporal coherence $\gamma_{11}(\tau)$ at a time delay τ, where the subscript 11 signifies the interference derived from the same point in space. Needless to say, when the path difference is zero, the visibility measure corresponds to $\gamma_{11}(0)$.

Since the property of the source affects the temporal coherence, the spectral width of the source would affect the temporal coherence of the source. The time interval Δt over which the wave is coherent can be approximated by

$$\Delta t = \frac{2\pi}{\Delta\omega}, \tag{1.36}$$

where $\Delta\omega$ is the spectral bandwidth of the source. Thus the *coherence length* of the light source can be defined as [Eq. (1.4)],

$$\Delta r = \Delta t c \cong \frac{2\pi c}{\Delta\omega}, \tag{1.37}$$

where c is the velocity of light.

1.4. TWO-BEAM INTERFEROMETRIC MEASUREMENT

In the 1930s Van Cittert [1.4] and Zernike [1.5] predicted a profound relationship between the spatial coherence and the intensity distribution of the light source. It was, however, the works of Thompson and Wolf [1.11, 1.12] that demonstrated a two-beam interference technique to measure the degree of partial coherence. They showed that, under quasi-monochromatic illumination, the degree of spatial coherence is dependent on the source size and the distance between two arbitrary points. The degree of temporal coherence is, however, dependent on the spectral bandwidth of the light source. They also illustrated several coherence measurements that are very consistent with Van Cittert and Zernike's predictions.

In this section we illustrate some results obtained by interferometric techniques, as described in the previous section, for the measurement of spatial and temporal coherence.

Figure 1.5 shows a sequence of interference fringe patterns obtained with a dual-beam technique (Fig. 1.3) where a circular extended source is used. This set of fringe patterns is obtained by increasing the separation d of the two pinholes. From this set of fringe patterns, we see that the visibility (i.e., the degree of spatial coherence) decreases from Fig. 1.5*a* to Fig. 1.5*d*, then increases to Fig. 15*e* and decreases again to Fig. 1.5*f* (as can be seen in Fig. 1.2). We also note that the phase of the fringe patterns is shifted a few times from Fig. 1.5*a* to 1.5*f*, which is due to the bipolar nature of the first-order Bessel function, as described in Eq. (1.33). In other words, Fig. 1.5*a* to 1.5*c* show the fringe patterns recorded in the first lobe of Fig. 1.2, while Fig. 1.5*d* and 1.5*e* are in the second lobe and Fig. 1.5*f* is in the third lobe. It is also apparent that the spatial frequency of the interference fringes is proportional to the separation d, as shown from Fig. 1.5*a* to 1.5*f*.

Figure 1.6 shows a set of interference fringe patterns that result when the separation d is held constant while the diameter of the circular source is increased. From this figure, we see that the visibility decreases as the source size increases, and that the spatial frequency of the fringe patterns remains unchanged.

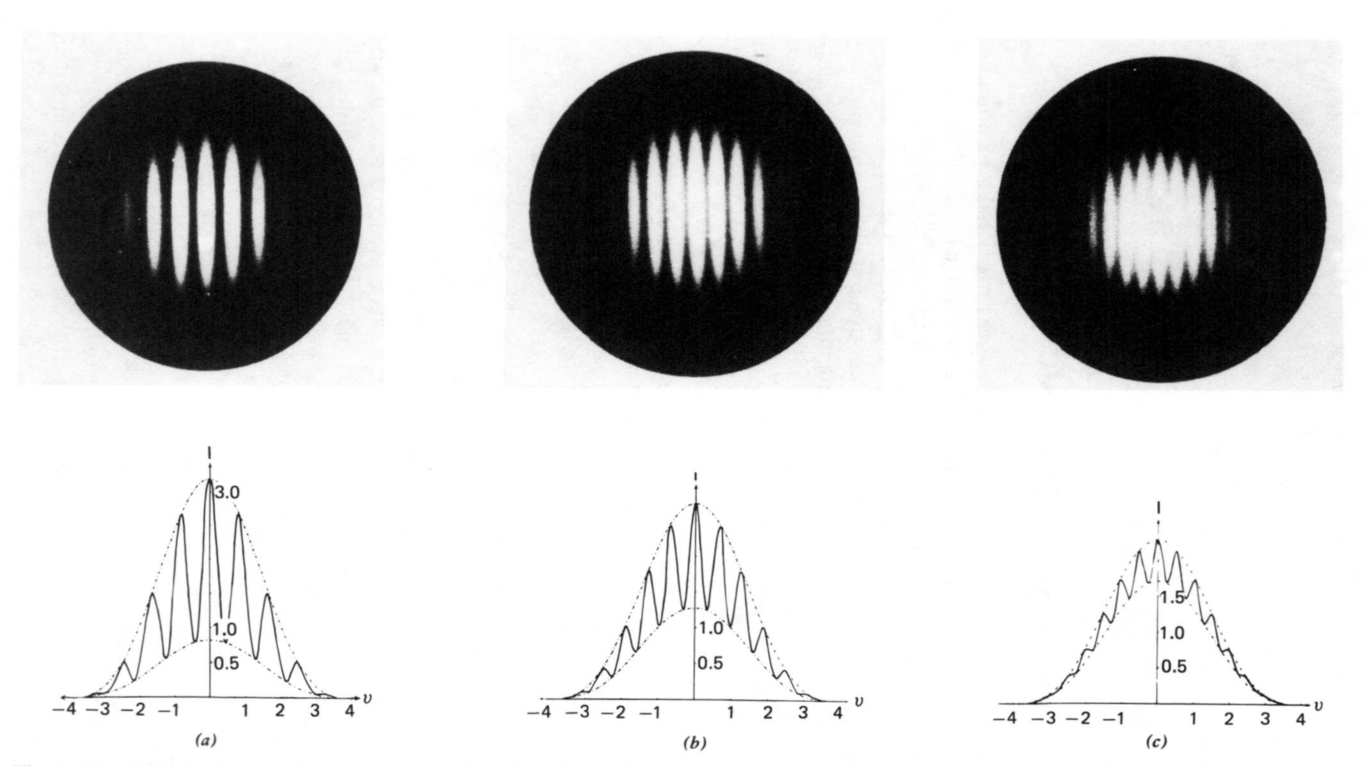

Figure 1.5. Spatial coherence measurement as a function of pinhole separation. d, the separation between pinholes; γ, the degree of coherence. (*a*) $d = 0.6$ cm, $|\gamma_{12}| = 0.593$. (*b*) $d = 0.8$ cm, $|\gamma_{12}| = 0.361$. (*c*) $d = 1$ cm, $|\gamma_{12}| = 0.146$. (Permission of B. J. Thompson.)

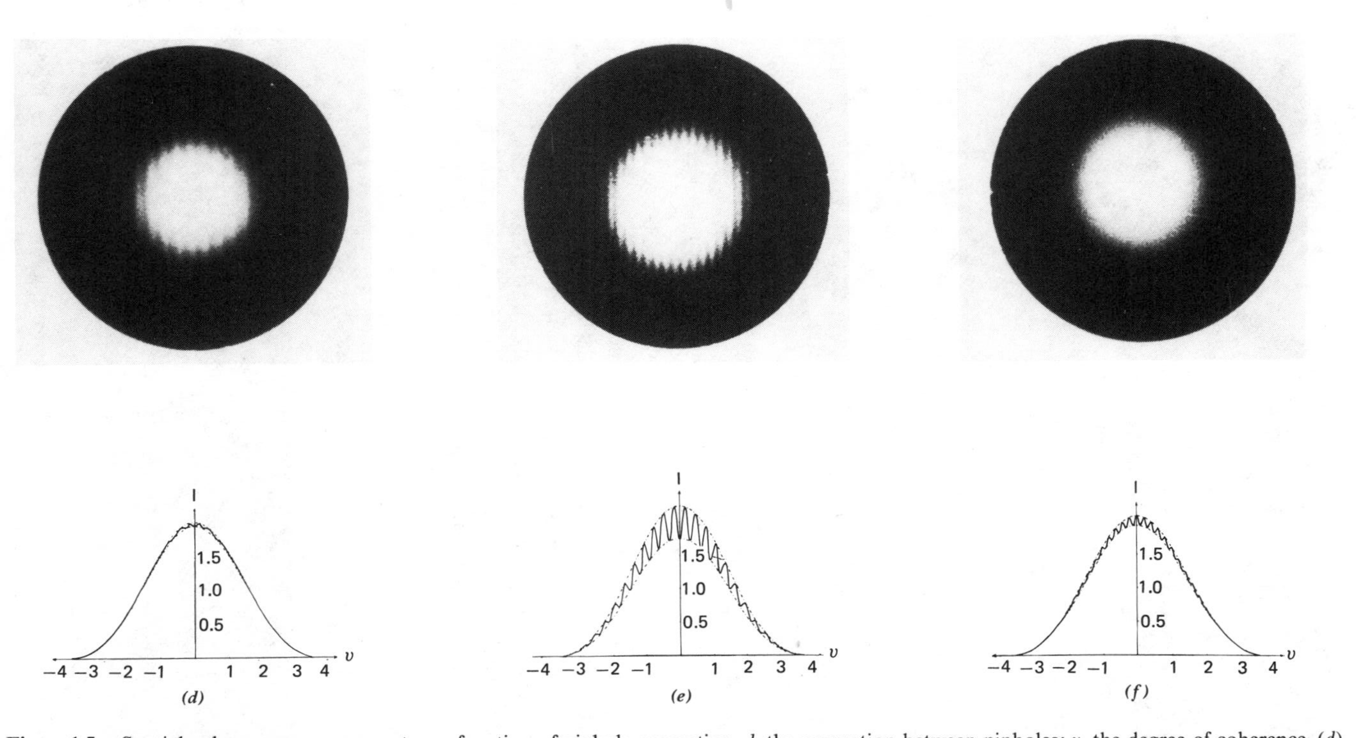

Figure 1.5. Spatial coherence measurement as a function of pinhole separation. d, the separation between pinholes; γ, the degree of coherence. (*d*) $d = 1.2$ cm, $|\gamma_{12}| = 0.015$. (*e*) $d = 1.7$ cm, $|\gamma_{12}| = 0.123$. (*f*) $d = 2.3$ cm, $|\gamma_{12}| = 0.035$. (Permission of B. J. Thompson.)

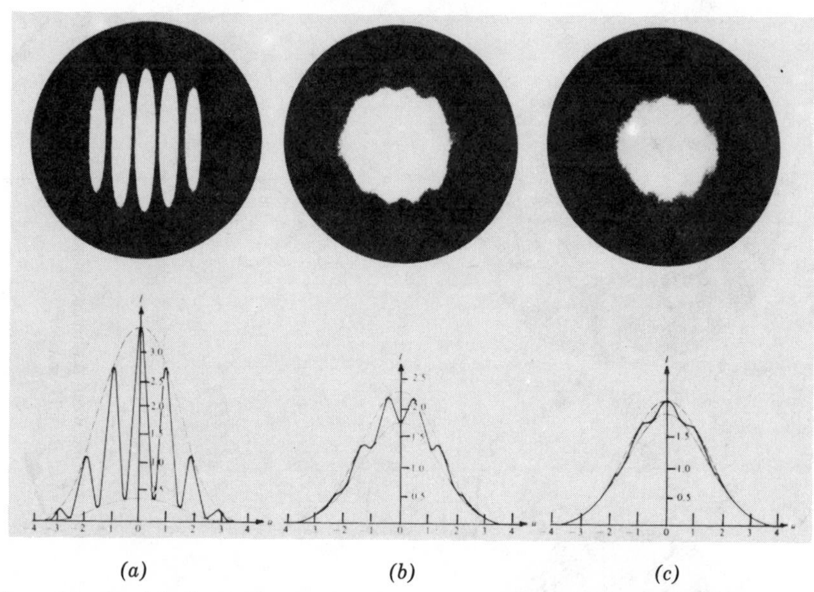

Figure 1.6. Spatial coherence measurement as a function of source size. γ, the degree of coherence; ϕ, the phase shift; and d, the separation between pinholes. (*a*) $\phi_{12} = 0$, $d = 0.5$ cm, $|\gamma_{12}| = 0.703$. (*b*) $\phi_{12} = \pi$, $d = 0.5$ cm, $|\gamma_{12}| = 0.132$. (*c*) $\phi_{12} = 0$, $d = 0.5$ cm, $|\gamma_{12}| = 0.062$. (Permission of B. J. Thompson.)

In the measurement of temporal coherence, however, we use the Michelson interferometer, as described previously. Figure 1.7 shows a sequence of interference fringe patterns that we have obtained with this technique. From this figure we see that the visibility of the fringe patterns decreases as the path length difference (i.e., τ) between the two waves increases. Thus the coherence length of a temporal partially coherent source can be measured with this technique.

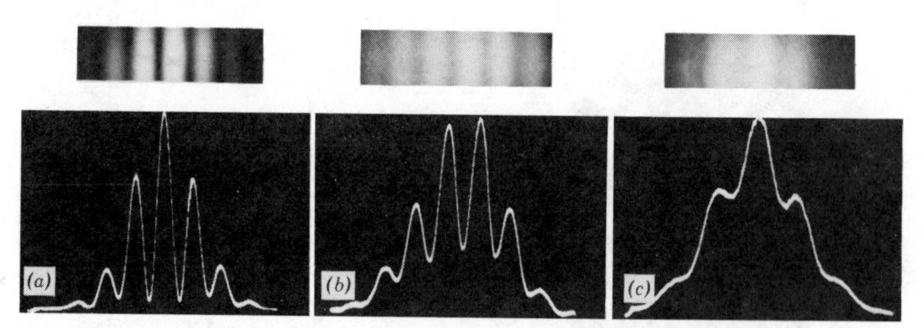

Figure 1.7. Temporal coherence measurement. The light source is a Spectra-physics, Model 120, He–Ne laser. Δd, the difference in optical path; γ; the degree of coherence. (*a*) $\Delta d \simeq 0$, $\gamma = 0.94$. (*c*) $\Delta d = 25$ cm, $\gamma = 0.67$. (*c*) $\Delta d = 43$ cm, $\gamma = 0.12$.

1.5. PROPAGATION OF MUTUAL COHERENCE

Equations that describe the manner in which the mutual coherence function is propagated can be found by first assuming that the field can be represented by a complex scalar function $u(t)$ that satisfies the scalar wave equation:

$$\nabla^2 u(t) = \frac{1}{c^2}\frac{\partial^2 u(t)}{\partial t^2}. \tag{1.38}$$

The mutual coherence function, as we have previously defined it, is

$$\Gamma_{12}(\tau) = \langle u_1(t+\tau)u_2^*(t)\rangle. \tag{1.39}$$

The Laplacian, taken at the point Q_1 of Eq. (1.39), may be written

$$\nabla_1^2\Gamma_{12}(\tau) = \langle \nabla_1^2 u_1(t+\tau)u_2^*(t)\rangle. \tag{1.40}$$

From Eq. (1.38) we can write

$$\nabla_1^2 u_1(t+\tau) = \frac{1}{c^2}\frac{\partial^2 u_1(t+\tau)}{\partial(t+\tau)^2}. \tag{1.41}$$

But

$$\frac{\partial^2 u_1(t+\tau)}{\partial(t+\tau)^2} = \frac{\partial^2 u_1(t+\tau)}{\partial\tau^2},$$

and therefore Eq. (1.41) becomes

$$\nabla_1^2\Gamma_{12}(\tau) = \left\langle \frac{\partial^2 u_1(t+\tau)}{c^2\partial\tau^2}\, u_2^*(t)\right\rangle. \tag{1.42}$$

The complex field $u_2(t)$ is independent of τ, and we can therefore make the second partial derivative apply to the whole time-average function:

$$\nabla_1^2\Gamma_{12}(\tau) = \frac{1}{c^2}\frac{\partial^2}{\partial\tau^2}\langle u_1(t+\tau)u_2^*(t)\rangle = \frac{1}{c^2}\frac{\partial^2\Gamma_{12}(\tau)}{\partial\tau^2}. \tag{1.43}$$

This may be regarded as the fundamental equation describing the propagation of the mutual coherence function.

In the same way, we can take the Laplacian of Eq. (1.39) with respect to the coordinates of the point Q_2, and it can be shown that it also reduces to

$$\nabla_2^2\Gamma_{12}(\tau) = \frac{1}{c^2}\frac{\partial^2\Gamma_{12}(\tau)}{\partial\tau^2}. \tag{1.44}$$

Each of the Eqs. (1.43) and (1.44) contains only four independent variables, while $\Gamma_{12}(\tau)$ contains seven (i.e., the three spatial coordinates of each of the two points plus the time delay τ). A complete propagation equation may be obtained by combining Eqs. (1.43) and (1.44):

$$\nabla_1^2\nabla_2^2\Gamma_{12}(\tau) = \frac{1}{c^4}\frac{\partial^4\Gamma_{12}(\tau)}{\partial\tau^4}. \tag{1.45}$$

Suppose that the source of radiation is a surface of finite extent. For each pair of points on the surface a mutual coherence function, $\Gamma_{12}(\tau)$, can be specified. We then apply Sommerfeld's radiation condition at infinity, which says that $u_1(t)$ and $u_2^*(t)$ behave asymptotically as point radiators:

$$\frac{A_1}{r_1}\exp(ikr_1) \qquad \text{and} \qquad \frac{A_2}{r_2}\exp(ikr_2), \tag{1.46}$$

as r_1 and r_2 (the distances from a pair of points on the source surface) approach infinity. These conditions may be applied to Eqs. (1.43) and (1.44), and then the equations can be solved as simultaneous wave equations. Since each equation contains four variables, their solutions can be written as four-dimensional Green's functions. The number of variables in each can, however, be reduced to three by first taking the Fourier transform of $\Gamma_{12}(\tau)$, which we designate by $\hat{\Gamma}_{12}(\omega)$. Now $\Gamma_{12}(\tau)$, as an analytic function, contains only posi-tive frequencies, and we can write

$$\Gamma_{12}(\tau) = \int_0^\infty \hat{\Gamma}_{12}(\omega)\exp(i\omega\tau)d\omega, \tag{1.47}$$

and

$$\hat{\Gamma}_{12}(\omega) = \begin{cases} \int_{-\infty}^{\infty} \Gamma_{12}(\tau)\exp(-i\omega\tau)d\tau, & \omega > 0, \\ 0, & \omega < 0. \end{cases} \tag{1.48}$$

By substituting Eq. (1.47) into Eqs. (1.43) and (1.44), and reversing the order of integrations and differentiations, we have the following integral equations:

$$\int_0^\infty [\nabla_1^2 + k^2(\omega)]\hat{\Gamma}_{12}(\omega)\exp(i\omega\tau)d\omega = 0, \tag{1.49}$$

and

$$\int_0^\infty [\nabla_2^2 + k^2(\omega)]\hat{\Gamma}_{12}(\omega)\exp(i\omega\tau)d\omega = 0. \tag{1.50}$$

These two equations must hold for every value of τ, and therefore

$$[\nabla_1^2 + k^2(\omega)]\hat{\Gamma}_{12}(\omega) = 0, \tag{1.51}$$

and

$$[\nabla_2^2 + k^2(\omega)]\hat{\Gamma}_{12}(\omega) = 0. \tag{1.52}$$

From these equations it can be seen that the Fourier transform of the mutual coherence function satisfies the scalar Helmholtz equation.

Now we set up a Green's function $G_1(P_1, P'_1, \omega)$ such that

$$[\nabla_1^2 + k^2(\omega)]G_1(P_1, P'_1, \omega) = -\delta(P_1 - P'_1), \tag{1.53}$$

with the following boundary condition:

$$G_1(P_1 P'_1, \omega)|_{P'_1 = S_1} = 0, \tag{1.54}$$

where P and S are the position coordinates, and δ is the Dirac delta function.

The solution of Eq. (1.51) may be put in terms of this Green's function:

$$\hat{\Gamma}(P_1, S_2, \omega) = -\int_{S_1} \hat{\Gamma}(S_1, S_2, \omega) \left.\frac{\partial G_1(P_1, P'_1, \omega)}{\partial n_{S_1}}\right|_{P'_1 = S_1} dS_1, \tag{1.55}$$

Using another Green's function, $G_2(P_2, P'_2, \omega)$, defined as in Eqs. (1.53) and (1.54), and noting the fact that Eq. (1.55) provides a boundary condition, we see that the solution of Eq. (1.52) becomes

$$\hat{\Gamma}(P_1, P_2, \omega) = -\int_{S_2} \hat{\Gamma}(P_1, S_2, \omega) \left.\frac{\partial G_2(P_2, P'_2, \omega)}{\partial n_{S_2}}\right|_{P'_2 = S_2} dS_2. \tag{1.56}$$

By combining Eqs. (1.55) and (1.56), we have

$$\hat{\Gamma}(P_1, P_2, \omega) = \int_{S_2}\int_{S_1} \hat{\Gamma}(S_1, S_2, \omega) \left.\frac{\partial G_1(P_1, P'_1, \omega)}{\partial n_{S_1}}\right|_{P'_1 = S_1} \cdot \left.\frac{\partial G_2(P_2, P'_2, \omega)}{\partial n_{S_2}}\right|_{P'_2 = S_2} dS_1\, dS_2. \tag{1.57}$$

Let us note that $\hat{\Gamma}(P_1, S_2, \omega)$ in Eq. (1.55) is the measure of coherence between a point on the source surface and an arbitrary point in space. The phrase *longitudinal direction* is used for space points that lie on perpendiculars to the surface points. It is evident, as we have discussed in the previous sections, that the so-called *longitudinal coherence* is quite different from that between points on the surface, namely the so-called spatial coherence.

The Fourier transform $\hat{\Gamma}(P_1, P_2, \omega)$ is a power spectrum, and if P_1 and P_2 are the same, it is the power spectrum of a single point in space. Equation (1.57) then shows that this power spectrum is not determined by the power spectrum of the whole source surface, but rather by $\hat{\Gamma}(S_1, S_2, \omega)$, which is the cross-correlation between two points of the surface.

According to the Sommerfeld radiation condition, Eq. (1.46), the behavior of $u_1(\omega)$ and $u_2^*(\omega)$ as r_1 and r_2 approach infinity should follow

$$\frac{a_1(\theta_1, \phi_1, \omega)}{r_1} \exp(ikr_1) \qquad \text{and} \qquad \frac{a_2(\theta_2, \phi_2, \omega)}{r_2} \exp(ikr_2). \tag{1.58}$$

If we assume that the statistics between $u_1(t)$ and $u_2(t)$ are those of a ergodic random process, then $\Gamma(P_1, P_2, \omega)$ may be expressed as a time average:

$$\hat{\Gamma}(P_1, P_2, \omega) = \lim_{T \to \infty} \frac{1}{2T} \int_{-T}^{T} u_1(\tau, \omega) u_2^*(\tau, \omega) d\tau, \tag{1.59}$$

and from Eqs. (1.58) and (1.59) we see that

$$\hat{\Gamma}(P_1, P_2, \omega) \to a_{12}(\theta_1, \theta_2, \phi_1, \phi_2, \omega) \frac{\exp[ik(r_1 - r_2)]}{r_1 r_2}, \tag{1.60}$$

as $r_1, r_2 \to \infty$.

With $\hat{\Gamma}(P_1, P_2, \omega)$ given by Eq. (1.57), the solution for $\Gamma(P_1, P_2, \tau)$ is

$$\Gamma(P_1, P_2, \tau) = \int_0^{\infty} \hat{\Gamma}(P_1, P_2, \omega) e^{i\omega\tau} \, d\omega. \tag{1.61}$$

1.6. SOME PHYSICAL CONSTRAINTS

The coherence functions have certain limits, which are discussed in this section. We have already, in discussing Eq. (1.23), called attention to a set of limits [0, 1] for the function $|\gamma_{12}(\tau)|$, with 0 representing complete incoherence and 1 representing complete coherence. We have pointed out that the degree of coherence depends upon the value of τ, but it is also true that it depends upon the particular pair of points chosen for comparison. Thus we may expect $|\gamma_{12}(\tau)|$ to be zero for some points and some delays, but we would not expect it to vanish in general. The question remains, however, whether it is possible for an extended field (which might be the source itself) to have the property that $|\gamma_{12}(\tau)| = 0$ or $|\gamma_{12}(\tau)| = 1$ for every pair of points in the field and for any time delay τ. If so, it would seem proper to call the entire field incoherent or coherent, respectively.

In this connection, we quote three well-known theorems, proofs of which the reader can find in the text by Beran and Parrent [1.13, pp. 47–52].

Theorem 1. A complex electromagnetic field has a unity degree of coherence (i.e., $|\gamma_{12}(\tau)| = 1$) for every pair of points in the field and every time delay τ, if and only if the field is monochromatic.

Theorem 2. A nonnull electromagnetic field, for which $|\gamma_{12}(\tau)| = 0$ for every pair of points in the field and for every time delay τ, cannot exist in free space. Conversely, if $|\gamma_{12}(\tau)| = 0$ for every pair of points on a continuous closed surface, then the surface does not radiate.

Theorem 3. If spectral filtering is used on the radiation from two source points whose degree of coherence is zero, the degree of coherence will remain zero.

With regard to Theorem 1, it must be said that strictly monochromatic fields do not exist in practice, since all fields have some finite spectral bandwidth. However, it is possible for a field to have a spectral bandwidth $\Delta\omega$ that is small compared with the center frequency ω_0 of the radiation. Such a field is called *quasi-monochromatic*. If path differences in the radiation are small, a theory for quasi-monochromatic fields may be developed. Of course the concept of quasi-monochromatic fields is intended to give a practical approach to real monochromaticity, but there are some aspects in which they differ widely. For example, although $|\gamma_{12}(\tau)| = 1$ is a requirement for all pairs of points in a strictly monochromatic field, this need not be the case for quasi-monochromaticity. In fact, there may be pairs of points for which $|\gamma_{12}(\tau)| = 0$. In order to help the reader obtain a more realistic feeling for the quasi-monochromatic field, we discuss it a little further.

We take as the condition for *quasi-monochromatic radiation*

$$\frac{\Delta\omega}{\omega_0} \ll 1. \tag{1.62}$$

It then follows that $\hat{\Gamma}_{12}(\omega)$ is essentially zero for all frequencies outside the band. That is, for an appreciable $\hat{\Gamma}_{12}(\omega)$ we must have

$$|\omega - \omega_0| < \Delta\omega. \tag{1.63}$$

From Eq. (1.47), the mutual coherence function can then be written

$$\Gamma_{12}(\tau) = \exp(-i\omega_0\tau)\int_0^\infty \hat{\Gamma}_{12}(\omega)\exp[-i(\omega-\omega_0)\tau]d\omega. \quad (1.64)$$

By considering only small values of τ, such that $\Delta\omega|\tau| \ll 2\pi$, Eq. (1.64) reduces to

$$\Gamma_{12}(\tau) = \Gamma_{12}(0)\exp(-i\omega_0\tau). \quad (1.65)$$

Instead of zero for the value of τ [giving $\Gamma_{12}(0)$], the standard point can be taken as any other value, which we call τ_0, making $\tau = \tau_0 + \tau'$, $\Delta\omega|\tau'| \ll 2\pi$. Then instead of Eq. (1.65), we write

$$\Gamma_{12}(\tau_0 + \tau') = \Gamma_{12}(\tau_0)\exp(-i\omega_0\tau'). \quad (1.66)$$

The condition $\Delta\omega|\tau'| \ll 2\pi$ is essential in the theory of quasi-monochromatic fields. We can then define coherence in a limited way, by saying that a field is coherent if a τ_0 can be found for every pair of points, such that $|\gamma_{12}(\tau_0)| = 1$ when $\Delta\omega|\tau'| \ll 2\pi$. For very narrow bandwidths ($\Delta\omega/\omega_0$ very small), the field may be considered monochromatic for all values of τ that are practical in a given problem.

In a strictly monochromatic field, the mutual coherence function may be stated as

$$\Gamma_{12}(\tau) = [\Gamma_{11}(0)\Gamma_{22}(0)]^{1/2}\exp[i\phi_{12}(\tau)], \quad (1.67)$$

and for $\Delta\omega|\tau'| \ll 2\pi$, this equation can be reduced to

$$\Gamma_{12}(\tau) = u(P_1)u^*(P_2)\exp(-i\omega_0\tau'), \quad (1.68)$$

where u is the field at a particular point (P_1 or P_2).

We do not want to give the impression that all fields with a narrow bandwidth are necessarily coherent, even when τ' is kept small. That is to say, that with $\Delta\omega/\omega_0 \ll 1$ and $\Delta\omega|\tau'| \ll 2\pi$, it still is not necessary that $|\gamma_{12}(\tau)| = 1$. In fact, this quantity can have any value between 0 and 1 under these conditions. However, if the quasi-monochromatic field at every point can be expressed as

$$u(t) = A(t)\exp\{-i[\omega_0 t + a(t)]\}, \quad (1.68)$$

where the functions $A(t)$ and $a(t)$ are slow in their variation with time, compared with $2\pi/\omega_0$, then the field can be said to behave in a coherent manner.

Finally, we note that the treatment of *partial coherence theory* in this chapter is by no means complete. For a more intensive treatment of this topic, the reader is referred to the excellent book *Theory of Partial Coherence*, by Beran and Parrent [1.13].

REFERENCES

1.1. M. E. Verdet, "Constitution de la lumiere non polarisee Et De la lumiere partiellement polarisee," *L'Ec. Norm. Super.* **2**, 291 (1869).

1.2. A. A. Michelson, "On the application of interference methods to astronomical measurements," *Phil. Mag.* **30**, (5), 1 (1890).

1.3. M. Berek, "Uber Koharenz und Konsonanz des Lichtes," *Z. Phys.* **36**, 824 (1926).

1.4. P. H. Van Cittert, "Die Wahrschcinlicke Schwingungs verteilung in einer von einer lichtquelle direkt Oden Mittels einer linse," *Physica* **1**, 201 (1934).

1.5. F. Zernike, "The concept of degree of coherence and its application to optical problems," *Physica* **5**, 785 (1938).

1.6. H. H. Hopkins, "The concept of partial coherence in optics," *Proc. Roy. Soc. A* **208**, 263 (1951).

1.7. H. H. Hopkins, "On the diffraction theory of optical image," *Proc. Roy. Soc. A* **217**, 408 (1953).

1.8. M. Born and E. Wolf, *Principles of Optics*, 2nd rev. ed., Pergamon Press, New York, 1964.

1.9. A. M. Yaglom, *An Introduction to the Theory of Stationary Random Functions*, Prentice-Hall, Englewood Cliffs, N.J., 1962.

1.10. J. M. Stone, *Radiation and Optics*, McGraw-Hill, New York, 1963.

1.11. B. J. Thompson and E. Wolf, "Two-beam interference with partially coherent light," *J. Opt. Soc. Am.* **47**, 895 (1957).

1.12. B. J. Thompson, "Illustration of the phase change in two-beam interference with partially coherent light," *J. Opt. Soc. Am.* **48**, 95 (1958).

1.13. M. J. Beran and G. B. Parrent, Jr., *Theory of Partial Coherence*, Prentice-Hall, Englewood Cliffs, N.J., 1964.

2

Partially Coherent Processing

Apparently signal processing was originated by a group of electrical engineers; their interest was originally focused at the electrical communication. Nonetheless, from the very beginning of the development of signal processing, the interest in its application to optical signal processing has never been totally absent. Recent advances in optical signal processing have made the relationship between electronic signal processing and optical signal processing more profound than ever.

During the past two and a half decades, optical signal processing has received increasing attention from a growing number of engineers and physicists. This increase in activity has stemmed in part from the discovery of a strong coherent source (i.e., the laser), and in part from a realization that optical configurations can be used to perform a wide variety of processing operations on signals. Although both coherent and noncoherent processing techniques have been employed for various analog processings, the former have generally been accepted as more versatile and useful. In coherent optical signal processings the phenomenon of diffraction is employed. If a lens is placed at a focal distance away from a transparency, the light at a focal length behind the lens is distributed according to a two-dimensional spectral analysis of the object transparency. By simple insertion of complex spatial filters in this spectral plane, a wide variety of signal processing operations can be performed. With a rather simple arrangement of the optical configuration, the output light distribution, affected by the spatial filter function, can be imaged.

We further note that optical configurations are capable of performing essentially any linear operation on a function of two variables. With the addition of cylindrical lenses to the configuration, the optical system can be converted to multichannel one-dimensional signal processing [2.1, 2.2].

2.1. MOTIVATION AND OBJECTIVE

Although coherent optical signal processors can perform a wide variety of complex signal processing operations, coherent processing systems are usually plagued with coherent artifact noise, which frequently limits their processing capability. As noted by the late D. Gabor, the Nobel Prize winner in physics in 1970 for his invention of holography, this coherent noise is the number one enemy of modern optical signal processing [2.3]. In addition to the problem of coherent artifact noise, there are further disadvantages arising from the expense of the coherent source and from the need for a special processing environment. For example, a heavy optical bench and a dust-free environment are generally required.

These difficulties have prompted us to look at optical processing from a new standpoint, and to consider whether it is necessary for all optical signal processing operations to be carried out by coherent sources. We have found that many optical signal processing operations can be carried out by a white-light source [2.4]. The basic advantages of white-light image processing are: (1) It can suppress the coherent artifact noise; (2) the white-light source is usually inexpensive; (3) the processing environment is generally very relaxed; (4) the white-light system is relatively easy and economical to operate; and (5) the white-light processor is particularly suitable for color signal processing.

One question that the reader may ask is why, since the white-light signal processor offers all these glamorous advantages, has it been ignored for so long? The answer is that it has been generally accepted that an incoherent source cannot process the optical wave field in complex amplitude. However, none of the practical light sources are strictly incoherent, not even a white-light source that includes the sunlight. In fact we have been able to utilize the partial coherence of a white-light source to perform a wide variety of complex amplitude processings. In other words, the white-light signal processor that we propose is capable of suppressing the coherent artifact noise like an incoherent processor, while still being able to process a signal in complex amplitude like a coherent processor.

2.2. SIGNAL PROCESSING UNDER STRICTLY COHERENT AND INCOHERENT REGIMES

Here we develop some equations for a general optical signal processing system under two different cases, namely completely incoherent and completely coherent illuminations [2.5, 2.6].

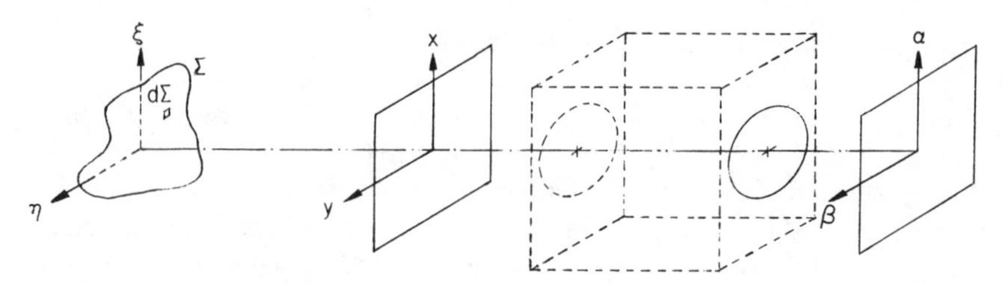

Figure 2.1. A hypothetical optical signal processing system. The black-box optical system is shown lying between the input or signal plane (x, y) and the output plane (α, β).

In Fig. 2.1, a hypothetical optical signal processing system is shown. Assume that the light emitted by the source Σ is monochromatic, and suppose that an output light distribution of the input signal is formed at the output plane of the optical system. In order to demonstrate the complex light distribution of the optical system, we let $u(x, y)$ be the complex light amplitude distribution at the input signal plane due to an incremental light source $d\Sigma$. If the complex amplitude transmittance of the input plane is $f(x, y)$, then the complex light field immediately behind the signal plane would be $u(x, y)f(x, y)$.

If it is assumed that the optical signal processor in the black box is linearly spatially invariant with a spatial impulse response of $h(x, y)$, then the complex light field at the output plane of the system due to $d\Sigma$ can be determined by the convolution equation

$$g(\alpha, \beta) = [u(x, y)f(x, y)] * h(x, y), \tag{2.1}$$

where the asterisk denotes the convolution operation.

At this point it may be emphasized that the assumption of linearity of the optical system is generally valid for small amplitude disturbances; however, the spatial-invariance condition may be applicable only over a small region of the signal plane.

From Eq. (2.1), the intensity distribution at the output plane due to $d\Sigma$ is

$$dI(\alpha, \beta) = g(\alpha, \beta)g^*(\alpha, \beta)d\Sigma, \tag{2.2}$$

where the superscript asterisk represents the complex conjugate. The overall output intensity distribution is therefore

$$I(\alpha, \beta) = \iint |g(\alpha, \beta)|^2 \, d\Sigma, \tag{2.3}$$

which can be written out in the following convolution integral:

$$I(\alpha, \beta) = \iiiint_{-\infty}^{\infty} \Gamma(x, y; x', y')h(\alpha - x, \beta - y)h^*(\alpha - x', \beta - y')$$

$$\cdot f(x, y)f^*(x', y')dx\, dy\, dx'\, dy', \tag{2.4}$$

where

$$\Gamma(x, y; x', y') = \iint_{\Sigma} u(x, y)u^*(x', y')d\Sigma, \tag{2.5}$$

is the spatial coherence function at input plane (x, y).

Let us now choose two arbitrary points Q_1 and Q_2 at the input signal plane. If r_1 and r_2 are the respective distances from Q_1 and Q_2 to $d\Sigma$, then the complex light disturbances at Q_1 and Q_2 due to $d\Sigma$ are

$$u_1(x, y) = \frac{[I(\xi, \eta)]^{1/2}}{r_1} \exp(ikr_1), \tag{2.6}$$

and

$$u_2(x', y') = \frac{[I(\xi, \eta)]^{1/2}}{r_2} \exp(ikr_2), \tag{2.7}$$

where $I(\xi, \eta)$ is the intensity distribution of the light source. By substituting Eqs. (2.6) and (2.7) in Eq. (2.5) we have

$$\Gamma(x, y; x', y') = \iint_{\Sigma} \frac{I(\xi, \eta)}{r_1 r_2} \exp[ik(r_1 - r_2)]d\Sigma. \tag{2.8}$$

In the paraxial case, $r_1 - r_2$ may be approximated by

$$r_1 - r_2 \simeq \frac{2}{r_1 + r_2}[\xi(x - x') + \eta(y - y')] \simeq \frac{1}{r}[\xi(x - x') + \eta(y - y')], \tag{2.9}$$

where r is the separation between the source plane and the signal plane. Then Eq. (2.8) can be reduced to

$$\Gamma(x, y; x', y') = \frac{1}{r^2} \iint I(\xi, \eta) \exp\left\{i\frac{k}{r}[\xi(x - x') + \eta(y - y')]\right\} d\xi\, d\eta, \tag{2.10}$$

which is known as the Van Cittert–Zernike theorem. Equation (2.10) is the inverse Fourier transform of the source intensity distribution.

Now one of the two extreme cases of the hypothetical optical signal processing system can be seen by letting the light source become infinitely large. If the intensity distribution of the light source is assumed uniform, that is if $I(\xi, \eta) \simeq K$, Eq. (2.10) becomes

$$\Gamma(x, y; x', y') = K_1 \delta(x - x', y - y'), \tag{2.11}$$

where K_1 is an appropriate positive constant. This equation describes a completely *incoherent* optical processing system.

On the other hand, if the light source is vanishingly small, that is, $I(\xi, \eta) \simeq K\delta(\xi, \eta)$, Eq. (2.10) becomes

$$\Gamma(x, y; x', y') = K_2, \tag{2.12}$$

where K_2 is a positive constant. This equation in fact describes a completely *coherent* optical processing system.

Referring to Eq. (2.4), for the completely incoherent case $[\Gamma(x, y; x', y') = K_1 \delta(x - x', y - y')]$, the intensity distribution at the output plane is

$$I(\alpha, \beta) = \iiiint_{-\infty}^{\infty} \delta(x' - x, y' - y) h(\alpha - x, \beta - y) \cdot h^*(\alpha - x', \beta - y') f(x, y) f^*(x', y') dx\, dy\, dx'\, dy', \tag{2.13}$$

which can be reduced to

$$I(\alpha, \beta) = \iint_{-\infty}^{\infty} |h(\alpha - x, \beta - y)|^2 |f(x, y)|^2\, dx\, dy. \tag{2.14}$$

It is clear from Eq. (2.14) that for the incoherent case the output intensity distribution is the convolution of the input signal intensity with respect to the impulse response irradiance. In other words, for the completely incoherent case, the optical signal processing system is linear in *intensity*, that is,

$$I(\alpha, \beta) = |h(x, y)|^2 * |f(x, y)|^2, \tag{2.15}$$

where the asterisk denotes the convolution operation.

By Fourier transformation, Eq. (2.15) can be expressed in the spatial frequency domain:

$$I(p, q) = [H(p, q) * H^*(p, q)][F(p, q) * F^*(p, q)], \tag{2.16}$$

where: $I(p, q)$, $H(p, q)$, and $F(p, q)$ are the Fourier transforms of $I(\alpha, \beta)$, $h(x, y)$, and $f(x, y)$, respectively; p and q are the spatial frequency coordinates, and the superscript asterisk represent the complex conjugate.

In a more convenient form, Eq. (2.15) can be written as

$$I(\alpha, \beta) = h_i(x, y) * f_i(x, y), \tag{2.17}$$

where $h_i(x, y) = |h(x, y)|^2$ and $f_i(x, y) = |f(x, y)|^2$ are the irradiances of the impulse response and the input signal, respectively. Then Eq. (2.16) may be written as

$$I(p, q) = H_i(p, q)F_i(p, q), \tag{2.18}$$

where $H_i(p, q)$ and $F_i(p, q)$ are the Fourier transforms of $h_i(x, y)$ and $f_i(x, y)$, respectively. In fact, $H_i(p, q)$ is given by

$$H_i(p, q) = \iint_{-\infty}^{\infty} h(x, y)h^*(x, y)\exp[-i(px + qy)]dx\, dy. \tag{2.19}$$

Then by the Fourier multiplication theorem, Eq. (2.19) becomes

$$H_i(p, q) = \frac{1}{4\pi^2} \iint_{-\infty}^{\infty} H(p', q')H^*(p' - p, q' - q)dp'\, dq', \tag{2.20}$$

which is the convolution of the complex transfer function with respect to its conjugate. On the other hand, for the completely coherent case $[\Gamma(x, y; x', y') = K_2]$, Eq. (2.4) becomes

$$I(\alpha, \beta) = g(\alpha, \beta)g^*(\alpha, \beta) = \iint_{-\infty}^{\infty} h(\alpha - x, \beta - y)f(x, y)dx\, dy$$

$$\cdot \iint_{-\infty}^{\infty} h^*(\alpha - x', \beta - y')f^*(x', y')dx'\, dy'. \tag{2.21}$$

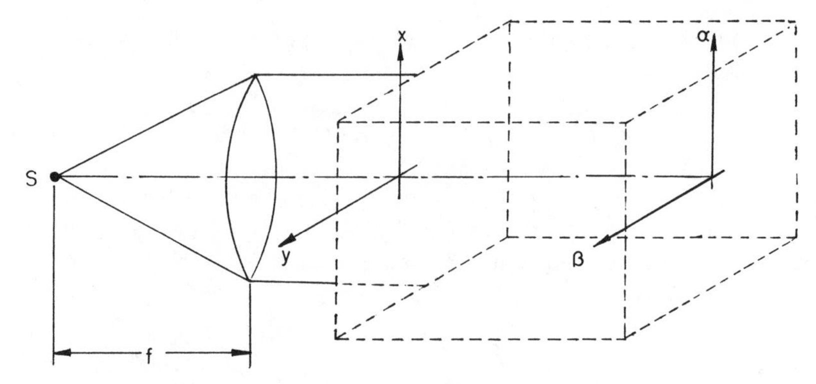

Figure 2.2. A simple optical configuration to achieve $\Gamma(x, y; x', y') = K$. A monochromatic point source is located at S.

From Eq. (2.21) it is apparent that the optical signal processing system is linear in *complex amplitude*, that is,

$$g(\alpha, \beta) = \iint_{-\infty}^{\infty} h(\alpha - x, \beta - y) f(x, y) dx\, dy. \tag{2.22}$$

Again by Fourier transformation, Eq. (2.22) becomes

$$G(p, q) = H(p, q) F(p, q), \tag{2.23}$$

where $G(p, q)$, $H(p, q)$, and $F(p, q)$ are the corresponding Fourier transforms of $g(\alpha, \beta)$, $h(x, y)$, and $f(x, y)$, respectively.

We further note that a coherence-preserving optical signal processing system that makes $\Gamma(x, y; x', y') = K$ (a constant) can be achieved by the configuration of Fig. 2.2. In this figure, a point source of monochromatic light is located at the front focal point of a positive lens. As a result, a collimated monochromatic plane wave illuminates the input signal plane (x, y) of the optical processing system.

2.3. SIGNAL PROCESSING UNDER PARTIALLY COHERENT REGIME

There are three fold limitations associated with strictly coherent optical signal processing. First, coherent optical signal processing requires that the dynamic range of a spatial light modulator be about 100000:1 for the input wave intensity, or a photographic density range of 4.0. Such a dynamic range is often quite difficult or impossible to achieve in practice. Second, coherent optical processing systems are susceptible to coherent artifact noise, which frequently limits their processing capabilities. Third, in coherent optical signal processing, the signal being processed is carried by a complex wave amplitude distribution. However, what is actually measured at the output plane of the optical signal processor is the output wave intensity. The loss of the output phase distribution may seriously limit the applicability of the optical processing technique in some applications.

To overcome the drawbacks of coherent optical signal processing it is probably effective to reduce either the temporal coherence or the spatial coherence, and there are several techniques available that use that approach. In the following sections we briefly mention three frequently used concepts. The first method is based on a spatially partially coherent source, and the other two methods utilize a broad spectral band light source (e.g., a white-light source) to perform complex signal processing operations.

2.3.1. Spatially Partially Coherent Processing

A technique of utilizing a spatially partially coherent illumination to perform complex data processing has been proposed by Lohmann [2.7], Rhodes [2.8], and Stoner [2.9]. These techniques share a basic concept: The optical processing system is characterized by the use of an impulse response or point spread function (PSF), which is not constrained to the class of nonnegative real functions used in conventional incoherent processing. The output intensity distribution can be controlled by changing the PSF, that is,

$$I(x', y') = \iint O(x, y)h(x', y'; x, y)dx\, dy, \tag{2.24}$$

where $h(x', y'; x, y)$ is the PSF, and $I(x', y')$ and $O(x, y)$ are the image and object intensity distributions, respectively. If the pupil function of an optical system is $P(\alpha, \beta)$, the PSF is equal to the square of the Fourier spectrum of the pupil function. Therefore, the output intensity distribution can actually be adjusted by selecting an adequate pupil function. A typical processing system, as proposed by Rhodes, is shown in Fig. 2.3. This optical processing system, which is characterized by an extended pupil region, has two input pupil functions, $P_1(\alpha, \beta)$ and $P_2(\alpha, \beta)$. With the optical path lengths of the two arms of the system being equal, the overall system pupil function is given by the sum of $P_1(\alpha, \beta)$ and $P_2(\alpha, \beta)$. If the path length in one arm is changed slightly,

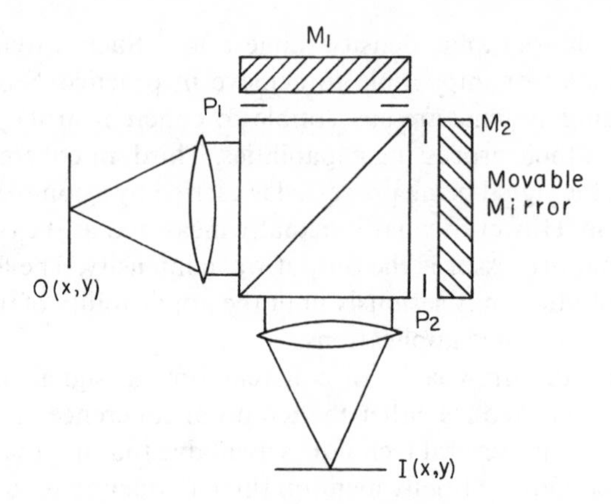

Figure 2.3. Two-pupil incoherent processing system.

however, for example, by moving mirror M_2 a small distance, a phase factor is introduced in one of the component pupil functions, with the result,

$$P(\alpha, \beta) = P_1(\alpha, \beta) + P_2(\alpha, \beta)\exp\{i\phi(\alpha, \beta)\}. \tag{2.25}$$

It is therefore evident that, if the pupil transparencies $P_1(\alpha, \beta)$ and $P_2(\alpha, \beta)$ are recorded holographically, arbitrary PSF's can then be synthesized using the 0°–180° phase switching operation. Thus we see that this optical system is capable of performing complex data processing under spatially partially coherent illumination.

2.3.2. Achromatic Optical Processing

An achromatic optical signal processing technique with broad-band source (e.g., white light) was proposed by Leith, Roth, and Swanson [2.10, 2.11], as shown in Fig. 2.4. Lens L_0 collimates the white-light point source S illumina-

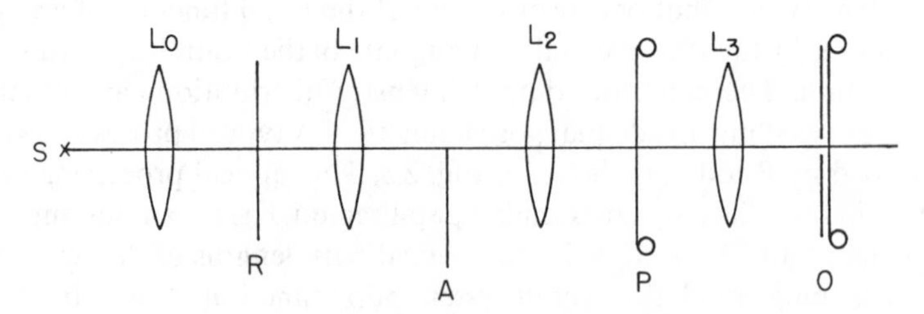

Figure 2.4. Achromatic optical processing system.

ting hologram R. The diffraction wavefront is spatially filtered by the aperture A, which removes the zero-order term and the lower sideband. The upper sideband is imaged onto the signal plane P; lens L_3 Fourier transforms the resulting wavefront; and observation in the transform plane is confined to the optical axis by a slit in the output plane O. The convolution of the demodulated input signal with the desired reference function is recorded by synchronously translating the signal and output films. Such a system, termed an achromatic system, thus has the flexibility of a coherent optical signal processor along with the potential for the noise immunity of an incoherent system.

2.3.3. Partially Coherent Processing with Grating Technique

We have introduced an achromatic partially coherent optical signal processing technique with a white-light source [2.12, 2.13]. Figure 2.5 illustrates this processing system, where all the transform lenses are achromatic. A high diffraction efficiency phase grating with an angular spatial frequency p_0 is used at the input plane to disperse the input signal spectrum into rainbow colors in the Fourier plane P_2, permitting a stripwise design of a complex spatial filter for each narrow spectral band in the Fourier plane. The achromatic output image irradiance can therefore be observed at the output plane P_3. The advantages of this technique are that each channel (e.g., each spectral band filter) behaves as a coherent processing channel, while the overall output noise performance is primarily due to the incoherent addition of each channel, which behaves as if under incoherent illumination.

2.3.4. Some Remarks

As we mentioned earlier, partially coherent optical processing systems generally exhibit a much better noise performance. Also, since a broad spectral band light source is utilized, the processor affords the advantage of being able to process color signals. On the other hand, a strictly coherent or incoherent

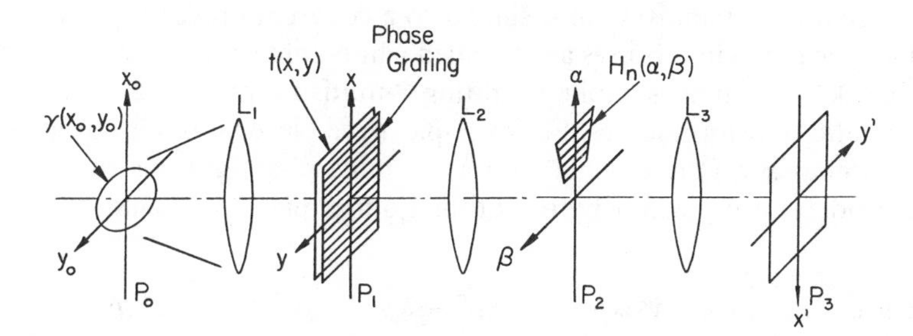

Figure 2.5. Partially coherent optical processing with grating technique.

optical field is a mathematical idealization. An optical field that occurs in practice consists of a very limited degree of coherence. However, the electromagnetic radiation from a real physical source is never strictly monochromatic. In reality, a physical source cannot be a point; rather it must have finite dimensions consisting of many elementary radiations. Thus a real optical signal processor actually performs various tasks under partially coherent illumination, rather than under completely coherent or completely incoherent illumination. Therefore, the investigation of the performance of a partially coherent optical signal processor should be quite important and interesting. There are generally three basic problems that we would investigate, as stated in the following.

First, do all the information processing operations require a strictly coherent source? In fact, there are several optical processing operations that can be carried out with reduced coherence, and thus we need to investigate the coherence requirements for partially coherent optical systems.

Second, the transfer function is usually used for evaluating the quality of an optical signal processor. It is necessary to determine the relationship between the transfer function and the degree of coherence. Consequently, the concept of the transfer function can also be adopted as a valuable criterion for selecting an appropriate partially coherent light source for a specific signal processing operation.

Third, the noise immunity of an optical signal processor is directly related to the partially coherent illumination. A fruitful study of the noise performance of a system can be made if we look into its relation to both the temporal and the spatial coherence illuminations.

2.4. PARTIALLY COHERENT PROCESSING WITH WHITE-LIGHT SOURCE

We now describe a partially coherent optical signal processing technique that can be carried out by a white-light source, as illustrated in Fig. 2.6. The white-light signal processing system is similar to a coherent processing system, except for the following: It uses an extended white-light source, a source encoding mask, a signal sampling grating, multispectral band filters, and achromatic transform lenses. For example, if we place an input object signal transparency $s(x, y)$ in contact with a sampling phase grating, the complex wave field, for every wavelength λ, at the Fourier plane P_2 would be

$$E(p, q; \lambda) = \iint s(x, y)\exp(ip_0 x)\exp[-i(px + qy)]dx\, dy = S(p - p_0, q), \tag{2.26}$$

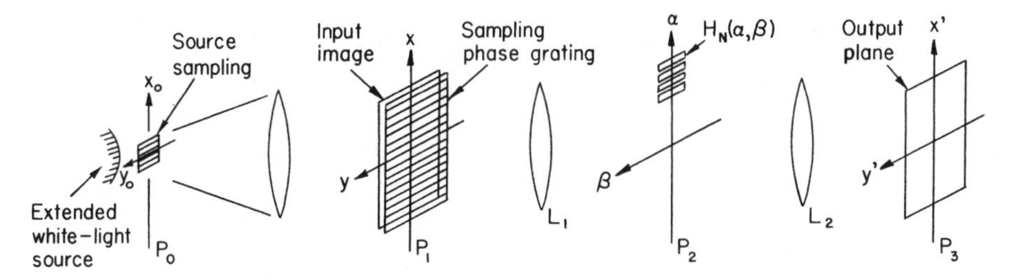

Figure 2.6. A white-light optical signal processor.

where we assume a small pinhole source encoding mask, the integral is over the spatial domain of the input plane P_1, (p, q) denotes the angular spatial frequency coordinate system, p_0 is the angular spatial frequency of the sampling phase grating, and $S(p, q)$ is the Fourier spectrum of $s(x, y)$. If we write Eq. (2.26) in the form of a spatial coordinate system (α, β), we have

$$E(\alpha, \beta; \lambda) = S\left(\alpha - \frac{\lambda f}{2\pi} p_0, \beta\right), \tag{2.27}$$

where $p = (2\pi/\lambda f)\alpha$, $q = (2\pi/\lambda f)\beta$, and f is the focal length of the achromatic transform lens. Thus we see that the Fourier spectra would disperse into rainbow color along the α-axis, and each Fourier spectrum for a given wavelength λ is centered at $\alpha = \pm(\lambda f/2\pi)p_0$.

In optical signal filtering, we assume that a set of narrow spectral band complex spatial filters is available. In practice, all the input signals are spatial frequency limited; the spatial bandwidth of each spectral band filter $H(p_n, q_n)$ is therefore

$$H(p_n, q_n) = \begin{cases} H(p_n, q_n), & \alpha_1 < \alpha < \alpha_2, \\ 0, & \text{otherwise}, \end{cases} \tag{2.28}$$

where $p_n = (2\pi/\lambda_n f)\alpha$, $q_n = (2\pi/\lambda_n f)\beta$, λ_n is the main wavelength of the filter, $\alpha_1 = (\lambda_n f/2\pi)(p_0 + \Delta p)$ and $\alpha_2 = (\lambda_n f/2\pi)(p_0 - \Delta p)$ are the upper and the lower spatial limits of $H(p_n, q_n)$, and Δp is the spatial bandwidth of the input image $s(x, y)$.

Since the limiting wavelengths of each $H(p_n, q_n)$ are

$$\lambda_h = \lambda_n \frac{p_0 + \Delta p}{p_0 - \Delta p}, \qquad \text{and} \qquad \lambda_l = \lambda_n \frac{p_0 - \Delta p}{p_0 + \Delta p}, \tag{2.29}$$

the spectral bandwidth of each $H(p_n, q_n)$ can be approximated by

$$\Delta\lambda_n = \lambda_n \frac{4p_0 \Delta p}{p^2 - (\Delta p)^2} \simeq \frac{4\Delta p}{p_0} \lambda_n. \tag{2.30}$$

If we place this set of spectral band filters side-by-side and position them properly over the smeared Fourier spectra, the intensity distribution of the output light field can be shown as

$$I(x, y) \simeq \sum_{n=1}^{N} \Delta\lambda_n |s(x, y; \lambda_n) * h(x, y; \lambda_n)|^2, \tag{2.31}$$

where $h(x, y; \lambda_n)$ is the spatial impulse response of $H(p_n, q_n)$ and $*$ denotes the convolution operation. Thus the proposed white-light processor is capable of processing the signal in complex amplitude. Since the output intensity is the sum of the mutually incoherent narrow-band spectral irradiances, the annoying coherent artifact noise can be suppressed. It is also apparent that the white-light source emanates all the visible wavelengths; thus the white-light processor is very suitable for color signal or image processing.

We should, however, emphasize that in white-light signal processing we would approach the processing operation from a different standpoint. For example, if signal filtering is two-dimensional (e.g., a two-dimensional correlation operation), we would synthesize a set of narrow spectral band filters for each λ_n for the entire smeared Fourier spectra, as illustrated in Fig. 2.7. On the other hand, if the image filtering is one-dimensional (e.g., deblurring due to linear motion), a broad-band fan-shaped spatial filter, to accommodate the scale variation due to wavelength, can be utilized as illustrated in Fig. 2.8.

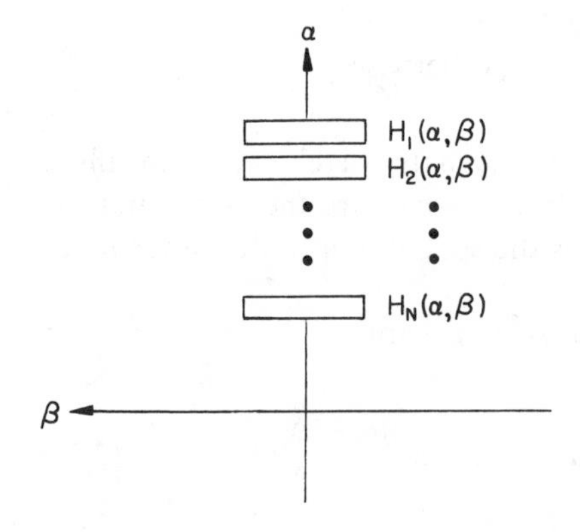

Figure 2.7. A set of narrow spectral band filters.

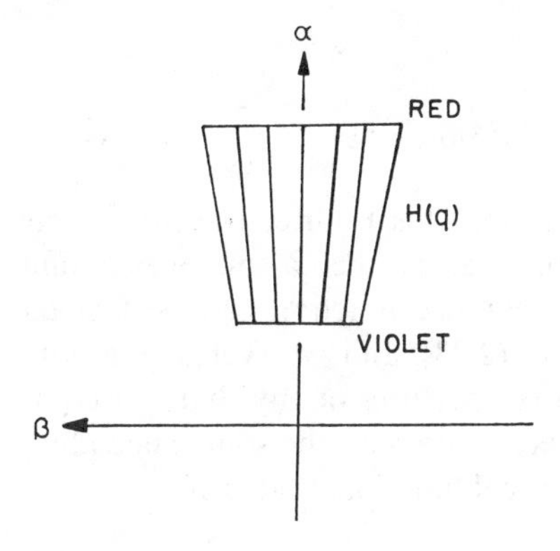

Figure 2.8. A broad-band fan-shaped filter.

There is, however, a temporal coherence requirement imposed on the signal filtering in the Fourier plane. Since the scale of the Fourier spectrum varies with wavelength, a temporal coherence requirement should be imposed on each narrow spectral band filter in the Fourier plane. Thus $H(p_n, q_n)$ should be satisfied by the following temporal coherence condition:

$$\frac{\Delta\lambda_n}{\lambda_n} = \frac{4\Delta p}{P_0} \ll 1. \tag{2.32}$$

Thus we see that a high degree of temporal coherence is attainable in the Fourier plane, by simply increasing the spatial frequency of the sampling grating. Needless to say, this temporal coherence requirement of Eq. (2.32) can also be applied for the broad-band fan-shaped filter of Fig. 2.8.

There is also a spatial coherence requirement imposed on the input plane of the white-light signal processor. The spatial coherence function at the input plane can be shown as [2.14, 2.15]

$$\Gamma(\mathbf{x} - \mathbf{x}') = \iint \gamma(\mathbf{x}_0) \exp\left[i2\pi \frac{\mathbf{x}_0}{\lambda f} (\mathbf{x} - \mathbf{x}') \right] d\mathbf{x}_0, \tag{2.33}$$

which essentially is the Van Cittert–Zernike theorem [2.16, 2.17] where $\gamma(\mathbf{x}_0)$ denotes the intensity distribution of the source encoding function.

From the above equation, we see that the spatial coherence and source encoding functions form a Fourier transform pair, that is,

$$\gamma(\mathbf{x}_0) = \mathscr{F}[\Gamma(\mathbf{x} - \mathbf{x}')], \tag{2.34}$$

and

$$\Gamma(\mathbf{x} - \mathbf{x}') = \mathscr{F}^{-1}[\gamma(\mathbf{x}_0)], \tag{2.35}$$

where $\mathscr{F}$ denotes the Fourier transformation. This Fourier transform pair implies that, if a spatial coherence function is given, then a source encoding function corresponding to this spatial coherence function can be formed through the Fourier transformation of Eq. (2.23), and vice versa. We note that source encoding function can consist of apertures of any shape or complicated gray-scale transmittance. However, in practice, the source encoding is only realizable if it is a bounded positive real function, that is, if

$$0 \leqq \gamma(\mathbf{x}_0) \leqq 1. \tag{2.36}$$

Furthermore, in white-light optical signal processing, we would search for a reduced spatial coherence function for the signal processing operation. With reference to the calculated reduced spatial coherence function, a source encoding function that satisfies the physical realizability condition of Eq. (2.36) can be obtained. One of the basic objectives of the source encoding is to alleviate the spatial coherence constraints of an extended source, so that signal processing can be carried out by a physical white-light source. In other words, it is possible to improve the effective utilization of the light source through a source encoding.

Before investigating the behavior of a white-light optical signal processor, it is necessary to establish a transformational relationship of the mutual coherence function (or correlation function), which determines the degree of coherence under partially coherent illumination. We use Wolf's theory [2.14] of partially coherent light to develop a transformational formula for the mutual coherence function propagating through an ideal thin lens. We then apply this transformational formula to derive an operational formula for the white-light optical signal processor that we have proposed.

2.5. FORMULATION OF PARTIALLY COHERENT FUNCTION

One of the most remarkable and useful properties of a converging lens is its inherent ability to perform the two-dimensional Fourier transformation of a complex wave field. For partially coherent illumination, we show that a four-dimensional Fourier transformation of the mutual coherence function can be established by an ideal thin lens.

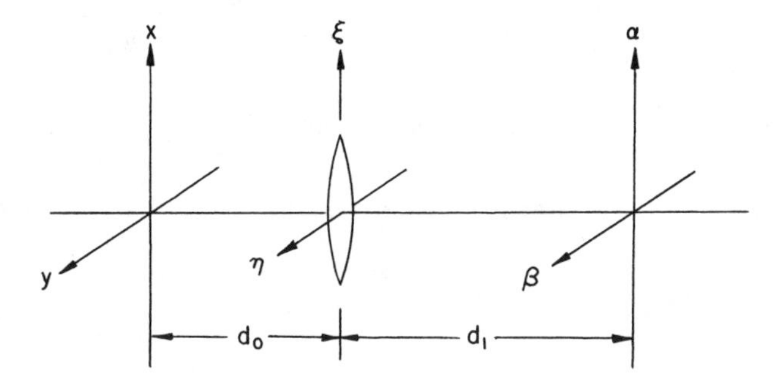

Figure 2.9. A partial coherence optical system.

It is obvious that a monochromatic plane wave passing through a thin lens suffers a phase delay transformation, such as [2.2]

$$T(\xi, \eta) = \exp(ik\eta\Delta_0)\exp\left[\frac{-ik(\xi^2 + \eta^2)}{2f}\right], \tag{2.37}$$

where η is the refractive index of the thin lens, f is the focal length, Δ_0 is the maximum thickness of the lens, $k = 2\pi/\lambda$, λ is the wavelength, and (ξ, η) is the spatial coordinate system of the thin lens.

Let us now consider an object transparency inserted at a distance d_0 in front of the lens and illuminated by a spatially partially coherent light. If the mutual coherence function at the object plane is $\Gamma_1(x_1, y_1; x_2, y_2)$ as depicted in Fig. 2.9, then the mutual coherence function at the output (α, β) plane can be evaluated. The mutual coherence function at the front of the lens can be written as

$$\Gamma_2(\xi_1, \eta_1; \xi_2, \eta_2)$$
$$= -\frac{1}{\lambda^2 d_0^2} \iiiint_{-\infty}^{\infty} \Gamma_1(x_1, y_1; x_2, y_2)\exp\left\{i\frac{k}{2d_0}[(\xi_1 - x_1)^2 + (\eta_1 - y_1)^2]\right\}$$
$$\cdot\exp\left\{-i\frac{k}{2d_0}[(\xi_2 - x_2)^2 + (\eta_2 - y_2)^2]\right\}dx_1\,dy_1\,dx_2\,dy_2, \tag{2.38}$$

so the mutual coherence function immediately behind the lens would be

$$\Gamma_3(\xi_1, \eta_1; \xi_2, \eta_2) = \Gamma_2(\xi_1, \eta_1; \xi_2, \eta_2) T(\xi_1, \eta_1) T^*(\xi_2, \eta_2). \tag{2.39}$$

It is apparent that the mutual coherence function at the output plane can therefore be written as

$$\begin{aligned}\Gamma_4(\alpha_1, \beta_1; \alpha_2, \beta_2) &= -\frac{1}{\lambda^2 d_1^2} \iiiint_{-\infty}^{\infty} \Gamma_3(\xi_1, \eta_1; \xi_2, \eta_2) \\ &\cdot \exp\left\{ i \frac{k}{2d_1} [(\alpha_1 - \xi_1)^2 + (\beta_1 - \eta_1)^2] \right\} \\ &\cdot \exp\left\{ -i \frac{k}{2d_1} [(\alpha_2 - \xi_2)^2 + (\beta_2 - \eta_2)^2] \right\} d\xi_1\, d\eta_1\, d\xi_2\, d\eta_2 .\end{aligned} \tag{2.40}$$

By further substituting Eqs. (2.37) and (2.39) into Eq. (2.40) we have

$$\begin{aligned}\Gamma_4(\alpha_1, \beta_1; \alpha_2, \beta_2) &= \frac{1}{4d_0^2 d_1^2} \exp\left\{ i \frac{k}{d_1} [(\alpha_1^2 + \beta_1^2) - (\alpha_2^2 + \beta_2^2)] \right\} \\ &\iiiint_{-\infty}^{\infty} \Gamma_1(x_1, y_1; x_2, y_2) \exp\left\{ i \frac{k}{d_0} [(x_1^2 + y_1^2) - (x_2^2 + y_2^2)] \right\} \\ &\left[\iiiint_{-\infty}^{\infty} \exp\left\{ ik \left[\left(\frac{1}{d_1} + \frac{1}{d_0} - \frac{1}{f} \right) (\xi_1^2 + \eta_1^2 - \xi_2^2 - \eta_2^2) - 2\xi_1 \left(\frac{\alpha_1}{d_1} + \frac{x_1}{d_0} \right) \right.\right.\right. \\ &- 2\eta \left(\frac{\beta_1}{d_1} + \frac{y_1}{d_0} \right) + 2\xi_2 \left(\frac{\alpha_2}{d_1} + \frac{x_2}{d_0} \right) \\ &\left.\left.\left. + 2\eta_2 \left(\frac{\beta_2}{d_1} + \frac{y_2}{d_0} \right) \right] \right\} d\xi_1\, d\eta_1\, d\xi_2\, d\eta_2 \right] dx_1\, dy_1 dx_2\, dy_2 .\end{aligned} \tag{2.41}$$

We further note that

$$\int_{-\infty}^{\infty} \exp\left\{ik\left[\left(\frac{1}{d_1}+\frac{1}{d_0}-\frac{1}{f}\right)\xi_1^2 - 2\xi_1\left(\frac{\alpha_1}{d_1}+\frac{x_1}{d_0}\right)\right]\right\}d\xi_1$$

$$= \exp\left\{\frac{-ik(\alpha_1/d_1 + x_1/d_0)}{1/d_1 + 1/d_0 - 1/f}\right\}\int_{\infty}^{\infty}\exp\left\{ik\left(\frac{1}{d_1}+\frac{1}{d_0}-\frac{1}{f}\right)\right.$$

$$\left.\cdot\left[\xi_1 - \frac{\alpha_1/d_1 + x_1/d_0}{1/d_1 + 1/d_0 - 1/f}\right]^2\right\}d\xi_1 = \frac{1}{2}\left[\frac{\pi}{ik(1/d_1 + 1/d_0 - 1/f)}\right]^{1/2}$$

$$\cdot\exp\left\{\frac{-ik(1/d_1 + x_1/d_0)^2}{1/d_1 + 1/d_0 - 1/f}\right\}, \tag{2.42}$$

and similar formulas can also be written for the η_1, ξ_2, and η_2 coordinates. Therefore the transformational formula for the mutual coherence function at the output plane is

$$\Gamma_4(\alpha_1, \beta_1; \alpha_2, \beta_2)$$

$$= C\exp\left\{i\frac{k}{2d_1}\left[1 - \frac{1}{\alpha_1(1/d_1 + 1/d_0 - 1/f)}\right][(\alpha_1^2 + \beta_1^2) - (\alpha_2^2 + \beta_2^2)]\right\}$$

$$\iiiint_{-\infty}^{\infty}\Gamma_1(x_1, y_1; x_2, y_2)$$

$$\cdot\exp\left\{i\frac{k}{2d_0}\left[1 - \frac{1}{d_0(1/d_1 + 1/d_0 - 1/f)}\right][(x_1^2 + y_1^2) - (x_2^2 + y_2^2)]\right\}$$

$$\cdot\exp\left\{-ik\frac{(\xi_1 x_1 + \eta_1 y_1) - (\xi_2 x_2 + \eta_2 y_2)}{d_0 d_1(1/d_1 + 1/d_0 - 1/f)}\right\}dx_1\,dy_1\,dx_2\,dy_2, \tag{2.43}$$

where C is an appropriate constant. This equation illustrates that the output mutual coherence function can be calculated by a four-dimensional integral function.

To this point we have neglected the finite extent of the lens aperture. Such an approximation is an accurate one if the distance d_0 is sufficiently small to place the input transparency deep within the region of Fresnel diffraction with respect to the lens aperture. This condition is satisfied in the vast majority of problems of interest, particularly for optical signal processing. The limitation of the effective input object by the finite lens aperture is known as a vignetting effect. The vignetting effect in the input space can be minimized if the input object is placed close to the lens or if the lens aperture is much larger than the input object. In practice it is often preferred to place the object transparency directly against the lens to minimize vignetting.

There are, however, two special cases of Eq. (2.43) worth mentioning:

1. For $d_1 = f$, the output plane is located at the back focal plane. Equation (2.43) reduces to

$$\Gamma_4(\alpha_1, \beta_1; \alpha_2, \beta_2) = C \exp\left\{i \frac{k}{2f}\left(1 - \frac{d_0}{f}\right)[(\alpha_1^2 + \beta_1^2) - (\alpha_2^2 + \beta_2^2)]\right\}$$

$$\cdot \iiiint_{-\infty}^{\infty} \Gamma_1(x_1, y_1; x_2, y_2)$$

$$\cdot \exp\left\{-i \frac{k}{2f}[(\alpha_1 x_1 + \beta_1 y_1) - (\alpha_2 x_2 + \beta_2 y_2)]\right\} dx_1\, dy_1\, dx_2\, dy_2. \tag{2.44}$$

Thus we see that, except for a quadratic phase factor, the output mutual coherence function is essentially the Fourier transform of the input mutual coherence function.

2. For $d_0 = d_1 = f$, both input and output plane are located at the front and back focal length of the lens. The quadratic phase factor vanishes and Eq. (2.44) reduces to

$$\Gamma_4(\alpha_1, \beta_1; \alpha_2, \beta_2) = C \iiiint_{-\infty}^{\infty} \Gamma_1(x_1, y_1; x_2, y_2)$$

$$\cdot \exp\left\{-i \frac{k}{2f}[(\alpha_1 x_1 + \beta_1 y_1) - (\alpha_2 x_2 + \beta_2 y_2)]\right\} dx_1\, dy_1\, dx_2\, dy_2, \tag{2.45}$$

which is exactly a four-dimensional Fourier transformation, between the input and output mutual coherence functions.

In partially coherent optical processing (e.g., white-light processing) an extended incoherent source is usually used at the front focal plane of a collimating lens to illuminate the input object transparency. Thus the mutual coherence function at the source plane (x_0, y_0) can be written as

$$\Gamma(x'_0, y'_0; x''_0, y''_0) = \begin{cases} \gamma(x_0, y_0), & \text{for } x'_0 = x''_0 = x_0, \, y'_0 = y''_0 = y_0, \\ 0, & \text{otherwise,} \end{cases} \tag{2.46}$$

where $\gamma(x_0, y_0)$ is the intensity distribution of the incoherent light source, and (x_0, y_0) is the spatial coordinate system of the source plane. If we assume that the input object transparency is located at the back focal plane of the collimator, then the mutual coherence function at the input plane reduces to

$$\Gamma_1(x_1 - x_2; y_1 - y_2)$$
$$= \iint_{\infty}^{\infty} \gamma(x_0, y_0) \exp\left\{ -i \frac{k}{2f} [(x_1 - x_2)x_0 + (y_1 - y_2)y_0] \right\} dx_0 \, dy_0, \tag{2.47}$$

which is essentially the Van Cittert–Zernike theorem [2.16, 2.17]. It is also evident that the mutual coherence function at the input plane, illuminated by an extended incoherent source, is a space invariant function.

2.6. COHERENCE MEASUREMENT FOR A WHITE-LIGHT OPTICAL PROCESSOR

Van Cittert and Zernike [2.16, 2.17] showed a profound Fourier transform relationship between the spatial coherence and the irradiance of an extended source. It is, however, the two-beam interferometric technique of Thompson and Wolf [2.18, 2.19] that demonstrates the degree of coherence measurement. In this section we devise a dual-beam achromatic interference technique to measure the coherence of a white-light signal processor [2.20]. We show that a high degree of coherence can be achieved in the white-light processor, such that the input signal can be processed in a complex wave field.

2.6.1. Measurement Technique

We now describe a visibility measurement to determine the degree of coherence in the β-direction at the Fourier plane of the white-light signal processor shown in Fig. 2.6. For simplicity we utilize a narrow slit as a one-dimensional

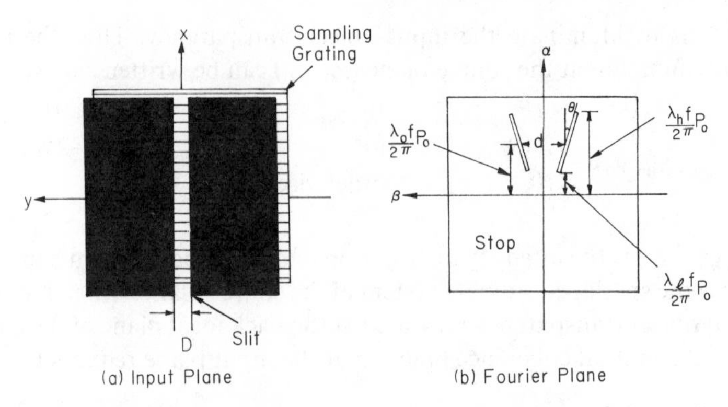

Figure 2.10. Visibility measurement along the β-direction. (*a*) Input plane. (*b*) Fourier plane. D, input slit width; d, mean slit separation; λ_0, mean wavelength of the light source; p_0, angular spatial frequency of the sampling grating.

object at the input plane, as shown in Fig. 2.10*a*. The complex light distribution at the Fourier plane can be written as

$$E(\alpha, \beta; \lambda) = C\left[\operatorname{sinc}\left(\frac{\pi D}{\lambda f}\beta\right)\right] * \delta\left(\alpha - \frac{\lambda f}{2\pi}p_0, \beta\right), \tag{2.48}$$

where C represents an appropriate complex constant, D is the slit width, f is the focal length of the achromatic transform lens, p_0 is the angular spatial frequency of the phase sampling grating, and $*$ denotes the convolution operation.

It is clear that Eq. (2.48) describes a fan-shaped smeared Fourier spectra of the input slit, for which the scale of the sinc factor increases as a function of the wavelength λ and decreases as the size of the object D (i.e., slit width) increases. To increase the efficiency of the visibility measurement along the β-direction, we would use a pair of slanted narrow slits at the smeared Fourier spectra, as illustrated in Fig. 2.10*b*. The inclination angle of this pair of slits should be adjusted with the separation of the slits, such as

$$\theta = \tan^{-1}\left(\frac{2\pi d}{\lambda_0 p_0}\right), \tag{2.49}$$

where d is the mean separation of the slits, λ_0 is the mean wavelength of the light source, and p_0 is the angular spatial frequency of the sampling grating. The transfer function of this pair of slanted slits can be written as

$$H(\alpha, \beta) = \left[\delta\left(\beta - \frac{d\lambda}{2\lambda_0}\right) + \delta\left(\beta + \frac{d\lambda}{2\lambda_0}\right)\right] * \delta\left(\alpha - \frac{\lambda f}{2\pi}p_0, \beta\right). \tag{2.50}$$

The output intensity distribution can be shown as

$$I(x', y') = K[1 + \cos(2\pi\, dx')], \tag{2.51}$$

where K is an appropriate proportionality constant. Equation (2.51) represents an achromatic fringe pattern of the entire spectral band of the light source.

Strictly speaking, all practical white-light sources are extended sources. For simplicity of illustration, we assume that an extended *square* source is used. Thus the output intensity distribution can be written as

$$I(x', y') = \int_{\lambda_l}^{\lambda_h} \iint_{-\infty}^{\infty} \operatorname{rect}\left(\frac{x_0}{a}\right) \cdot \operatorname{rect}\left(\frac{y_0}{a}\right)$$

$$\cdot \left| \iint_{-\infty}^{\infty} \operatorname{sinc}\left[\frac{\pi D}{\lambda f}(\beta + y_0)\right] * \delta\left(\alpha + x_0 - \frac{\lambda f}{2\pi} p_0, \beta\right)\right.$$

$$\left. \cdot H(\alpha, \beta) \exp\left[-i\frac{2\pi}{\lambda f}(x'\alpha + y'\beta)\right] d\alpha\, d\beta \right|^2 dx_0\, dy_0\, d\lambda, \tag{2.52}$$

where $H(\alpha, \beta)$ is given by Eq. (2.50),

$$\operatorname{rect}\left(\frac{x_0}{a}\right) \triangleq \begin{cases} 1, & |x_0| \leq a/2, \\ 0, & |x_0| > a/2, \end{cases}$$

and a is the dimension of the extended light source. From this equation, we see that the visibility (i.e., degree of coherence) is dependent upon the source size a, the object size D (i.e., slit width), and the angular spatial frequency p_0 of the sampling grating.

To investigate the degree of coherence variation in the α-direction, again we insert a narrow slit aperture as an input object, but position it parallel to the sampling direction of the sampling grating, as shown in Fig. 2.11*a*. The smeared Fourier spectra can be written as

$$E(\alpha, \beta; \lambda) = \operatorname{sinc}\left(\frac{\pi D}{\lambda f}\alpha\right) * \delta\left(\alpha - \frac{\lambda f}{2\pi} p_0, \beta\right), \tag{2.53}$$

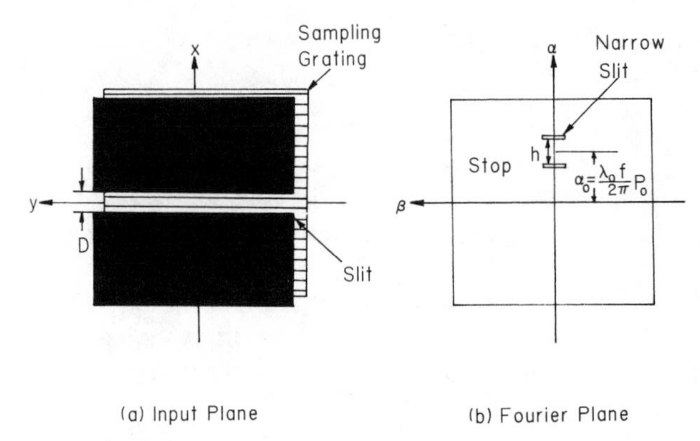

Figure 2.11. Visibility measurement along the α-direction. (*a*) Input plane. (*b*) Fourier plane. *D*, input slit width; *h*, slit separation; α_0, center location of the pair of narrow slits; λ_0, mean wavelength of the light source; p_0, angular spatial frequency of the sampling grating.

which describes a narrow smeared rainbow color spectra along the α-axis. For the visibility measurement, a pair of narrow slits is inserted at the Fourier plane perpendicular to the α-axis and centered at $\alpha = \lambda_0 f p_0/(2\pi)$, as shown in Fig. 2.11*b*, where λ_0 is the center wavelength of the light source and f is the focal length. The transfer function of this pair of slits can be written as

$$H(\alpha, \beta) = \delta\left(\alpha - \frac{h}{2} - \frac{\lambda_0 f}{2\pi} p_0, \beta\right) + \delta\left(\alpha + \frac{h}{2} - \frac{\lambda_0 f}{2\pi} p_0, \beta\right), \tag{2.54}$$

where h is the separation between the slits. Again with the assumption of a *square* light source, the output intensity distribution can be shown as

$$I(x', y') = \int_{\lambda_l}^{\lambda_h} \iint_{-\infty}^{\infty} \operatorname{rect}\left(\frac{x_0}{a}\right)\operatorname{rect}\left(\frac{y_0}{a}\right) \cdot \left| \iint_{-\infty}^{\infty} \operatorname{sinc}\left[\frac{\pi D}{\lambda f}(\alpha + x_0)\right] * \delta\left(\alpha + \frac{h}{2} - \frac{\lambda f}{2\pi} p_0, \beta + y_0\right) \cdot H(\alpha, \beta)\exp\left[-i\frac{2\pi}{\lambda f}(x'\alpha + y'\beta)\right] d\alpha\, d\beta \right|^2 dx_0\, dy_0\, d\lambda, \tag{2.55}$$

where $H(\alpha, \beta)$ is defined by Eq. (2.54). From this equation, again we see that the degree of coherence in the α-direction is dependent on the source size a, the object size D, and the angular spatial frequency p_0 of the sampling grating.

The optical setup for the measurement of the degree of coherence at the Fourier plane of a white-light optical image processor is shown in Fig. 2.12. This setup utilizes the principle of a dual-beam interference technique (see Thompson and Wolf [2.18]) for coherence measurement. The output interference fringe pattern can be traced by a linear scanning photometer and displayed on an oscilloscope for the visibility measurement. In the experiment, the photometer is made by mounting a photomultiplier on the top of a motor-driven linear translator. Since it is a one-dimensional fringe pattern, a narrow slit can be utilized at the input end of the photomultiplier for the visibility measurement, that is,

$$V \triangleq \frac{I_{max} - I_{min}}{I_{max} + I_{min}}, \tag{2.56}$$

where I_{max} and I_{min} are the maximum and minimum irradiances of the fringes. And the visibility of the fringes is in fact a measure of the degree of coherence measurement (see Section 1.1), that is,

$$V = |\gamma|, \tag{2.57}$$

where γ is the complex degree of coherence.

In the following we illustrate the visibility (i.e., coherence) measurement in the β- and in the α-directions, at the Fourier plane.

2.6.2. Coherence Measurement in the β -Direction

We now describe the coherence measurement in the β-direction, as illustrated in Fig. 2.12. We show that the visibility varies as a function of the mean separation of the slanted slits d, the source size a, the object size D (i.e., slid width), and the spatial frequencies of the sampling grating. Figure 2.13 shows the variation of the degree of coherence as a function of the main separation d for various values of source sizes a. From this figure we see that the degree of coherence decreases as the separation d increases. Further increase in d increases the reappearance of the coherence. Still further increase in d causes the repeated fluctuation of visibility. In this figure, we also see that the degree of coherence increases as the source size a decreases. There is an interesting phenomenon in this coherence measurement. We see that as the source size a decreases further, the reappearance of the visibility is higher. However, the

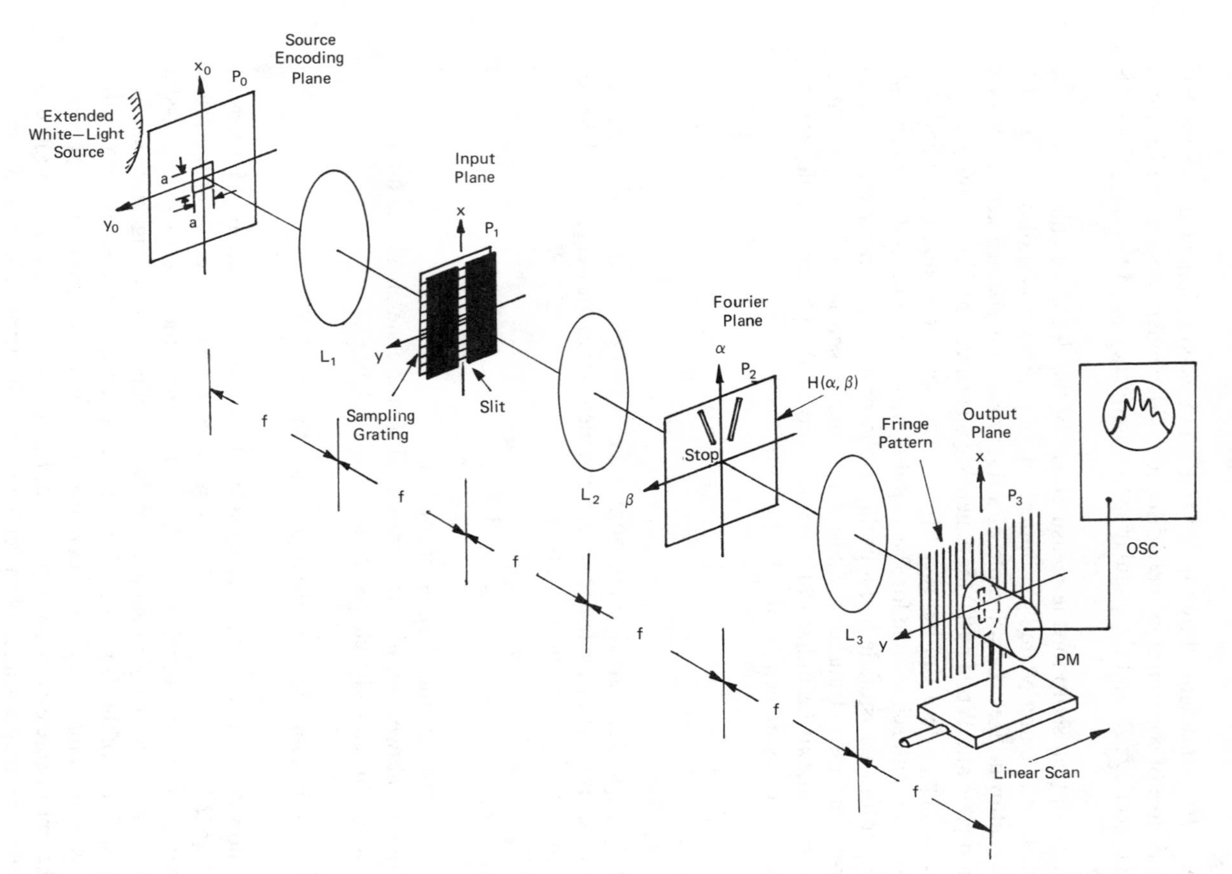

Figure 2.12. An optical setup for the coherence measurement along the β-direction. $\gamma(x_0, y_0)$, source encoding mask; a, source size; L, achromatic transform lenses; $H(\alpha, \beta)$, pair of slant slits; PM, photometer; OSC, oscilloscope.

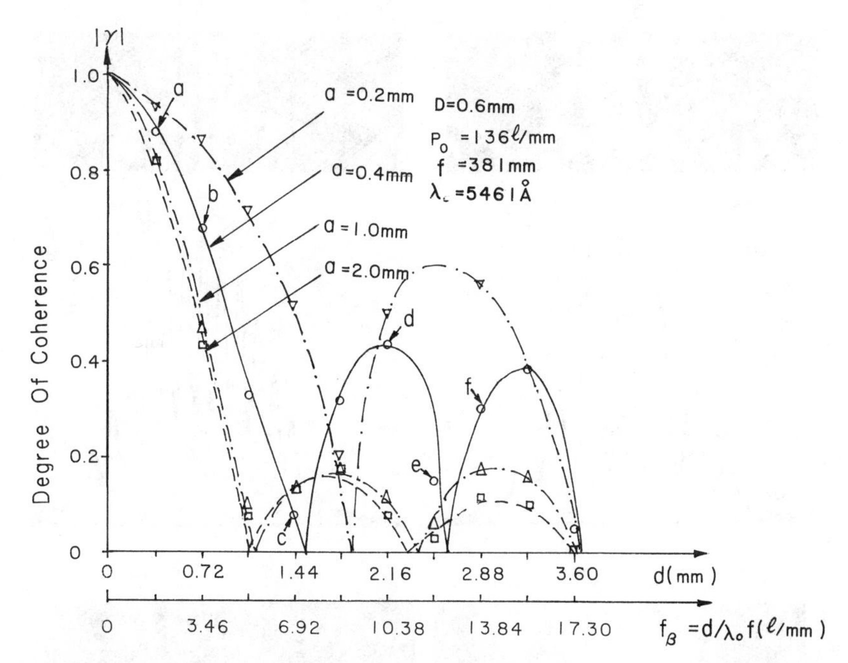

Figure 2.13. Plots of degree of coherence along the β-direction as a function of mean slit separation d for various values of source size a, $f_\beta = d/(\lambda_0 f)$, the mean slit separation in spatial frequency.

overall irradiance of the smeared Fourier spectra decreases. This phenomenon is primarily due to the finite object size under a uniform source size illumination. Furthermore, if the source size is further increased, for example if it exceeds 1.0 mm in Fig. 2.13, the decrease in coherence due to the source size is not apparent. This is primarily due to a comparable broader object size D (e.g., $D = 0.6$ mm) as compared with a narrower sinc factor derived from source size a. Nevertheless, if the input slit size decreases further, the changes in degree of coherence for large source sizes can be seen.

We now provide a set of output fringe patterns with a set of normalized photometer traces, as shown in Fig. 2.14. The fringe patterns of Fig. 2.14*a* through *c* were taken at visibilities of 0.88, 0.68, and 0.08, which corresponded to the mean separations $d = 0.36$, 0.72, and 1.44 mm at points *a*, *b*, and *c* indicated on the main lobe of the plot $a = 0.4$ mm in Fig. 2.13. Figure 2.14*d* and *e* were taken on the second lobe of the visibility reappearance, which corresponded to points *d* and *e* in Fig. 2.13. The degrees of coherence at these two points are 0.42 and 0.15, respectively. And the corresponding mean slit separations are 2.16 mm and 2.52 mm. Figure 2.14*f* was taken on the third

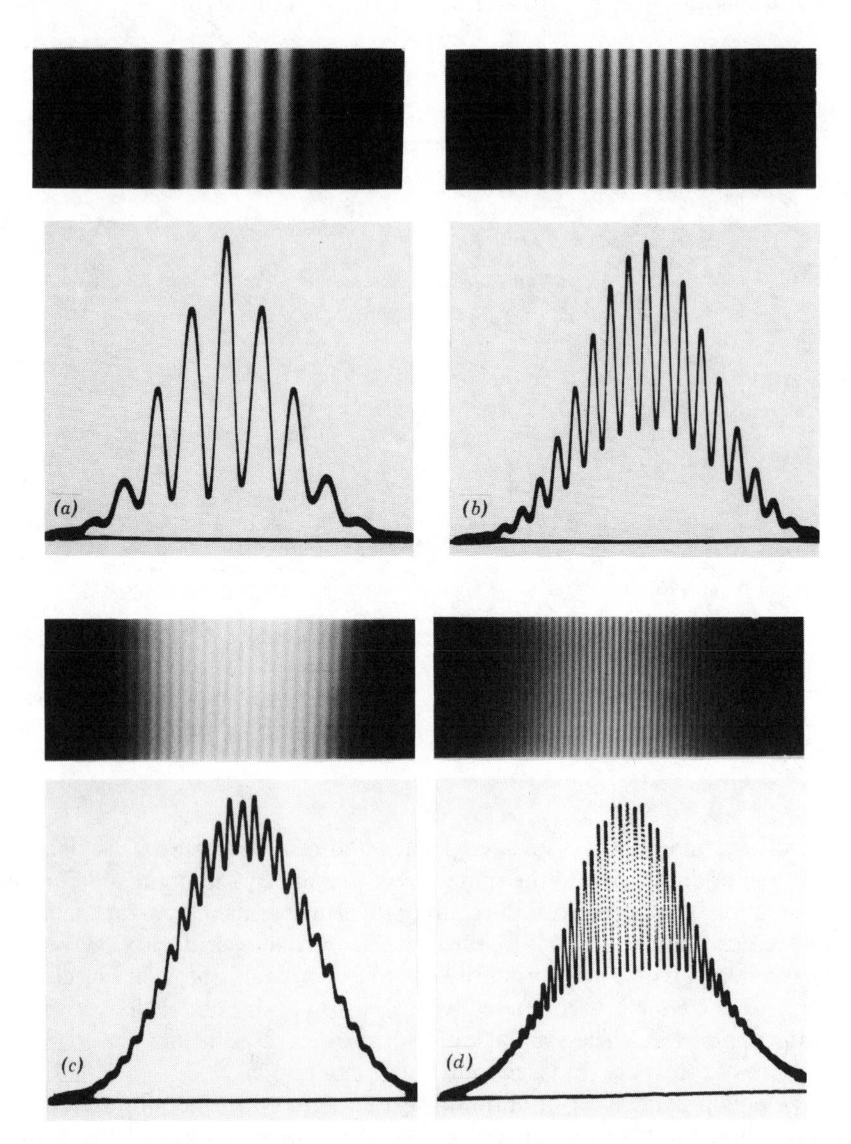

Figure 2.14. Samples of fringe visibility patterns at the output plane. The upper portions of (*a*) through (*f*) show the fringe visibility patterns. The lower portions show the corresponding photometer traces. In these experiments, the fringe visibility and the spatial frequency were varied by changing the mean separation distance between the two slanted slits. (*a*) through (*c*) were obtained at points *a*, *b*, and *c* on the first lobe of Fig. 2.13. These figures show the decrease in fringe visibility and corresponding increase in spatial frequency as the separation *d* increases. (*d*) and (*e*) were obtained at points *d* and *e* on the second lobe of Fig. 2.13. These two figures also show the increase of spatial frequency as *d* further increases. (*f*) was obtained at point *f* on the third lobe of Fig. 2.13.

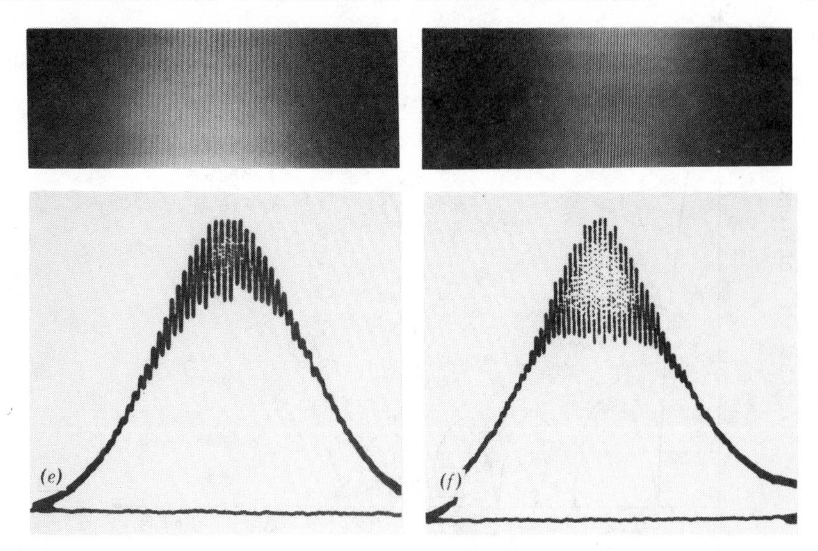

Figure 2.14 (*Continued*)

lobe visibility at point *f*. The degree of coherence is 0.3 and the mean slit separation is 2.88 mm.

Let us now investigate the degree of coherence as a function of mean separation d for various input object sizes D (i.e., slid widths), as plotted in Fig. 2.15. Again, we see that the degree of coherence decreases as d increases. Further increase in d also causes side lobes to reappear. In this figure, we also see that the degree of coherence increases as object size D decreases. Finally, Fig. 2.16 shows the visibility measure as a function of mean separation d for two values of sampling grating frequencies. From this figure we see that the degree of coherence is dramatically improved in the Fourier plane by the insertion of the sampling grating.

2.6.3. Coherence Measurement in the α-Direction

We now measure the degree of coherence in the α-direction at the Fourier plane. The measurement technique is essentially identical to that of Fig. 2.12, except that the input object is replaced by a pair of horizontal slits, as shown in Fig. 2.11. In coherence measurement, we centered the pair of slits at the center of the smeared Fourier spectra, corresponding to $\lambda = 5461$ Å.

Figure 2.17 shows plots of degree of coherence as a function of slit separation h, for various values of source size a. From this figure, we see that the degree of coherence decreases as h increases. Further increase in h again causes the reappearance of the visibility side lobes. However, the degree of

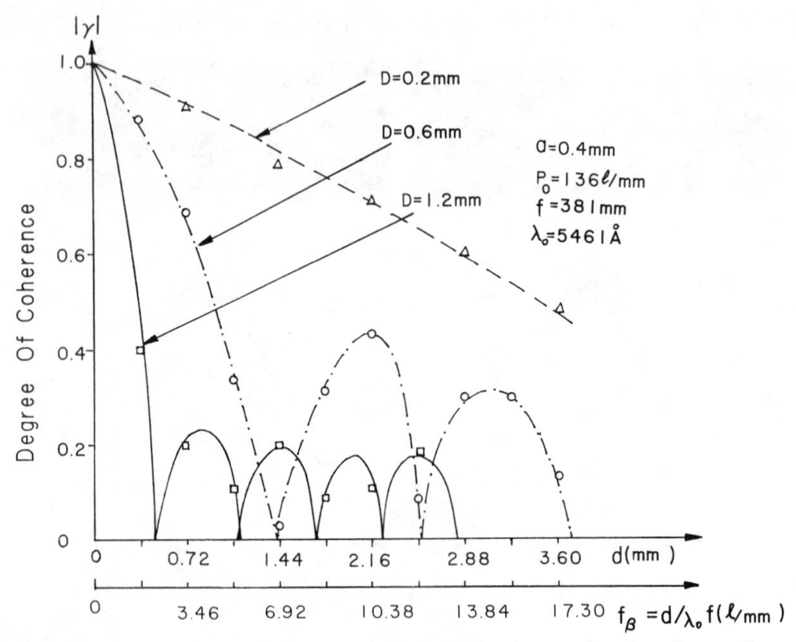

Figure 2.15. Plots of degree of coherence along the β-direction as a function of mean slit separation d for various values of input slit width D.

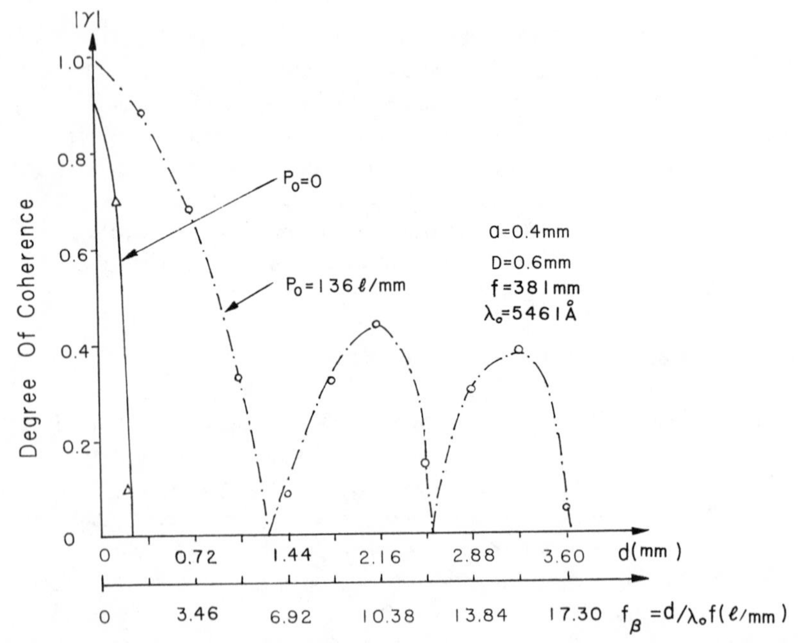

Figure 2.16. Plots of degree of coherence along the β-direction as a function of mean slit separation d for two values of sampling grating frequency p_0.

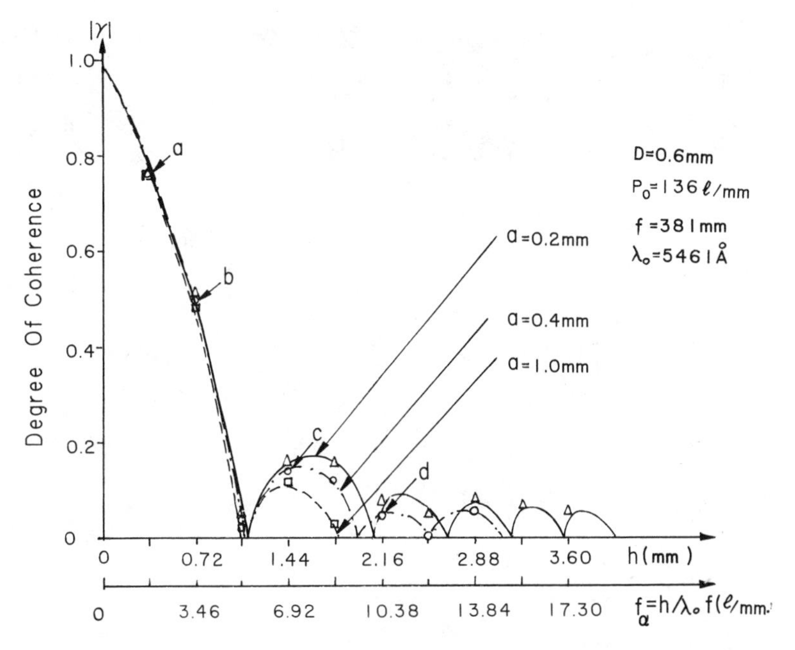

Figure 2.17. Plots of degree of coherence along the α-direction as a function of slit separation h for various values of source size a. $f_\alpha = h/\lambda_0 f$, corresponding slit separation in spatial frequency.

coherence is generally not affected by the variation of the source size a. Figure 2.18 shows a set of the visibility fringe patterns that we have obtained at the output image plane. These pictures were taken at points a, b, c, and d as shown in Fig. 2.17. The corresponding degrees of coherence are 0.76, 0.50, 0.14, and 0.05. The respective separations are $h = 0.36$, 0.72, 1.44, and 2.16 mm.

We now plot the degree of coherence as a function of h for various values of object sizes D (i.e., slit widths D), as shown in Fig. 2.19. From these plots, we see that the degree of coherence decreases as the object size D increases. Figure 2.20 shows the variation of coherence due to spatial frequency of the sampling grating, as a function of slit separation h. From this figure, we see that a higher degree of coherence is achievable with the insertion of a high spatial frequency grating at the input plane.

Let us now briefly discuss the overall effect of coherence in the (α, β) spatial frequency plane. By comparing the visibility measurements of Figs. 2.13 and 2.17, we see that the degree of coherence substantially increases in the β-direction as the source size decreases. There is, however, no significant improvement in the α-direction for smaller source sizes. Although both cases

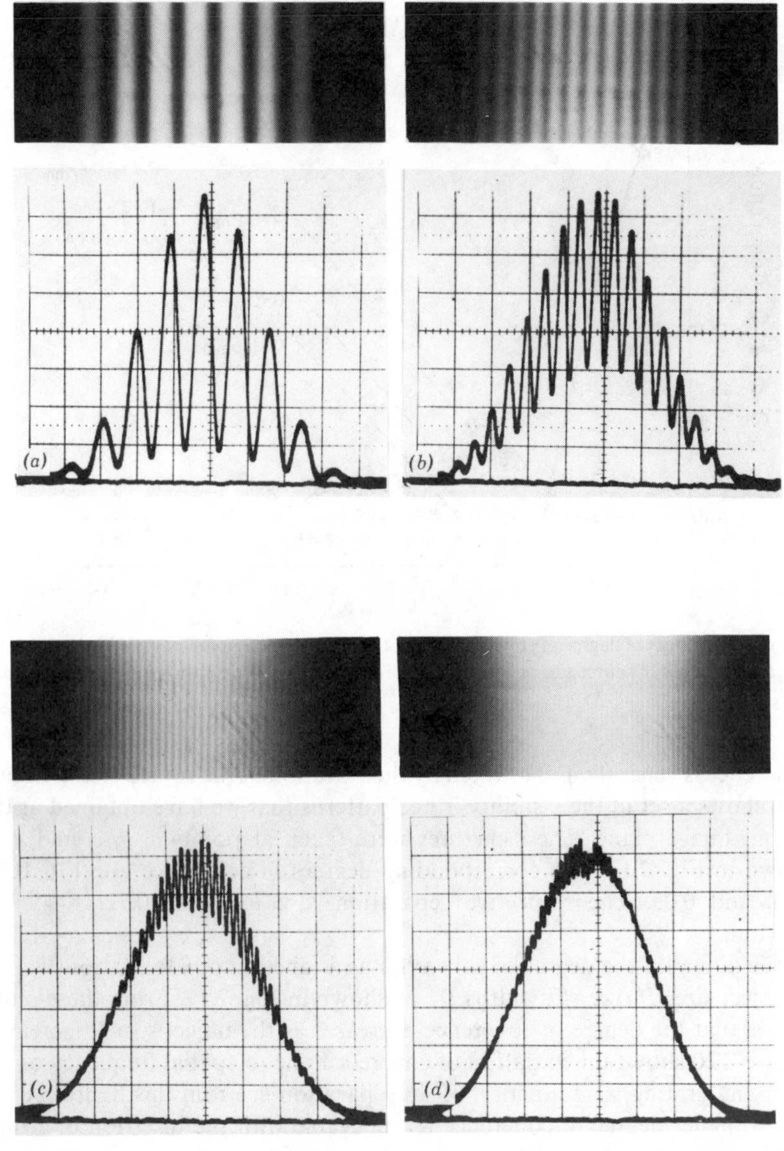

Figure 2.18. Samples of fringe visibility patterns. The upper portions of (*a*) through (*d*) show fringe visibility patterns; the lower portions show the corresponding intensity profiles. (*a*) and (*b*) were obtained at points *a* and *b* on the first lobe of Fig. 2.17. (*c*) was obtained at point *c* on the second lobe of Fig. 2.17, and (*d*) was obtained at point *d* on the third lobe of Fig. 2.17.

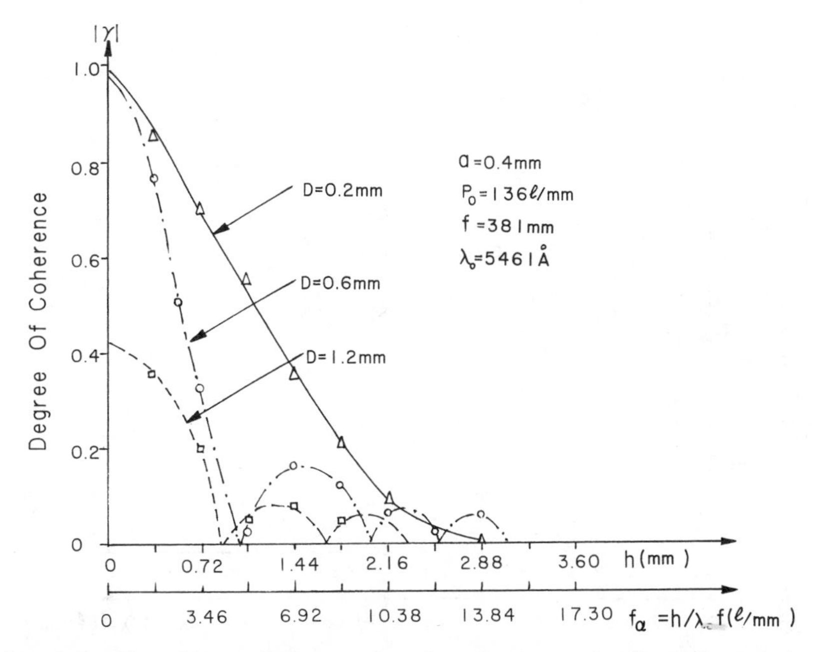

Figure 2.19. Plots of degree of coherence along the α-direction as a function of slit separation h for various values of object size D.

show the increase in coherence for smaller object sizes, the increase in coherence is higher in the β-direction than in the α-direction, as shown in Figs. 2.15 and 2.19. With reference to the plots of Figs. 2.16 and 2.20, both cases show significant improvement in the degree of coherence with the insertion of a sampling grating. However, the improvement in coherence in the β-direction is somewhat higher than in the α-direction. This is primarily due to overlapping of the smeared rainbow Fourier spectra.

2.6.4. Remarks

We have devised a dual-beam interference technique to measure the degree of coherence in the Fourier plane of a white-light optical image processor. The effect of coherence variation due to source size, input object size, and the spatial frequency of the sampling grating have been plotted as functions of distance in the β- and in the α-directions of the Fourier plane. We have shown that the degree of coherence increases as the spatial frequency of the sampling grating increases. Although the improvement in degree of coherence at the Fourier plane is quite evident, the improvement in the β-direction (i.e., the direction perpendicular to the light dispersion) is somewhat more effective

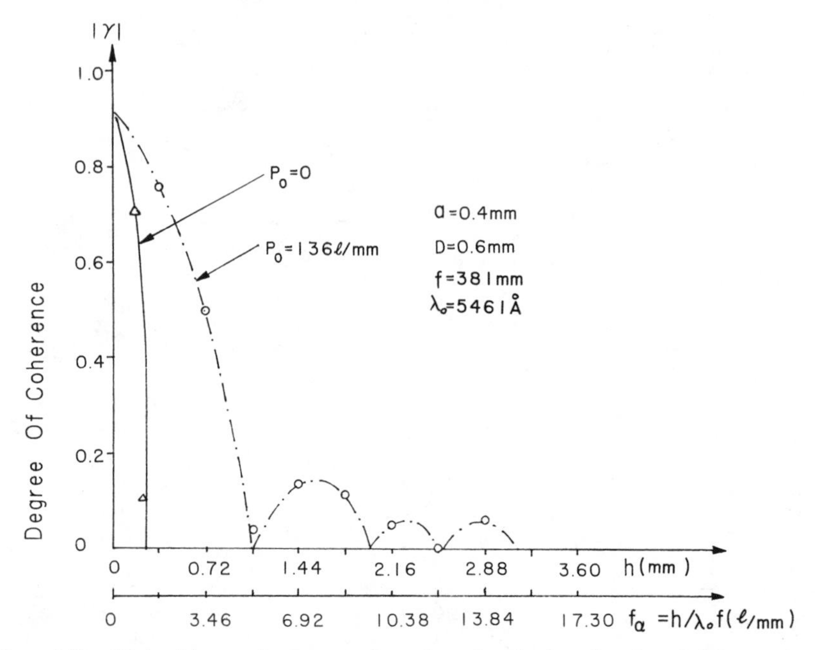

Figure 2.20. Plots of degree of coherence along the α-direction as a function of slit separation h for two values of sampling grating frequency p_0.

than that in the α-direction. The results indicate that this white-light optical signal processing technique is somewhat more effective in the β-direction than in the α-direction. Nevertheless, the existence of the high degree of coherence in Fourier plane allows us to process the signal in complex amplitude rather than in intensity. And the white-light processing technique is very suitable for color signal or image processing.

REFERENCES

2.1. L. J. Cutrona et al., "Optical data processing and filtering system," *IRE Trans. Inform. Theory* **IT-6**, 386 (1960).

2.2. F. T. S. Yu, *Optical Information Processing*, Wiley-Interscience, New York, 1983, ch. 6.

2.3. D. Gabor, "Laser speckle and its elimination," *IBM J. Res. Develop.* **14**, 509 (1970).

2.4. F. T. S. Yu and J. L. Horner, "Optical processing of photographic images," *Opt. Eng.* **20**, 666 (1981).

2.5. H. H. Hopkins, "The concept of partial coherence in optics," *Proc. Roy. Soc. A* **208**, 263 (1961).

2.6. H. H. Hopkins, "On the diffraction theory of optical images," *Proc. Roy. Soc. A* **217**, 408 (1953).

2.7. A. W. Lohmann, "Incoherent optical processing of complex data," *Appl. Opt.* **16**, 261 (1977).

2.8. W. T. Rhodes, "Bipolar pointspread function by phase switching," *Appl. Opt.* **16**, 265 (1977).

2.9. W. Stoner, "Incoherent optical processing via spatially offset pupil masks," *Appl. Opt.* **17**, 2454 (1978).

2.10. E. Leith and J. Roth, "White-light optical processing and holography," *Appl. Opt.* **16**, 2565 (1977).

2.11. E. Leith and G. Swanson, "Achromatic interferometers for white-light optical processing and holography," *Appl. Opt.* **19**, 638 (1980).

2.12. F. T. S. Yu, "A new technique of incoherent complex signal detection," *Opt. Commun.* **27**, 23 (1978).

2.13. F. T. S. Yu, "Restoration of smeared photographic image by incoherent optical processing," *Appl. Opt.* **17**, 3571 (1978).

2.14. M. Born and E. Wolf, *Principles of Optics*, 2nd rev. ed., Pergamon Press, New York, 1964.

2.15. F. T. S. Yu, *Optical Information Processing*, Wiley-Interscience, New York, 1983, p. 294.

2.16. P. H. Van Cittert, "Die Wahrschcinlicke Schwingungs verteilung in einer von einer lichtquelle direkt Oden Mittels einer linse," *Physica* **1**, 201 (1934).

2.17. F. Zernike, "The concept of degree of coherence and its application to optical problems," *Physica* **5**, 785 (1938).

2.18. B. J. Thompson and E. Wolf, "Two-beam interference with partially coherent light," *J. Opt. Soc. Am.* **47**, 895 (1957).

2.19. B. J. Thompson, "Illustration of the phase change in two-beam interference with partially coherent light," *J. Opt. Soc. Am.* **48**, 95 (1958).

2.20. F. T. S. Yu, F. K. Hsu, and T. H. Chao, "Coherence measurement of a grating-based white-light optical signal processor," *Appl. Opt.* **23**, 333 (1984).

3

Coherence Requirement

Optical systems perform myriad sophisticated information processing operations using coherent light. However, coherent systems are susceptible to coherent artifact noise, which frequently limits their processing capability. The use of incoherent light alleviates the noise problem [3.1–3.4], usually at the cost of a lowered signal-to-noise ratio due to dc bias. Various workers have studied optical systems that employ (1) a totally incoherent source [3.5–3.7] or (2) a broad-band source with reduced coherence [3.8, 3.9]. The question to be addressed in this chapter is: To what degree can we relax the coherence requirement without sacrificing the overall results of the processing system? We use Wolf's theory of partially coherent light [3.10] to develop the necessary coherence criteria for a white-light or partially coherent optical signal processor. We use the results to discuss the temporal and spatial coherence requirements for white-light optical processor, as applied to specific processing operations.

3.1. GENERAL FORMULATION

We now develop a general formulation for the propagation of mutual coherence functions through a white-light optical signal processor.

With reference to the white-light signal processor of Fig. 3.1, the mutual coherence function at the input plane P_1 due to the source irradiance $\gamma(x_0, y_0; \lambda)$ can be shown as [see Eq. (2.46)]

$$\Gamma(x_1, y_1; x_2, y_2; \lambda) = \iint \gamma(x_0, y_0; \lambda) \cdot \exp\left\{ -i \frac{2\pi}{\lambda f} [(x_1 - x_2)x_0 + (y_1 - y_2)y_0] \right\} dx_0 \, dy_0, \tag{3.1}$$

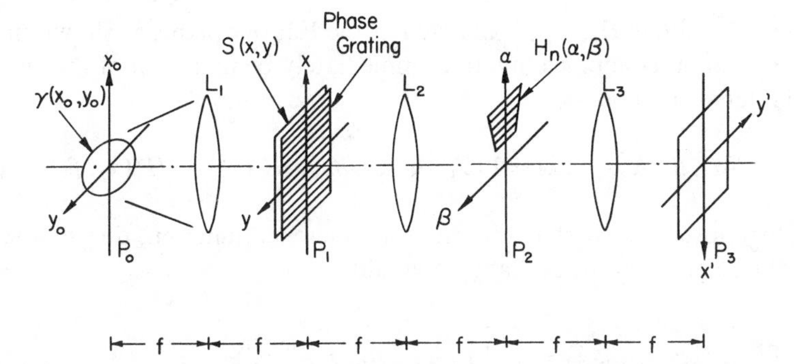

Figure 3.1. A partially coherent optical processing system: P_0, source plane; P_1, input plane; P_2, Fourier plane; P_3, output plane; L, acromatic lenses.

where the integration is over the source plane P_0. The mutual coherence function immediately behind the sampling phase grating can be written as

$$\Gamma'(x_1, y_1; x_2, y_2; \lambda) = \Gamma(x_1, y_1; x_2, y_2; \lambda)s(x_1, y_1)s^*(x_2, y_2) \cdot \exp(i2\pi\nu_0 x_1)\exp(-i2\pi\nu_0 x_2), \tag{3.2}$$

where the superscript $*$ denotes the complex conjugate, and ν_0 is the spatial frequency of the sampling grating. Similarly, the mutual coherence function at the Fourier plane P_2 can be written as

$$\Gamma(\alpha_1, \beta_1; \alpha_2, \beta_2; \lambda) = \iiiint \Gamma'(x_1, y_1; x_2, y_2; \lambda) \cdot \exp\left\{-i\frac{2\pi}{\lambda f}[\alpha_1 x_1 + \beta_1 y_1 - \alpha_2 x_2 - \beta_2 y_2]\right\}dx_1\, dy_1\, dx_2\, dy_2, \tag{3.3}$$

where the integration is over the input plane P_1. By substituting Eqs. (3.1) and (3.2) into Eq. (3.3), we have

$$\Gamma(\alpha_1, \beta_1; \alpha_2, \beta_2; \lambda) = \iint \gamma(x_0, y_0; \lambda)S(x_0 + \alpha_1 - \lambda f\nu_0, y_0 + \beta_1) \cdot S^*(x_0 + \alpha_2 + \lambda f\nu_0, y_0 + \beta_2)dx_0\, dy_0, \tag{3.4}$$

where the integration is over the source plane P_0, and $S(\alpha, \beta)$ is the Fourier spectrum of the input image $s(x, y)$. If we assume that a set of narrow spectral band spatial filters $H_n(\alpha, \beta)$ is inserted at the Fourier plane, as shown in Fig. 3.1, the mutual coherence function immediately behind each of the spectral band filters would be

$$\Gamma'(\alpha_1, \beta_1; \alpha_2, \beta_2; \lambda) = \Gamma(\alpha_1, \beta_1; \alpha_2, \beta_2; \lambda) H_n(\alpha_1, \beta_1) H_n^*(\alpha_2, \beta_2). \quad (3.5)$$

It is therefore evident that the mutual coherence function, due to the nth spatial filter, at the output plane P_3 would be

$$\Gamma(x_1', y_1'; x_2', y_2'; \lambda) = \iiiint \Gamma'(\alpha_1, \beta_1; \alpha_2, \beta_2; \lambda) \cdot \exp\left[-i \frac{2\pi}{\lambda f}(x_1'\alpha_1 + y_1'\beta_1 - x_2'\alpha_2 - y_2'\beta_2)\right] d\alpha_1\, d\beta_1\, d\alpha_2\, d\beta_2. \quad (3.6)$$

Since the interest is usually centered at the output intensity distribution, by letting $x_1' = x_2' = x'$ and $y_1' = y_2' = y'$ we can write the output signal irradiance due to nth spectral band filter as

$$I_n(x', y'; \lambda) = \iint \gamma(x_0, y_0; \lambda) dx_0\, dy_0 \iint S(x_0 + \alpha_1 - \lambda f v_0, y_0 + \beta_1) H_n(\alpha_1, \beta_1) \cdot \exp\left[-i \frac{2\pi}{\lambda f}(\alpha_1 x' + \beta_1 y')\right] d\alpha_1\, d\beta_1 \cdot \iint S^*(x_0 + \alpha_2 - \lambda f v_0, y_0 + \beta_2) \cdot H_n^*(\alpha_2, \beta_2) \exp\left[i \frac{2\pi}{\lambda f}(\alpha_2 x' + \beta_2 y')\right] d\alpha_2\, d\beta_2,$$
$$\text{for } \lambda_{ln} \leqq \lambda \leqq \lambda_{hn}, \quad (3.7)$$

where λ_{ln} and λ_{hn} are the lower and upper wavelength limits of $H_n(\alpha, \beta)$. By changing the variables (α_1, β_1) and (α_2, β_2), to (α, β), Eq. (3.7) reduces to

$$I_n(x', y'; \lambda) = \iint \gamma(x_0, y_0; \lambda) \left| \iint S(x_0 + \alpha - \lambda f v_0, y_0 + \beta) H_n(\alpha, \beta) \cdot \exp\left[-i \frac{2\pi}{\lambda f}(\alpha x' + \beta y')\right] d\alpha\, d\beta \right|^2 dx_0\, dy_0,$$
$$\text{for } \lambda_{ln} \leqq \lambda \leqq \lambda_{hn}. \quad (3.8)$$

Let us denote $S(\lambda)$ as the relative spectral intensity of the light source and $C(\lambda)$ as the relative spectral response of the output detector or recording material; the output signal irradiance resulting from each narrow spectral band filter would be

$$I_n(x', y') = \int_{\lambda_n - \Delta\lambda_n/2}^{\lambda_n + \Delta\lambda_n/2} \iint \gamma(x_0, y_0; \lambda) S(\lambda) C(\lambda) \left| \iint S(x_0 + \alpha - \lambda f v_0, y_0 + \beta) \cdot H_n(\alpha, \beta) \exp\left[-i \frac{2\pi}{\lambda f}(x'\alpha + y'\beta) \right] d\alpha\, d\beta \right|^2 dx_0\, dy_0\, d\lambda, \quad \text{for } n = 1, 2, \ldots, N, \tag{3.9}$$

where λ_n and $\Delta\lambda_n$ are the center wavelength and the bandwidth of the nth narrow spectral band filter $H_n(\alpha, \beta)$. It is therefore apparent that the overall output signal irradiance would be the incoherent addition of these signal irradiances of Eq. (3.9), that is,

$$\begin{aligned} I(x', y') &= \sum_{n=1}^{N} I_n(x', y') \\ &= \sum_{n=1}^{N} \int_{\lambda_n - \Delta\lambda_n/2}^{\lambda_n + \Delta\lambda_n/2} \iint \gamma(x_0, y_0; \lambda) S(\lambda) C(\lambda) \\ &\quad \cdot \left| \iint S(x_0 + \alpha - \lambda f v_0, y_0 + \beta) \cdot H_n(\alpha, \beta) \right. \\ &\quad \left. \cdot \exp\left[-i \frac{2\pi}{\lambda f}(x'\alpha + y'\beta) \right] d\alpha\, d\beta \right|^2 dx_0\, dy_0\, d\lambda, \end{aligned} \tag{3.10}$$

where N is the total number of the spectral band filters. Furthermore, if the signal processing is a one-dimensional operation, then the fan-shaped spatial filter of Fig. 2.8 (see Section 2.4) can be utilized; thus Eq. (3.10) can be reduced to the following form:

$$I(x', y') = \iiint \gamma(x_0, y_0; \lambda) S(\lambda) C(\lambda) \left| \iint S(x_0 + \alpha - \lambda f v_0, y_0 + \beta) H(\alpha, \beta) \cdot \exp\left[-i \frac{2\pi}{\lambda f}(x'\alpha + y'\beta) \right] d\alpha\, d\beta \right|^2 dx_0\, dy_0\, d\lambda. \tag{3.11}$$

Now we are in a position to ask a fundamental question: To what degree can the coherence requirements be relaxed without sacrificing the overall output signal quality of the processing system? The answer is that the nature of the signal processing operation governs the degree of temporal and spatial coherence necessary to obtain satisfactory results. We apply the general formulation of Eqs. (3.10) or (3.11) to suit the specific need of the processing operation under consideration. In the following we discuss the temporal and spatial coherence requirements for some specific cases, for example, for image deblurring, image subtraction, and signal correlation [3.11, 3.12]. For other specific processing operations, the coherence requirement can also be deduced by a similar approach. However, such a lengthy illustration is beyond the scope of this chapter and is therefore left for the readers to investigate.

3.2. COHERENCE REQUIREMENTS FOR IMAGE DEBLURRING

Since linear smeared image deblurring is primarily a one-dimensional processing operation and since the inverse filtering takes place with every image point, the spatial coherence requirement depends on the smeared length of the object. Thus this optical signal processing operation is limited by the source size and the spectral bandwidth of each of the narrow spectral band filters $H_n(\alpha, \beta)$; that is, it is limited by the spatial and temporal coherence requirements.

3.2.1. Temporal Coherence Requirement

There is a temporal coherence requirement imposed on white-light image processing for photographic image deblurring. This temporal coherence requirement is a measure of the output deblurred image quality due to the spectral bandwidth of each of the narrow spectral band inverse filters $H_n(\alpha, \beta)$.

For the analysis of the temporal coherence requirement, we assume that the white-light signal processor utilizes a broad spectral band point source, [i.e., $I(x_0, y_0) = \delta(x_0, y_0)$] and that its spectral distribution is uniform throughout the spectral bandwidth [i.e., $S(\lambda) = K$]. Without loss of generality, we also assume that the relative spectral response of the output detector is uniform [i.e., $C(\lambda) = K$]. If the smeared length w is known a priori, the output deblurred image irradiance of Eq. (3.10) can be written as

$$I(x', y') = \sum_{n=1}^{N} \int_{\lambda_{ln}}^{\lambda_{hn}} |A_n(x', y'; \lambda)|^2 \, d\lambda, \tag{3.12}$$

where: the proportionality constant has been ignored; λ_{ln} and λ_{hn} are the lower and upper wavelength limits of the nth narrow spectral band filter $H_n(\alpha, \beta)$;

$$A_n(x', y'; \lambda) = \iint S(\alpha - \lambda f v_0, \beta) H_n(\alpha, \beta) \cdot \exp\left[-i \frac{2\pi}{\lambda f}(\alpha x' + \beta y')\right] d\alpha \, d\beta, \qquad \text{for } \lambda_{ln} \leqq \lambda \leqq \lambda_{hn}, \tag{3.13}$$

is the complex light distribution of the deblurred image, resulting from the nth narrow spectral band filter; and $S(\alpha - \lambda f v_0, \beta)$ is the blurred image spectrum.

Thus it is apparent that the temporal coherence requirement for the image deblurring operation is set by the spectral bandwidth of each of the spectral band filters. If we denote by $\Delta\alpha_n$ the spatial width of the narrow spectral band filter, the spectral bandwidth $\Delta\lambda_n$ of $H_n(\alpha, \beta)$ can be written as

$$\Delta\lambda_n = \frac{\Delta\alpha_n}{f v_0}, \tag{3.14}$$

which is proportional to $\Delta\alpha_n$ and inversely proportional to v_0. It is now apparent that the temporal coherence (i.e., $\Delta\lambda_n$) of each $H_n(\alpha, \beta)$ is controlled by both the filter width $\Delta\alpha_n$ and the sampling grating frequency v_0.

Since each $H_n(\alpha, \beta)$ is designed to fit a specific wavelength λ_n, and since the filter width $\Delta\alpha_n$ is determined by the spatial frequency content of the input blurred image $s(x, y)$, the higher the spatial frequency content of the input image, the wider the filter $H_n(\alpha, \beta)$ required. In other words, to maintain a higher degree of temporal coherence for high spatial frequency image, a higher sampling grating frequency v_0 is needed.

Since the image deblurring takes place at every image point, the image deblurring operation can be evaluated with respect to every blurred image point, that is,

$$s(x, y) = \operatorname{rect}\left(\frac{y}{w}\right), \tag{3.15}$$

where w denotes the smeared length, and

$$\operatorname{rect}\left(\frac{y}{w}\right) \triangleq \begin{cases} 1, & \text{for } |y| < w/2, \\ 0, & \text{otherwise.} \end{cases}$$

The corresponding Fourier spectrum for every λ is

$$S(\alpha - \lambda f v_0, \beta) = \operatorname{sinc}\left(\frac{\pi w}{\lambda f}\beta\right), \tag{3.16}$$

where

$$\operatorname{sinc}(\chi) \triangleq \frac{\sin \chi}{\chi}.$$

Thus the nth spectral band deblurring filter is

$$H_n(\alpha, \beta) = \frac{1}{S(\alpha - \lambda f v_0, \beta)} = \frac{1}{\operatorname{sinc}[(\pi w/\lambda_n f)\beta]}, \tag{3.17}$$

where λ_n denotes the central wavelength of the nth deblurring filter $H_n(\alpha, \beta)$. From Eq. (3.16), it becomes evident that the smeared image deblurring is a one-dimensional processing operation. For simplicity, we adopt a one-dimensional notation in the following: By substituting Eqs. (3.16) and (3.17) into Eq. (3.13), we have

$$A_n(y'; \lambda) = \operatorname{rect}\left(\frac{y'}{w}\right) * \int \frac{1}{\operatorname{sinc}[(\pi w/\lambda_n f)\beta]} \cdot \exp\left(-i\frac{2\pi}{\lambda f}\beta y'\right) d\beta, \qquad \text{for } \lambda_{ln} \leqq \lambda \leqq \lambda_{hn},\ n = 1, 2, \ldots, N. \tag{3.18}$$

With reference to Appendix A, Eq. (3.18) can be further reduced to the following form:

$$A_n(y'; \lambda) = \begin{cases} \dfrac{4f\lambda}{\pi} \displaystyle\sum_{m=1}^{2M} (-1)^n \sin\frac{\pi m \lambda_n}{\lambda} \sin\frac{2\pi m \lambda}{\lambda w}, & \text{for } |y'| > \dfrac{w}{2}, \\ -\dfrac{4f\lambda_n}{\pi} \displaystyle\sum_{m=1}^{2M} (-1)^n \cos\frac{\pi m \lambda_n}{\lambda} \cos\frac{2\pi m \lambda_n}{\lambda w} y', & \text{for } |y'| \leqq \dfrac{w}{2}, \end{cases} \tag{3.19}$$

for $\lambda_{ln} \leqq \lambda \leqq \lambda_{hn}$, $n = 1, 2, \ldots, N$, where M is an arbitrary large positive integer. By substituting Eq. (3.19) in Eq. (3.12), the output deblurred image irradiance can be shown as

$$I^{(1)}(y') = \sum_{n=1}^{N} \frac{2f^2\lambda_n^2}{\pi^2} \sum_{m=1}^{2M} \sum_{m'=1}^{2M} \sum_{l=1}^{8} (-1)^{m+m'+l} \Big\{ 3\lambda_a^{-3} \cos\Big[a_{mm'}^l(y')\lambda_a \Big]$$

$$- 3\lambda_b^{-3} \cos[a_{nn'}^l(y')\lambda_b] - 6\lambda_a^{-2} a_{mm'}^l(y') \sin[a_{mm'}^l(y')\lambda_a]$$

$$- 6\lambda_b^{-2}[a_{mm'}^l(y')] \sin[a_{mm'}^l(y')\lambda_b] - 6[a_{mm'}^l(y')]^2[\lambda_a^{-1}]$$

$$\cdot \cos[a_{mm'}^l(y')\lambda_a] - 6\lambda_b^{-1}[a_{mm'}^l(y')] \cos[a_{mm'}^l(y')\lambda_b]$$

$$- 6[a_{mm'}^l(y')]^3[S_i(\lambda_a) - S_i(\lambda_b)] \Big\}, \qquad \text{for } |y'| > \frac{w}{2}, \tag{3.20}$$

and

$$I^{(2)}(y') = \sum_{n=1}^{N} \frac{2f^2\lambda_n^2}{\pi^2} \sum_{m=1}^{2M} \sum_{m'=1}^{2M} \sum_{l=1}^{8} (-1)^{m+m'} \Big\{ 6[a_{mm'}^l(y')]^3[S_i(\lambda_a) - S_i(\lambda_b)]$$

$$+ 6[a_{mm'}^l(y')]^2\lambda_a^{-1} \cos[a_{mm'}^l(x')\lambda_a] - 6\lambda_b^{-1}[a_{mm'}^l(y')]^2$$

$$\cdot \cos[a_{mm'}^l(x')\lambda_b] + 6[a_{mm'}^l(y')]\lambda_a^{-2} \sin[a_{mm'}^l(x')\lambda_a] - 6\lambda_b^{-2}[a_{mm'}^l(y')]$$

$$\cdot \sin[a_{mm'}^l(y')\lambda_b] - 3\lambda_a^{-3} \sin[a_{mm'}^l](y')\lambda_a - 3\lambda_b^{-3}$$

$$\cdot \cos[a_{mm'}^l(y')\lambda_b] \Big\}, \qquad \text{for } |y'| \leqq \frac{w}{2}, \tag{3.21}$$

where

$$\lambda_a = \frac{\lambda_n}{\lambda_n + \Delta\lambda_n},$$

$$\lambda_b = \frac{\lambda_n}{\lambda_n - \Delta\lambda_n},$$

$$S_i(\lambda') \triangleq \int_0^{\lambda'} \frac{\sin \chi}{\chi} \, d\chi,$$

$$a_{mm'}^{(1)}(y') = \pi(m - m')\left(1 - \frac{2y'}{w}\right),$$

$$a_{mm'}^{(3)}(y') = \pi(m - m')\left(1 + \frac{2y'}{w}\right),$$

$$a_{mm'}^{(5)}(y') = \pi(m + m')\left(1 - \frac{2y'}{w}\right),$$

$$a_{mm'}^{(7)}(y') = \pi(m + m')\left(1 + \frac{2y'}{w}\right),$$

and

$$a_{mm'}^{(2)}(y') = \pi\left[m\left(1 - \frac{2y'}{w}\right) - m'\left(1 + \frac{2y'}{w}\right)\right],$$

$$a_{mm'}^{(4)}(y') = \pi\left[m\left(1 + \frac{2y'}{w}\right) - m'\left(1 - \frac{2y'}{w}\right)\right],$$

$$a_{mm'}^{(6)}(y') = \pi\left[m\left(1 - \frac{2y'}{w}\right) + m'\left(1 + \frac{2y'}{w}\right)\right],$$

$$a_{mm'}^{(8)}(y') = \pi\left[m\left(1 + \frac{2y'}{w}\right) + m'\left(1 - \frac{2y'}{w}\right)\right].$$

Equations (3.20) and (3.21) provide the mathematical basis for the evaluation of the temporal coherence requirement of a smeared image deblurring with a white-light processor. Plots of the normalized deblurred image irradiance, defined as a function of y' and the bandwidth $\Delta\lambda_n$, are shown in Fig. 3.2. $\lambda_n = 5461$ Å was used for the calculation. It is evident that, if the light source is strictly coherent (i.e., $\Delta\lambda_n = 0$), the deblurred point image is infinitesimal. Also note that the degree of deblurring decreases as the temporal coherence of the deblurred filter is reduced. The deblurred length Δw re-

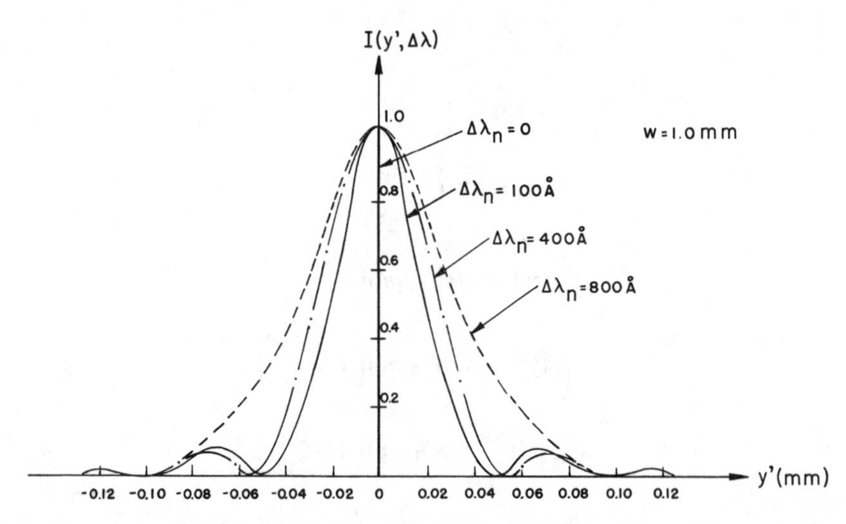

Figure 3.2. Output intensity distribution of the deblurred image. $\Delta\lambda_n$, spectral bandwidth of the narrow spectral band deblurring filter $H_n(\alpha, \beta)$; w, smeared length.

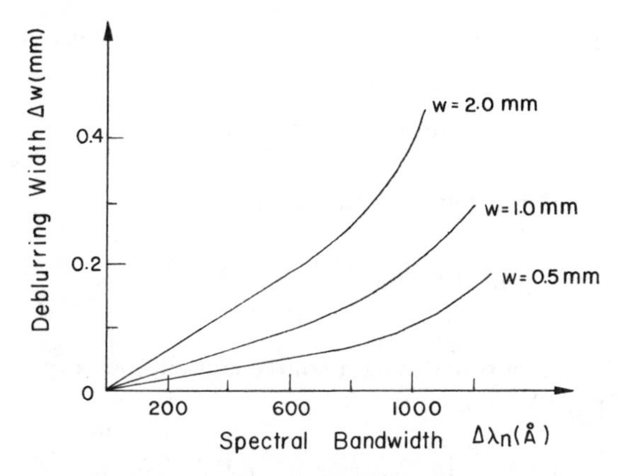

Figure 3.3. Plots of the deblurring width Δw as a function of the spectral bandwidth $\Delta\lambda_n$ of the narrow spectral band deblurring filter for various values of smeared length.

presents the spread of the deblurred image irradiance. It is formally defined as the separation between the 10% points of $I(y')$. The deblurred length Δw can be shown to decrease monotonically with $\Delta\lambda_n$. Plots of Δw as a function of the spectral bandwidth $\Delta\lambda_n$ for various values of smeared length w are shown in Fig. 3.3. It may be shown that the deblurred length Δw is linearly proportional to the smear length w for a given value of $\Delta\lambda_n$. The greater the smear, the more difficult the deblurring process. In principle this may be corrected by decreasing the spectral bandwidth of each of the deblurred filters $H_n(\alpha, \beta)$. Table 3.1 numerically summarizes the preceding analysis. The value of $\Delta\lambda_n$ can be regarded as the temporal coherence requirement for each $H_n(\alpha, \beta)$ for the deblurring process.

3.2.2. Spatial Coherence Requirement

The relationship between the source size and the image intensity distribution is the key factor in the calculation of the degree of spatial coherence requirement for a white-light image deblurring system. For simplicity in notation, we again use a one-dimensional representation in the following.

Table 3.1. Effect of Temporal Coherence Requirement

$\Delta w/w$	1/20	1/15	1/10	1/8	1/5
$\Delta\lambda_n$(Å)	270	400	640	750	990

Assume the intensity distribution of the light source is given by

$$\gamma(x_0, y_0) = \operatorname{rect}\left(\frac{y_0}{\Delta s}\right)\delta(x_0), \tag{3.22}$$

where $\delta(x_0)$ is the Dirac delta function, and

$$\operatorname{rect}\left(\frac{y_0}{\Delta s}\right) \triangleq \begin{cases} 1, & |y_0| \leqq \Delta s/2, \\ 0, & \text{otherwise.} \end{cases}$$

If an extended monochromatic light source is used, the output intensity distribution of Eq. (3.9) becomes

$$I_n(y') = \int_{-\infty}^{\infty} \operatorname{rect}\left(\frac{y_0}{\Delta s}\right) |A_n(y', y_0; \lambda)|^2 \, dy_0, \tag{3.23}$$

where

$$A_n(y', y_0; \lambda) = \int_{-\infty}^{\infty} \operatorname{sinc}\left[\frac{\pi w}{\lambda f}(\beta + y_0)\right] \exp\left(-i\frac{2\pi}{\lambda f}\beta y'\right) d\beta$$
$$* \int_{-\infty}^{\infty} \frac{1}{\operatorname{sinc}[(\pi w/\lambda f)\beta]} \exp\left(-i\frac{2\pi}{\lambda f}\beta y'\right) d\beta.$$

and $*$ denotes the convolution operation.

Equation (3.23) can be reduced to the form of

$$A_n(y', x_0; \lambda) = \begin{cases} i2f\lambda_n \sum_{m=1}^{2M} (-1)^m m \exp\left(\frac{i2\pi\sigma y'}{w}\right) \exp\left(i\frac{2\pi m}{w} y'\right) \operatorname{sinc}[\pi(\sigma + m)] \\ \quad - \exp\left(-i\frac{2\pi m}{w} y'\right) \operatorname{sinc}[\pi(\sigma - m)], \qquad \text{for } |y'| > \frac{w}{2}, \\ \frac{2f\lambda_n}{\pi} \sum_{m=1}^{M} (-1)^m m \exp\left(\frac{i2\pi\sigma y'}{w}\right) \\ \quad \cdot \left(\frac{1}{\sigma - m}\left\{1 - \exp\left[\frac{i2\pi(\sigma - m)y'}{w}\right]\cos[\pi(\sigma - m)]\right\}\right) \\ \quad + \frac{1}{\sigma + m}\left\{1 - \exp\left[\frac{i2\pi(\sigma + m)y'}{w}\right]\cos[\pi(\sigma + m)]\right\}, \\ \quad \text{for } |y'| \leqq \frac{w}{2}, \end{cases} \tag{3.24}$$

where $\sigma = y_0 w/(\lambda f)$.

By substituting Eq. (3.24) into Eq. (3.12), the deblurred image irradiance is found to be

$$I_n^{(1)}(y', \Delta s) = \frac{16 f^3 \lambda_n^3}{w\pi^2} \sum_m \sum_{m'} \left\{ \Phi_{mm'}^{(1)}(\Delta s) \cos\left[\frac{2\pi(m - m')y'}{w}\right] + \Phi_{mm'}^{(2)}(\Delta s) \cos\left[\frac{2\pi(m + m')y'}{w}\right] \right\}, \qquad \text{for } |y'| > \frac{w}{2}, \tag{3.25}$$

and

$$I_n^{(2)}(y', \Delta s) = \frac{16 \lambda_n^3 f^3}{w\pi^2} \sum_m \sum_{m'} \left\{ \Phi_{mm'}^{(3)}(\Delta s) + \Phi_{mm'}^{(4)}(\Delta s) \sin\left[\frac{2\pi(m' - m)y'}{w}\right] + \Phi_{mm'}^{(5)}(\Delta s) \cos\left[\frac{2\pi(m' - m)y'}{w}\right] + \Phi_{mm'}^{(6)}(\Delta s) \cos\left[\frac{2\pi(m + m')y'}{w}\right] \right\},$$
$$\text{for } |y'| \leqq \frac{w}{2}, \tag{3.26}$$

where

$$\Phi_{mm'}^{(1)}(\Delta s) = \frac{mm'}{m' - m} \left\{ m\left[\frac{m + \bar{\sigma}(\Delta s)}{m - \bar{\sigma}(\Delta s)}\right] - [(-1)^m C_j(m, \Delta s) - (-1)^{m'} C_j(m', \Delta s)] \right\},$$

$$\Phi_{mm'}^{(2)}(\Delta s) = \frac{mm'}{m' + m} \left\{ m\left[\frac{m + \bar{\sigma}(\Delta s)}{m - \bar{\sigma}(\Delta s)}\right] - [(-1)^m C_j(m, \Delta s) - (-1)^{m'} C_j(m', \Delta s)] \right\},$$

$$\Phi_{mm'}^{(3)}(\Delta s) = (-1)^m \frac{2m}{m^2 - m'^2} \left\{ m\left[\frac{m - \bar{\sigma}(\Delta s)}{m + \bar{\sigma}(\Delta s)}\right] + (-1)^m \frac{m}{m^2 - m'^2} [C_k^+(m', \Delta s) + C_k^-(m', \Delta s)] \right\},$$

$$\Phi_{mm'}^{(4)}(\Delta s) = \frac{m'}{m^2 - m'^2} [S_k^+(m, \Delta s) + S_k^-(m, \Delta s)],$$

$$\Phi^{(5)}_{mm'}(\Delta s) = (-1)^{m+1}\frac{m'}{m^2 - m'^2}[C^+_k(m, \Delta s) + C^-_k(m, \Delta s)]$$

$$+ \frac{2}{m' - m}\left\{m\left[\frac{m + \bar{\sigma}(\Delta s)}{m - \sigma(\delta s)}\right] + [C_j(m, \Delta s) - C_j(m', \Delta s)]\right\},$$

$$\Phi^{(6)}_{mm'}(\Delta s) = \frac{1}{m + m'}\left\{m\left[\frac{m - \bar{\sigma}(\Delta s)}{m + \bar{\sigma}(\Delta s)}\right] - [C_j(m, \Delta s) + C_j(m', \Delta s)]\right\},$$

$$S_j(m, \Delta s) \triangleq S_i\{2\pi[\bar{\sigma}(\Delta s) - m]\} - S_i\{2\pi[-\bar{\sigma}(\Delta s) - m]\},$$

$$C_j(m, \Delta s) \triangleq C_i[2\pi|\bar{\sigma}(\Delta s) + m|] - C_i[2\pi|\bar{\sigma}(\Delta s) - m|],$$

$$S_i(u) \triangleq \int_0^u \frac{\sin v}{v}\,dv,$$

$$C_i(u) \triangleq \int_u^\infty \frac{\cos v}{v}\,dv,$$

$$\bar{\sigma}(\Delta s) \triangleq \frac{\Delta s w}{(2\lambda f)},$$

$$S^{\pm}_k(m, \Delta s) \triangleq S_i\left\{\pi\left(\frac{2y'}{w \pm 1}\right)[\bar{\sigma}(\Delta s) + m]\right\}$$

$$- S_i\left\{\pi\left(\frac{2y'}{w \pm 1}\right)[-\bar{\sigma}(\Delta s) + m]\right\},$$

and

$$C^{\pm}_k(m, \Delta s) \triangleq C_i\left[\pi\left|\frac{2y'}{w \pm 1}\right||\bar{\sigma}(\Delta s) + m|\right]$$

$$- C_i\left[\pi\left|\frac{2y'}{w \pm 1}\right||-\bar{\sigma}(\Delta s) + m|\right].$$

Thus the spatial coherence requirement of an image deblurring process may be evaluated by using Eqs. (3.25) and (3.26). The output intensity distribution of this process is plotted in Fig. 3.4. Recall the definition of the deblurred length Δw stated previously (i.e., the separation between the 10% points of the output image irradiance I_n); the plots of Δw as a function of the source

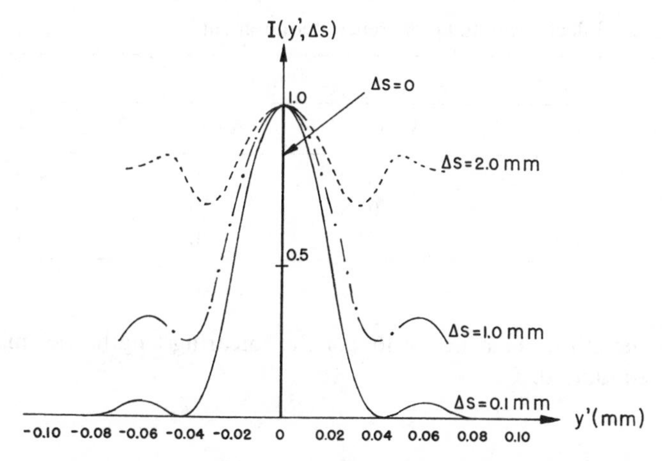

Figure 3.4. Output intensity distribution of the deblurred image for various values of the source size Δs.

size Δs and the smear length w are shown in Fig. 3.5. From this figure we see that when the spatial width of the light source increases beyond a critical size Δs_c, the deblurred length becomes independent of Δs and equal to w. Table 3.2 provides a brief numerical summary of the key parameters for the determination of the spatial coherence in image deblurring. From this table it is evident that the spatial coherence requirement is inversely proportional to

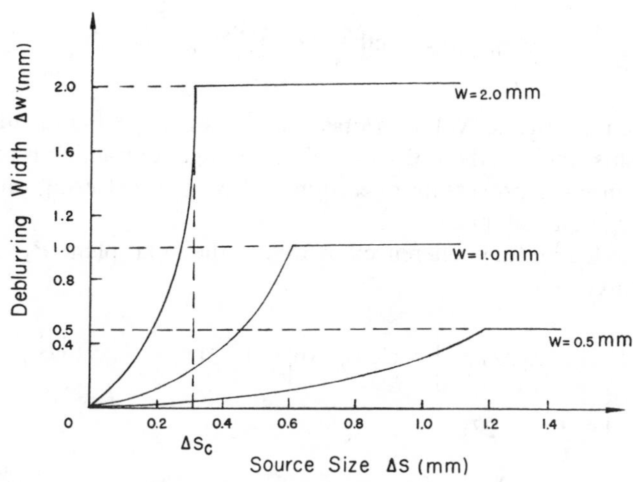

Figure 3.5. Plots of the deblurring width as a function of the source size for various values of the smeared length w.

Table 3.2. Effect of Spatial Coherence Requirement

	Δs(mm)			
w(mm)	$\Delta w/w = 1/20$	$\Delta w/w = 1/15$	$\Delta w/w = 1/10$	$\Delta w/w = 1/5$
0.5	0.2	0.38	0.6	0.92
1	0.1	0.18	0.26	0.40
2	0.05	0.08	0.12	0.18

the smear length w. That is, the longer the smearing length, the smaller the source size required.

3.3. COHERENCE REQUIREMENTS FOR IMAGE SUBTRACTION

We have reported an image subtraction procedure employing an extended incoherent source and a source encoded mask [3.13]. Let us now discuss the temporal and spatial coherence requirements of the image subtraction process. We note that the output intensity distribution can be derived by calculating the propagation of the mutual coherence function through an optical system that utilizes an encoded extended source. With reference to the image subtraction processor of Fig. 3.6, the encoded source intensity distribution may be described by [3.14]

$$\gamma(x_0, y_0) = \mathrm{rect}\left(\frac{x_0}{w}\right) * \sum_{n=-N}^{N} \delta(x_0 - nd), \qquad (3.27)$$

where $*$ denotes the convolution operation, d is the spacing of the $2N + 1$ encoding slits, and w is the width of each slit. Image subtraction is essentially a one-dimensional processing operation and is analyzed using one-dimensional notation for simplicity.

The encoded mutual coherence function at the input plane P_1 can therefore be written as

$$\Gamma(x_1 - x_2; \lambda) = \int \gamma(x_0, y_0) \exp\left[-i\frac{2\pi}{\lambda f}(x_1 - x_2)x_0\right] dx_0, \qquad (3.28)$$

or, in regard to Eq. (3.27),

$$\Gamma(x_1 - x_2; \lambda) = \sum_{n=-N}^{N} \mathrm{sinc}\left[\frac{\pi w(x_1 - x_2)}{\lambda f}\right] \exp\left[-i\frac{2\pi}{\lambda f}(x_1 - x_2)nd\right]. \qquad (3.29)$$

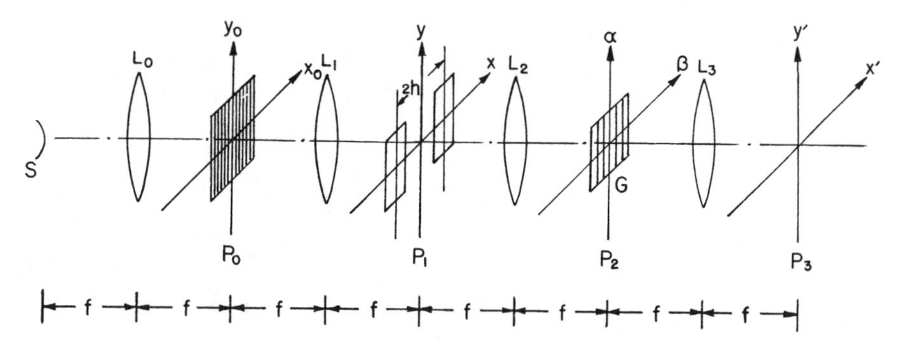

Figure 3.6. A partially coherent optical processing system for image subtraction: S, extended narrow-band incoherent source; P_0, encoded extended source plane; P_1, input plane; P_2, Fourier plane; P_3, output plane; L, achromatic lenses.

As shown in Fig. 3.6 two object transparencies, $A(x)$ and $B(x)$, are placed at the input plane P_1. These can be represented by

$$t(x) = A(x - h) + B(x + h), \tag{3.30}$$

where the separation between transparencies is $2h$. The mutual coherence function in the P_2 plane is simply the Fourier transform of $\Gamma(x_1 - x_2; \lambda)$ convolved with the spectrum of the amplitude transmittance function $t(x)$. The image subtraction of Lee [3.15] requires that a sinusoidal grating of spatial frequency v_0 be placed in the Fourier plane. Thus the mutual coherence function behind this grating is given by

$$\Gamma(\alpha_1, \alpha_2; \lambda) = [\Gamma(x_1 - x_2; \lambda)t(x_1)t^*(x_2)][1 + C\sin(2\pi v_0 \alpha_1)] \cdot [1 + C\sin(2\pi v_0 \alpha_2)]. \tag{3.31}$$

A final Fourier transform operation due to wavelength λ will give the intensity distribution at the output plane P_3:

$$I(x'; \lambda) = \iint \Gamma(\alpha_1, \alpha_2; \lambda)\exp\left[-i\frac{2\pi}{\lambda f}(\alpha_1 - \alpha_2)x'\right]d\alpha_1\, d\alpha_2, \tag{3.32}$$

$$\begin{aligned} I(x'; \lambda) = N[|A(x' - h)|^2 + |B(x' + h)|^2] + \frac{NC^2}{4}\{2\sin(2\pi w v_0) \\ \cdot \mathrm{Re}[A(x' - h + \lambda f v_0)B(x' + h - \lambda f v_0)] - |A(x' - h + \lambda f v_0)|^2 \\ - |B(x' + h - \lambda f v_0)|^2 - |A(x' - h - \lambda f v_0)|^2 \\ - |B(x' + h + \lambda f v_0)|^2\}, \end{aligned} \tag{3.33}$$

where Re denotes the real part of [ ]. From the above equation we see that there are six diffracted image terms at the output plane. If we only consider the diffracted images around the optical axis of the output plane, from Eq. (3.33) we have

$$\begin{aligned} I^{(0)}(x'; \lambda) = {} & |A(x' - h + \lambda f v_0)|^2 - 2\,\mathrm{sinc}(2\pi w v_0) \\ & \cdot \mathrm{Re}[A(x' - h + \lambda f v_0) B(x' + h - \lambda f v_0)] \\ & + |B(x' + h - \lambda f v_0)|^2. \end{aligned} \tag{3.34}$$

where $v_0 = h/\lambda f$.

We note that, if the slit size d equals zero, the analysis reduces to the case of strictly spatial coherence with the intensity distribution given by

$$I^{(0)}(x'; \lambda)|_{w=0} = [A(x') - B(x')]^2. \tag{3.35}$$

Equations (3.34) and (3.35) show that a high-contrast subtracted image requires a strictly spatial coherence system, and that the image quality will decrease as the slit width w increases.

To analyze the case of partial coherence we assume that the light source has a uniform spectral bandwidth and that the spectral response of the detector is also uniform [i.e., $S(\lambda) = K$; $C(\lambda) = K$]. Then the image intensity distribution at the output plane may be given by

$$\begin{aligned} I^{(0)}(x') = \int_{-\Delta\lambda/2}^{\Delta\lambda/2} & \{|A(x' + \lambda' f v_0)|^2 - 2\,\mathrm{sinc}(2\pi w v_0) \\ & \cdot \mathrm{Re}[A(x' + \lambda' f v_0) \cdot B(x' - \lambda' f v_0)] \\ & + |B(x' - \lambda' f v_0)|^2\} d\lambda', \end{aligned} \tag{3.36}$$

where $\lambda' = \lambda - \lambda_0$, λ_0 is the center frequency, and $\Delta\lambda$ is the spectral bandwidth of the source*; v_0 is the spatial frequency of the grating. Equation (3.36) may be simplified by using a Taylor series expansion for the input object functions. Thus for

$$\begin{aligned} A(x' + \lambda' f v_0) &\cong A(x') + \sum_{m=1}^{\infty} \frac{1}{m!} A^{(m)}(x')(\lambda' f v_0)^m, \\ B(x' - \lambda' f v_0) &\cong B(x') + \sum_{m=1}^{\infty} \frac{1}{m!} B^{(m)}(x')(\lambda' f v_0)^m, \end{aligned} \tag{3.37}$$

* Note: λ_0 and $\Delta\lambda$ are equivalent to the main wavelength λ_n and the bandwidth $\Delta\lambda_n$ of the narrower spectral band filter $H_n(\alpha, \beta)$.

where

$$A^{(m)}(x') = \frac{d^m A(x')}{dx'^m},$$
$$B^{(m)}(x') = \frac{d^m B(x')}{dx'^m}. \tag{3.38}$$

we find that

$$\begin{aligned}
I^{(0)}(x') = {} & [|A(x')|^2 - 2\,\mathrm{sinc}(2\pi w \nu_0) A(x')B(x') + |B(x')|^2]\Delta\lambda \\
& + \sum_{\substack{m=\text{even} \\ m \neq 0}} \frac{1}{2^{m-1}(m+1)!} (f\omega_0)^m [A^{(m)}(x') - \mathrm{sinc}(2\pi w \nu_0) B^{(m)}(x')] \\
& \cdot [A(x') + B(x')](\Delta\lambda)^{m+1} \\
& + \sum_{m}\sum_{m'}_{\substack{m+m'=\text{even} \\ m \neq 0 \\ m' \neq 0}} \frac{1}{2^{(m+m'-1)}(m+m'+1)m!m'!} \\
& \cdot (f\omega_0)^{m+m'} [A^{(m)}(x')A^{(m')}(x') \\
& - (-1)^{m'}\,\mathrm{sinc}(2\pi w \nu_0) A^{(m)(x')B(m')}(x') \\
& + (-1)^{m+m'} A^{(m)}(x')B^{(m')}(x')](\Delta\lambda)^{m+m'+1}.
\end{aligned} \tag{3.39}$$

This equation shows that a high-constrast subtracted image can be obtained with object transparencies of moderately low spatial frequency content. In addition this equation may be used to compute the spectral requirement (i.e., temporal coherence) of the light source.

An example analysis is presented to develop the modulation transfer function (MTF) equations. These are used to determine the temporal and spatial coherence requirements of the image subtraction process. Assume that the input object transparencies are given by

$$A(x) = 1, \qquad B(x) = \frac{1}{2[1 + C_0 \cos(2\pi \nu x)]}, \tag{3.40}$$

where C_0 is the contrast of the sinusoidal grating.

In the case of strict coherence (i.e., $\Delta\lambda = w = 0$), the subtracted image produces a contrast reversed image with intensity distribution

$$\bar{I}_0(x') = [A(x') - B(x')]^2, \tag{3.41}$$

or

$$\bar{I}_0(x') = \frac{1}{4} - \frac{C_0^2}{8} - \frac{C_0}{2}\cos(2\pi\nu x') - \frac{C_0^2}{8}\cos[2\pi(2\nu)x']. \tag{3.42}$$

For the partially coherent case due to Eq. (3.36), the intensity of the subtracted image can be shown to be

$$I_0(x') = \left[\frac{5}{4} - \frac{C_0^2}{8} - \operatorname{sinc}(2\pi w\nu_0)\right]\Delta\lambda + C_0\left[\frac{1}{2} - \operatorname{sinc}(2\pi w\nu_0)\right]$$
$$\cdot \operatorname{sinc}(f\pi\nu\nu_0\Delta\lambda)\Delta\lambda\cos(2\pi\nu x')$$
$$+ \frac{C_0^2}{8}\operatorname{sinc}(2\pi f\nu\nu_0\Delta\lambda)\Delta\lambda\cos[2\pi(2\nu)x']. \tag{3.43}$$

Note the addition of a second harmonic term $\{\cos[2\pi(2\nu)x']\}$ to the basic frequency. The MTF is defined as the ratio of the contrasts of the input and output sinusoidal objects. These are given by

$$MTF(\nu) = \frac{[1 - 2\operatorname{sinc}(2\pi w\nu_0)](2 - C_0^2)\operatorname{sinc}(f\pi\nu\nu_0\Delta\lambda)}{10 - C_0^2 - 8\operatorname{sinc}(2\pi w\nu_0)}, \tag{3.44}$$

$$MTF(2\nu) = \frac{(2 - C_0^2)\operatorname{sinc}(2\pi f\nu\nu_0\Delta\lambda)}{10 - C_0^2 - 8\operatorname{sinc}(2\pi w\nu_0)}. \tag{3.45}$$

As previously stated Eqs. (3.44) and (3.45) allow the evaluation of the temporal and spatial coherence requirements for image subtraction.

3.3.1. Temporal Coherence Requirement

The case of strictly spatial coherence is discussed first. This requires that the slit width w approach zero. Equations (3.44) and (3.45) then become

$$MTF(\nu) = \operatorname{sinc}(\pi f\nu\nu_0\Delta\lambda), \tag{3.46}$$

$$MTF(2\nu) = \operatorname{sinc}(2\pi f\nu\nu_0\Delta\lambda). \tag{3.47}$$

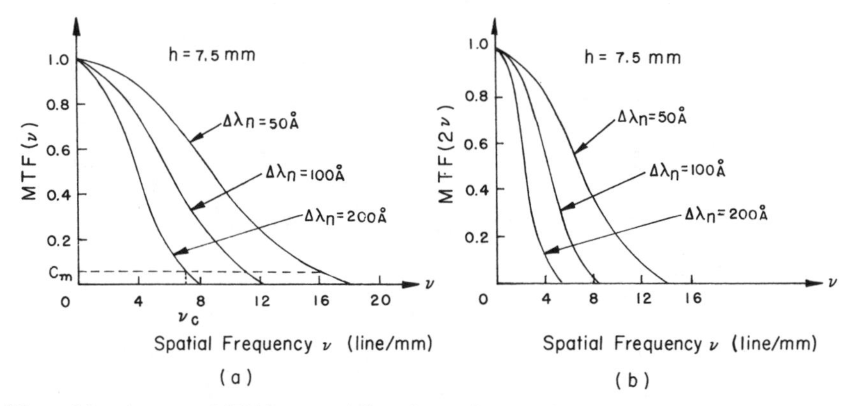

Figure 3.7. Apparent MTF for a partially coherent image subtraction. (*a*) Basic frequency. (*b*) Second harmonic.

The normalized *MTF* curves of the basic and harmonic frequencies are shown in Fig. 3.7. It is apparent that the contrast of the subtracted image decreases monotonically as a function of the object spatial frequency. However, the *MTF* of the subtracted image decreases as the spectral bandwidth of the light source increases. In other words, the quality of the subtracted image improves as the spectral bandwidth of light source and the spatial frequency of the object decrease.

Let ν_c be the cut-off spatial frequency where the *MTF* decreases to a minimum value C_m, as shown in Fig. 3.7. The value C_m depends on the maximum resolution of the output recording material. Figure 3.8 shows the functional

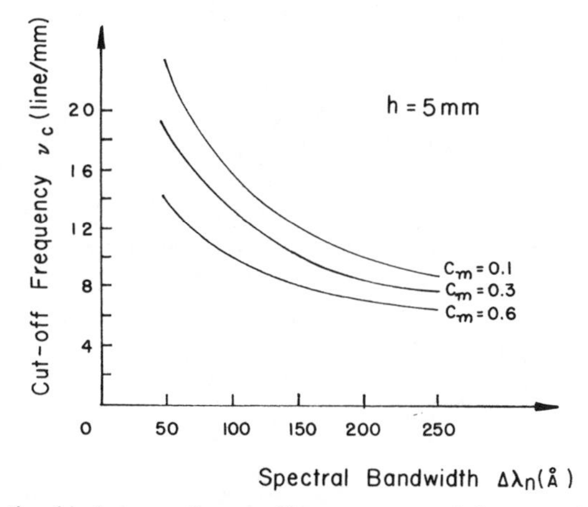

Figure 3.8. Relationship between the cut-off frequency ν_c and the spectral bandwidth $\Delta\lambda_n$ of $H_n(\alpha, \beta)$ for different minimum desirable contrasts C_m.

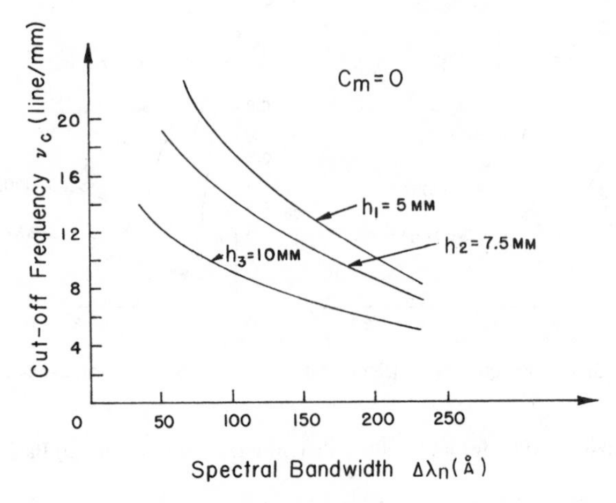

Figure 3.9. Relationship between the cut-off frequency ν_c and the spectral bandwidth $\Delta\lambda_n$ for various values of separation h. $2h$ is the main separation between the input object transparencies.

relationship of the cut-off frequency ν_c and the spectral width $\Delta\lambda_n$ for various values of C_m. It is possible to determine the spectral bandwidth requirement $\Delta\lambda_n$ from this figure. The relationship between the cut-off frequency ν_c, the spectral width $\Delta\lambda_n$, and the separation between two input transparencies h is shown in Fig. 3.9. Note that the spectral bandwidth required for a given cut-off frequency decreases with increasing separation h. Table 3.3 illustrates the dependence of $\Delta\lambda_n$ on ν_c and h. The focal length of the Fourier transform lens is assumed to be $f = 300$ mm for the calculation. It is clear from the table that, as the spatial frequency and object separation increase, the spectral bandwidth of the light source must decrease.

3.3.2. Spatial Coherence Requirement

Consider the case of perfect temporal coherence (i.e., $\Delta\lambda_n = 0$) and partially spatial coherence, where Eqs. (3.44) and (3.45) are of the forms

$$MTF_1 = \frac{(2 - C_0^2)[1 - 2\,\mathrm{sinc}(2\pi w\nu_0)]}{10 - C_0^2 - 8\,\mathrm{sinc}(2\pi w\nu_0)}, \tag{3.48}$$

and

$$MTF_2 = \frac{2 - C_0^2}{10 - C_0^2 - 8\,\mathrm{sinc}(2\pi w\nu_0)}. \tag{3.49}$$

Table 3.3. Temporal Coherence Requirement for Different ν_c and h.

	$\Delta\lambda$(Å)				
h(mm)	ν = 4 l/mm	ν = 8 l/mm	ν = 12 l/mm	ν = 16 l/mm	ν = 18 l/mm
5.0	240	135	92	67	52
7.5	176	92	62	45	32
10.0	130	70	45	32	17

Note that the *MTF*s are independent of the object's spatial bandwidth. The above equations are, however, dependent on the slit width w. This requires that the grating be precisely designed to match the separation of the object transparencies, that is

$$\nu_0 = \frac{h}{\lambda f}. \tag{3.50}$$

The plots of the *MTF* versus the separation h for Eqs. (3.48) and (3.49) are shown in Fig. 3.10. It is therefore apparent that to obtain a high-contrast subtraction image the separation h must be reduced. However, decreasing the separation between the object transparencies limits the size of the input objects to be processed. The relationship between the *MTF*, the object transparency separation h, and the slit width w is numerically presented in Table 3.4, where the focal length of the Fourier transform lens was again assumed to be 300 mm.

Note that this table indicates the necessity of a very narrow source size to achieve an adequate *MTF*. However, a high-intensity narrow source size is

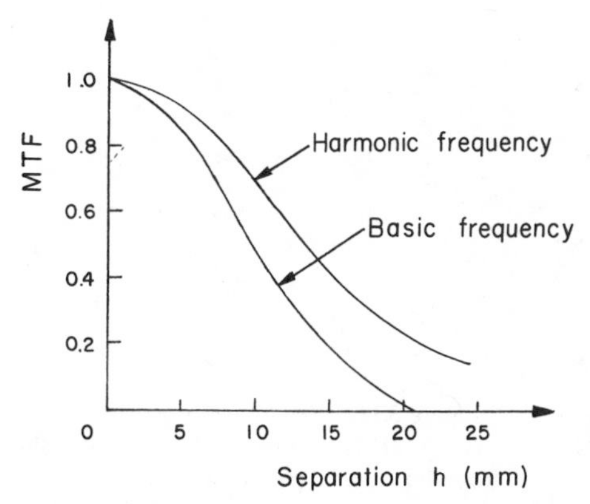

Figure 3.10. Apparent MTF versus the separation h.

Table 3.4. Source Size for Image Subtraction Under Different MTF, and Separation h

	w(mm)		
MTF	$h = 5.0$ mm	$h = 7.5$ mm	$h = 10.0$ mm
0.1	0.0076	0.005	0.0038
0.3	0.0052	0.0035	0.0026
0.6	0.0031	0.0021	0.0015

Table 3.5. Spatial Coherence Requirement for Various w/d

w/d	0.05	0.10	0.20	0.30
MTF	0.85	0.57	0.18	0.006

difficult to achieve in practice. Considerable progress toward solving the problem can be made if source encoding techniques for image subtracting are used [3.13, 3.14].

A multislit source encoding mask is used for this illustration. The spatial period of the encoding mask should be precisely equal to that of the diffraction grating G (i.e., $d = 1/\nu_0$). The spatial coherence requirement, although independent of the slit size, is governed by the ratio of the slit width to the spatial period of the encoding mask, that is, w/d. The ratio w/d must be relatively small to achieve a high degree of spatial coherence. The dependence of the MTF on w/d is shown in Fig. 3.11. It is obvious that the subtraction effect ceases when the MTF approaches zero, that is, when $w/d = 0.3$. A few numerical examples are presented in Table 3.5.

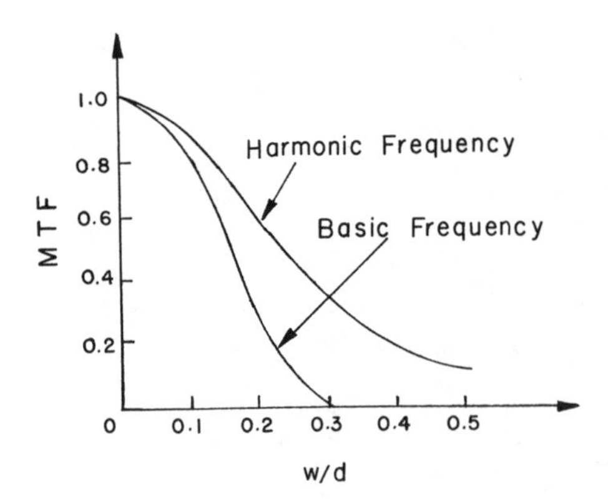

Figure 3.11. Relationship between MTF (ν), MTF (2ν), and the ratio of the slit width to spatial period, w/d.

3.4. COHERENCE REQUIREMENTS FOR SIGNAL CORRELATION

Recent investigations of Morris and George [3.9] have led to a relationship between the correlation intensity and the wavelength spread of the light source, using an argon laser and a spectrally broad-band light source. Prior to their work, Watrasiewicz [3.16] had evaluated the effect of spatial coherence on the correlation intensity with a rectangular function. In this section, we quantitatively evaluate the temporal and spatial coherence requirements for a white-light optical correlation detector that we proposed [3.17].

3.4.1. Temporal Coherence Requirement

We now discuss the temporal coherence requirement for a partially coherent optical correlator. With reference to the white-light optical signal processor of Fig. 3.1, we assume that a white-light point source [i.e., $\gamma(x_0, y_0) = \delta(x_0, y_0)$] is utilized for the temporal coherence investigation. Again for simplicity we let $S(\lambda)$ and $C(\lambda)$ be constant. The output intensity distribution of Eq. (3.10) becomes [see Eq. (3.12)]

$$I(x', y') = \sum_{n=1}^{N} \int_{\lambda_{ln}}^{\lambda_{hn}} |A_n(x', y'; \lambda)|^2 \, d\lambda, \tag{3.51}$$

where N is the total number of the spectral band filters, λ_{ln} and λ_{hn} are the lower and upper wavelength limits of the nth narrow spectral band filter $H_n(\alpha, \beta)$, $\Delta\lambda_n = |\lambda_{hn} - \lambda_{ln}|$ is the spectral bandwidth of the wavelength spread over $H_n(\alpha, \beta)$,

$$A_n(x', y'; \lambda) = \iint S(\alpha - \lambda f v_0, \beta) H_n(\alpha, \beta) \cdot \exp\left[-i \frac{2\pi}{\lambda f}(\alpha x' + \beta y')\right] d\alpha \, d\beta, \quad \text{for } \lambda_{ln} \leqq \lambda \leqq \lambda_{hn}, \tag{3.52}$$

$S(\alpha - \lambda f v_0, \beta)$ is the smeared signal spectra,

$$H_n(\alpha, \beta) = S^*(\alpha - \lambda_n f v_0, \beta) \tag{3.53}$$

is the nth narrow spectral band matched filter, λ_n is the center wavelength of the filter, and the superscript $*$ denotes the complex conjugate.

By virture of Eq. (3.53), we further note that Eq. (3.52) can be approximated by the following equation:

$$A_n(x', y'; \lambda) \cong \frac{\lambda_n}{\lambda} s(x', y') * s(x', y'), \tag{3.54}$$

where $*$ denotes the correlation operation.

As a special example to illustrate the temporal coherence requirement, let us consider a one-dimensional spatially Gaussian target, that is,

$$s(x, y) = \exp(-a^2 y^2), \tag{3.55}$$

where a is an arbitrary constant. The corresponding signal spectrum at the spatial frequency plane is

$$S(\alpha - \lambda_n f \nu_0, \beta) = \frac{\sqrt{\pi}}{a} \delta(\alpha - \lambda_n f \nu_0) \exp\left[-\left(\frac{\pi\beta}{a\lambda f}\right)^2 \right], \tag{3.56}$$

where $\delta(\ )$ is the Dirac delta function, λ is the wavelength of the light source, and f is the focal length of the achromatic transform lens. If a set of narrow spectral band matched filters [i.e., Eq. (3.53)] is properly placed in the Fourier plane of the partially coherent processor of Fig. 3.1, then the output irradiance would be

$$I(x', y') \cong \sum_{n=1}^{N} \frac{2\pi}{a^2} \int_{\lambda_{ln}}^{\lambda_{hn}} \frac{\lambda_n^2}{\lambda^2 + \lambda_n^2} \exp\left[-2a^2 \left(\frac{\lambda_n^2}{\lambda^2 + \lambda_n^2}\right)(y')^2 \right] d\lambda. \tag{3.57}$$

From this equation, we see that the output intensity is again Gaussian distributed. The correlation peak occurs at $y' = 0$, that is,

$$I(x', 0) = 2N\lambda_n \left[\tan^{-1}\left(\frac{1 + \Delta\lambda_n}{2\lambda_n}\right) - \tan^{-1}(1) \right]. \tag{3.58}$$

Since the spatially Gaussian target essentially has a low spatial frequency content, the normalized correlation peak, that is, $I(x', 0)/N\Delta\lambda_n$, is relatively independent of the spectral bandwidth, as shown in Fig. 3.12.

To investigate the effect of the spatial frequency on the temporal coherence

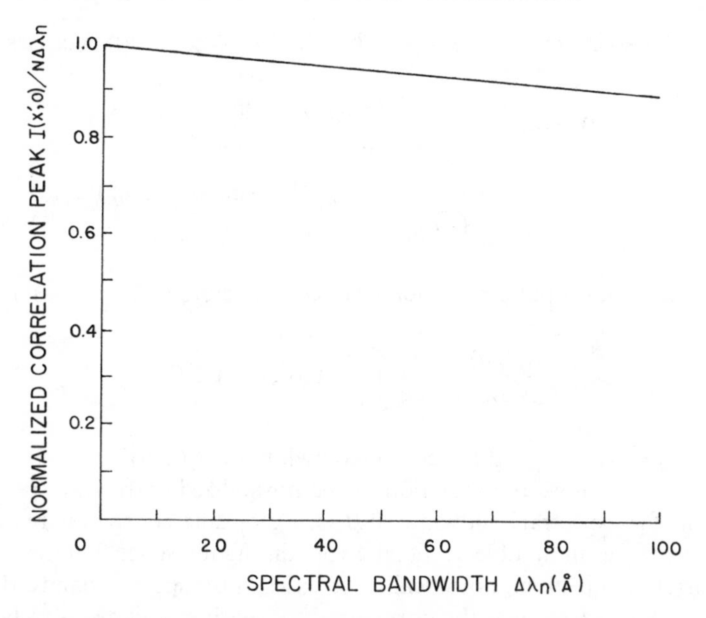

Figure 3.12. Normalized correlation peak as a function of spectral bandwidth $\Delta\lambda_n$ of $H_n(\alpha, \beta)$ for a Gaussian spatial signal.

requirement, we now assume a one-dimensional sinusoidal function as the input object, that is,

$$s(x, y) = \operatorname{rect}\left(\frac{y}{w}\right)[1 + \cos(q_0 y)] \tag{3.59}$$

where q_0 is the angular spatial frequency of the input sinusoid,

$$\operatorname{rect}\left(\frac{y}{w}\right) \triangleq \begin{cases} 1, & |y| \leqq \dfrac{w}{2}, \\ 0, & \text{otherwise,} \end{cases}$$

and w is the spatial extension of the input object. By substituting Eq. (3.59) into Eq. (3.54), we have

$$A_n(x', y'; \lambda) \cong \frac{\lambda_n}{\lambda} \int \operatorname{rect}\left(\frac{\eta}{w}\right)[1 + \cos(q_0 \eta)] \cdot \operatorname{rect}\left(\frac{\eta - y'}{w_0}\right)\{1 + \cos[q_0'(\eta - y')]\} d\eta, \tag{3.60}$$

where $w_0 = w\lambda/\lambda_n$ *and* $q_0' = q_0\lambda/\lambda_n$. Thus at $y' = 0$, Eq. (3.60) becomes

$$A_n(x', 0; \lambda) = \frac{\lambda_n}{\lambda}\left\{w + \frac{\sin(wq_0'/2)}{wq_0'/2} + \frac{\sin(wq_0/2)}{wq_0/2} + \frac{\sin[w(q_0 - q_0')/2]}{[w(q_0 - q_0')/2]} + \frac{\sin[w(q_0 + q_0')/2]}{w(q_0 + q_0')/2}\right\}. \tag{3.61}$$

The normalized output correlation peak can therefore be written as

$$\frac{I(x', 0)}{N\Delta\lambda_n} \cong \frac{1}{\Delta\lambda_n}\int_{\lambda_{ln}}^{\lambda_{hn}} |A_n(x', 0; \lambda)|^2 \, d\lambda, \tag{3.62}$$

where $\Delta\lambda_n = |\lambda_{hn} - \lambda_{ln}|$, the spectral bandwidth of $H_n(\alpha, \beta)$.

Figure 3.13 shows the variation of the normalized correlation peak as a function of the spectral bandwidth of $H_n(\alpha, \beta)$, for different spatial frequencies of the target (i.e., input object) signal. From this figure we see that the normalized correlation peak monotonically decreases as the spectral bandwidth $\Delta\lambda_n$ increases. We also see that the normalized correlation peak drops rather rap-

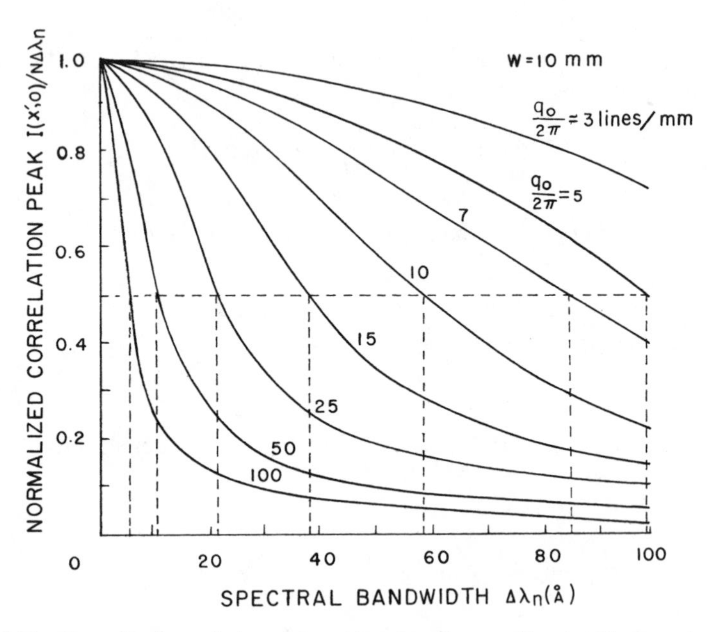

Figure 3.13. Normalized correlation peak as a function of spectral bandwidth, for various input spatial frequencies.

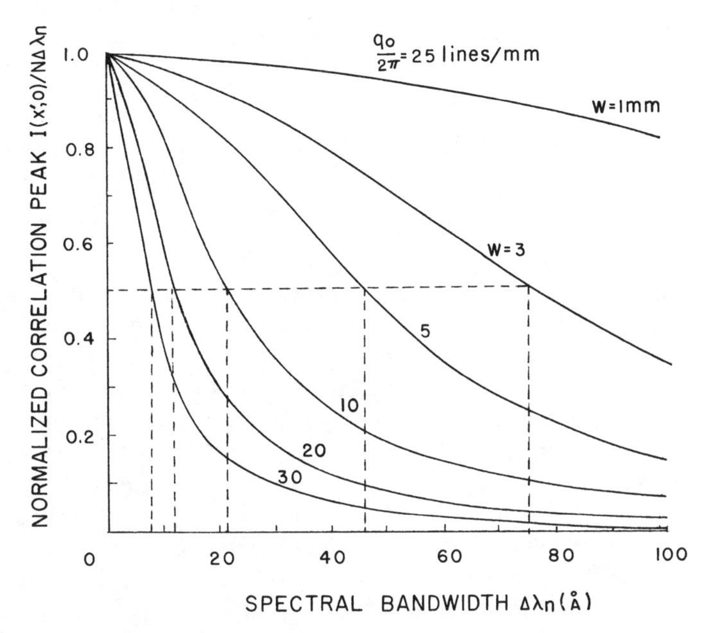

Figure 3.14. Normalized correlation peaks as a function of spectral bandwidth, for different target extensions.

idly as the spatial frequency of the target increases. On the other hand, Fig. 3.14 shows the variation of the normalized correlation peak as a function of the spectral bandwidth $\Delta\lambda_n$ for various target extensions w. Again we see that the normalized correlation peak decreases monotonically as $\Delta\lambda_n$ increases. And the normalized peak decreases very rapidly as the target extension w increases. Thus from Figs. 3.13 and 3.14 we see that the normalized correlation peak strongly depends on the spatial frequency and the extension of the target.

If we use the half-power-point criterion, the spectral bandwidth $\Delta\lambda_n$ (i.e., the temporal coherence requirement) of the filter $H_n(\alpha, \beta)$ can be determined from Figs. 3.13 and 3.14. Some of the numerical results are tabulated in Tables 3.6 and 3.7. From these we see that for a relatively small and lower

Table 3.6. Temporal Coherence Requirement as a Function of Spatial frequency $q_0/2\pi$ where $w = 10$ mm and $\lambda_n = 5461$ Å

$q_0/2\pi$(lines/mm)	1	3	5	7	10	25	50	100
$\Delta\lambda_n$(Å)	540	180	99	85	58	22	11	6

Table 3.7. Temporal Coherence Requirement as a Function of the Extension of target w, where $q_0/2\pi = 25$ lines/mm and $\lambda_n = 5461$ Å

w(mm)	1	2	3	5	10	20	30	60
$\Delta\lambda$(Å)	220	99	75	46	22	11	8	4.7

spatial frequency target, a relatively broader spectral bandwidth of spatial filter $H_n(\alpha, \beta)$ can be used. In other words, the higher the spatial frequency of the target or the larger the extension of the detecting signal, the higher the degree of temporal coherence required for each of the narrow spectral band filters. It is apparent that the degree of temporal coherence in the Fourier plane can be increased simply by increasing the sampling frequency ν_0 of the grating. However, when ν_0 is increased, larger achromatic transform lenses may be needed. This is the price we pay to improve the temporal coherence. Furthermore, from Tables 3.6 and 3.7, a plot of the space bandwidth product as a function of the required spectral bandwidth is shown in Fig. 3.15. From this figure, we see that the space bandwidth product exponentially decreases, as the spatial bandwidth of the filter $H_n(\alpha, \beta)$ increases. Thus for a large space

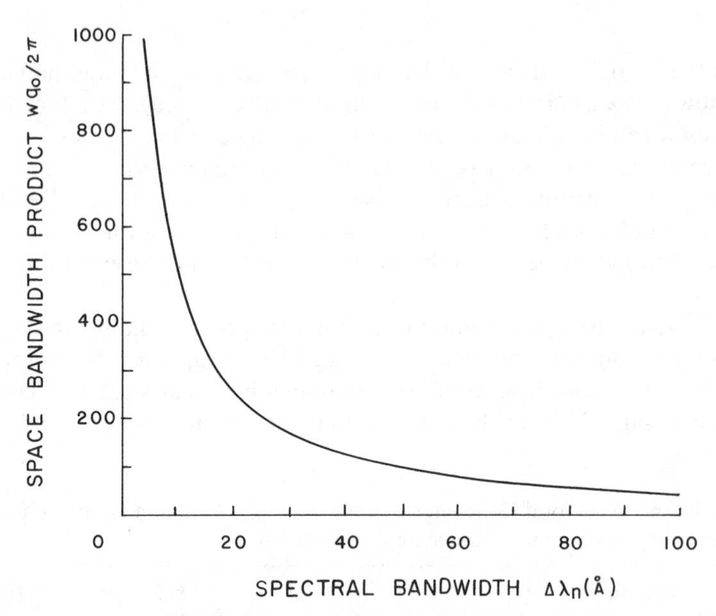

Figure 3.15. Space bandwidth product as a function of the required spectral bandwidth $\Delta\lambda_n$ of $H_n(\alpha, \beta)$.

bandwidth product, an extremely high degree of temporal coherence for the narrow spectral band filter $H_n(\alpha, \beta)$ is required.

3.4.2. Spatial Coherence Requirement

We now discuss the spatial coherence requirement for the correlation detection. Again, for simplicity, we assume that the light source $\gamma(x_0, y_0)$ is a one-dimensional extended monochromatic source, that is,

$$\gamma(x_0, y_0) = \operatorname{rect}\left(\frac{y_0}{\Delta s}\right)\delta(x_0), \tag{3.63}$$

where

$$\operatorname{rect}\left(\frac{y_0}{\Delta s}\right) \triangleq \begin{cases} 1, & |y_0| \leqq \dfrac{\Delta s}{2}, \\ 0, & \text{otherwise}, \end{cases} \tag{3.64}$$

and Δs is the size of the extended source. Similarly, the output plane intensity distribution can be shown as

$$I(x', y') = \int_{-\Delta s/2}^{\Delta s/2} \int_{-\infty}^{\infty} |A(x', y'; x_0, y_0)|^2 \, dx_0 \, dy_0, \tag{3.65}$$

where

$$A(x', y'; x_0, y_0) = \iint s^*(\xi, \eta) s(\xi - x', \eta - y') \exp\left[i \frac{2\pi}{\lambda_0 f}(x_0 \xi + y_0 \eta) \right] d\xi \, d\eta, \tag{3.66}$$

and λ_0 is wavelength of the monochromatic source.

We again consider a spatially Gaussian target of Eq. (3.55), where we obtain

$$A(x', y'; x_0, y_0) = \frac{1}{a}\sqrt{\frac{\pi}{2}} \exp\left[-\left(\frac{\pi y_0}{a\sqrt{2}\lambda_0 f} \right)^2 \right] \exp\left[-\frac{a^2}{2}(y')^2 \right]. \tag{3.67}$$

The corresponding output intensity distribution can be shown as

$$I(x', y') = \frac{\pi}{a^2} \exp[-a^2 (y')^2] \int_0^{\Delta s/2} \exp\left[-\left(\frac{\pi y_0}{a f \lambda_0} \right)^2 \right] dy_0. \tag{3.68}$$

Since the integral is independent of y', it can be shown that the output intensity, as before, has a Gaussian-like distribution. To determine the correlation peak, we let $y' = 0$, that is,

$$I(x; 0) = \frac{\pi}{a^2} \int_0^{\Delta s/2} \exp\left[-\left(\frac{\pi y_0}{af\lambda_0}\right)^2 \right] dy_0. \tag{3.69}$$

It is evident from this equation that $I(x; 0)$ depends on the size of the extended monochromatic source ΔS. Figure 3.16 shows the normalized correlation peak $I(x; 0)/\Delta S$ of the Gaussian target. We see that the correlation peak drops quickly with increasing size of the source.

We now consider the effect of spatial frequency on the correlation peak. Again we use a spatially limited sinusoidal grating of (3.59) as the input signal. For simplicity of notation we let $q_0 = 2\pi y_0/(x_0 f)$ and $g = \pi/(\lambda_0 f)$. At $y' = 0$, Eq. (3.66) becomes

$$\begin{aligned} A(0; y_0) = {} & \frac{\sin(gwy_0)}{gy_0} + \frac{\sin[w(gy_0 - 2q_0)]}{gy_0 - 2q_0} \\ & + \frac{\sin[w(gy_0 + 2q_0)]}{gy_0 + 2q_0} + \frac{1}{gy_0 + q_0} \\ & \cdot \left[\sin(wgy_0) \cdot \cos^2(2wq_0) \right. \\ & \left. + \frac{2q_0 \sin[w(gy_0 - q_0)]}{gy_0 - q_0} + \frac{2q_0 \sin(wgy_0)}{gy_0} \right], \end{aligned} \tag{3.70}$$

where a_0 is the angular spatial frequency of the input sinusoidal grating, and w is its spatial extension. And the output irradiance can be determined by the following equation:

$$I(x', 0) = 2w \int_0^{\Delta s/2} |A(0; y_0)|^2 \, dy_0. \tag{3.71}$$

Since g is not related to x_0, the spatial coherence requirement is evidently independent of the spatial frequency of the input object, as expected. However the spatial coherence strongly depends on the extension of the target, as can be seen in Fig. 3.17. From this figure, we see that the normalized correlation peak monotonically decreases with increasing source size. The rate of the decrease is rapidly increased for larger extension of the target. Again by using the half-power-point criterion, the spatial coherence requirement for various

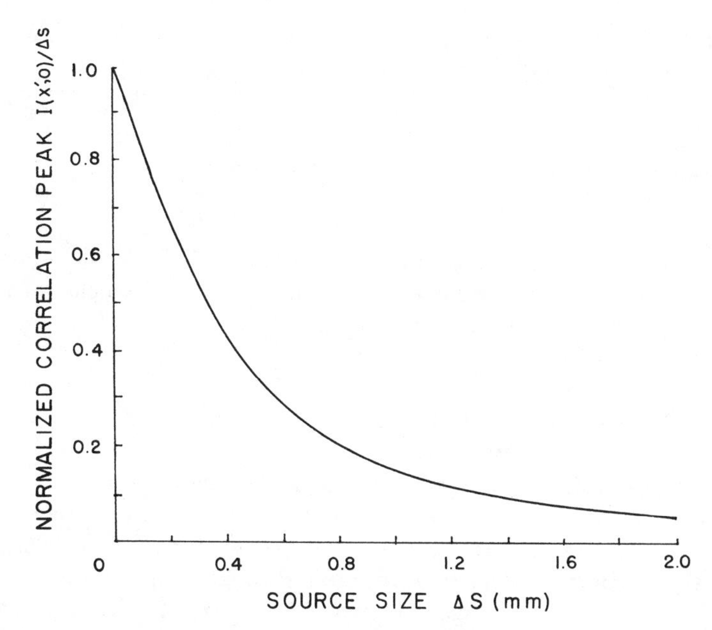

Figure 3.16. Normalized correlation peak as a function of source size, for a Gaussian spatial signal.

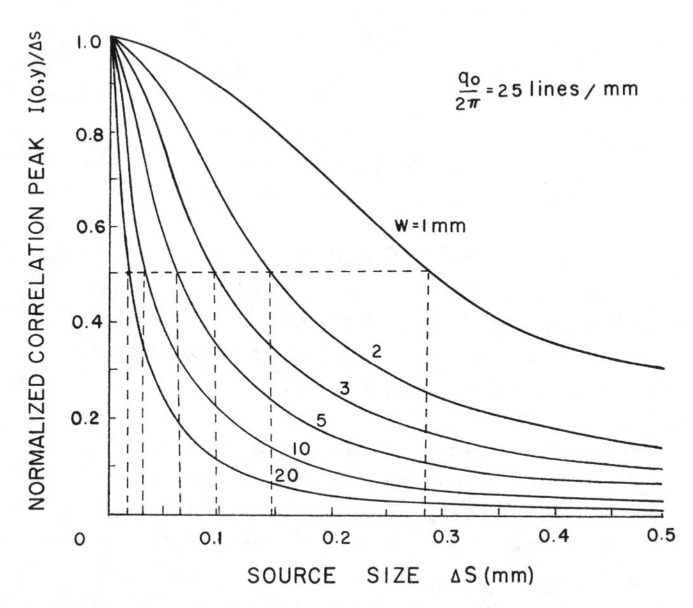

Figure 3.17. Normalized correlation peak as a function of source size, for different target extensions.

Table 3.8. Spatial Coherence Requirement as a Function of the Extension of Target, where $\lambda_0 = 5461$ Å

w(mm)	1	2	3	5	10	20	50
Δs(mm)	0.28	0.14	0.09	0.07	0.03	0.015	0.006

values of the target extension are tabulated in Table 3.8. The table shows that the spatial coherence requirement (i.e., source size requirement) increases rapidly, as the target extension increases.

3.5. SUMMARY AND REMARK

In summary, we have shown that the temporal and spatial coherence requirements for some partially coherent optical processing operations, namely, image deblurring, image subtraction, and signal correlation, can be determined in terms of the output intensity distribution through propagation of mutual coherence function. For image deblurring the temporal coherence requirement depends on the ratio of the deblurring width to the smeared length of the blurred image. To obtain a higher degree of deblurring a narrower spectral width of the spatial filter $H_n(\alpha, \beta)$ is required. For example, if the deblurring ratio $\Delta w/w$ is 0.1, the spectral width $\Delta\lambda_n$ should be < 640 Å.

For the spatial coherence requirement, the image deblurring depends on both the deblurring ratio $\Delta w/w$ and the smeared length w. If the deblurring ratio $\Delta w/w = 1/10$ and $w = 1$ mm, a slit source < 0.26 mm should be used. For a smeared image deblurring operation the constraints of the temporal and spatial coherence requirements are not critical, and can be achieved in practice.

For image subtraction, we have utilized an extended narrow-band source instead of a white-light source for the evaluation. The temporal coherence requirement is determined by the highest spatial frequency and the separation of the input object transparencies. If the separation and spatial frequency of the input transparencies are high, a narrower spectral bandwidth of the light source is required.

For the spatial coherence requirement, the MTF, which determines the contrast of the subtracted image, depends on the ratio of the slit width to the spatial period of the encoding mask, that is, w/d. If the ratio w/d is low, a higher contrast subtracted image can be obtained. For example, with $w/d = 0.05$, a relatively higher MTF $= 0.85$ can be obtained. Compared with the image deblurring operations, the coherence requirements are more stringent for the subtraction process.

We have also developed equations for the coherence requirements of correlation detection for partially coherent optical processing. In addition, we have calculated numerical estimates for these requirements using the half-power-point criterion for some special cases. We have shown that the temporal coherence requirement strongly depends on the spatial frequency and the extension (i.e., space bandwidth product) of the target signal to be detected. However, the spatial coherence requirement depends only upon the extension of the target. We have shown the coherence requirement for a spatially Gaussian signal as a special case. For this case, the normalized correlation peak is relatively independent of the spectral bandwidth of the light source, but it decreases rather rapidly as the source size increases.

In concluding this chapter, we remark that the solution to the coherence requirement for the white-light or partially coherent processing is not restricted to the examples that are presented; it can be applied to any other partially coherent signal processing problems.

REFERENCES

3.1. G. L. Rogers, "Non-coherent optical processing," *Opt. Laser Technol.* **7**, 153 (1975).

3.2. K. Bromley, "An optical incoherent correlation," *Opt. Acta* **21**, 35 (1974).

3.3. M. A. Monahan, K. Bromley, and R. P. Bocker, "Incoherent optical correlations," *Proc. IEEE* **65**, 121 (1977).

3.4. G. L. Rogers, *Noncoherent Optical Processing*, John Wiley, New York, 1977.

3.5. S. Lowenthal and P. Chavel, "Noise problems in optical image processing," in *Proceedings, ICO Jerusalem* 1976 *Conference on Holography and Optical Processing*, E. Marom, A. Friesem, and E. Wiener-Avnear, Eds., Pergamon, New York, 1977.

3.6. A. Lohmann, "Incoherent optical processing of complex data," *Appl. Opt.* **16**, 216 (1977).

3.7. W. T. Rhodes, "Bipolar point spread function by phase switching," *Appl. Opt.* **16**, 265 (1977)

3.8. E. N. Leith and J. Roth, "White-light optical processing and holography," *Appl. Opt.* **16**, 2565 (1977).

3.9. G. M. Morris and N. George, "Space and wavelength dependence of a dispersion-compensated matched filter," *Appl. Opt.* **19**, 3843 (1980).

3.10. M. Born and E. Wolf, *Principle of Optics*, 2nd rev. ed., Pergamon Press, New York, 1964.

3.11. S. L. Zhuang and F. T. S. Yu, "Coherence requirements for partially coherent optical processing," *Appl. Opt.* **21**, 2587 (1982).

3.12. F. T. S. Yu, Y. W. Zhang, and S. L. Zhuang, "Coherence requirement for partially coherent correlation detection," *Appl. Phys.* **B30**, 23 (1983).

3.13. S. T. Wu and F. T. S. Yu, "Image subtraction with encoded extended incoherent source," *Appl. Opt.* **20**, 4082 (1981).

3.14. F. T. S. Yu, "Source encoding, signal sampling and spectral band filtering for partially coherent optical signal processing," *J. Opt.* **14**, 173 (1983).

3.15. S. H. Lee, S. K. Yao, and A. G. Milnes, "Optical image synthesis (complex amplitude addition and subtraction) in real-time by a diffraction-grating interferometric method," *J. Opt. Soc. Am.* **60**, 1037 (1970).

3.16. B. Watrasiewicz, "Effect of spatial coherence on the correlation spot intensity," *Opt. Acta* **16**, 321 (1969).

3.17. F. T. S. Yu, "A new technique of incoherent complex signal detection," *Opt. Commu.* **27**, 23 (1978).

4

Apparent Transfer Function

The description of a transfer function for a linear spatially invariant optical image processing system is an important concept in image evaluation. The techniques of using amplitude and intensely sinusoidal objects as input signals to determine the transfer characteristic of a coherent and an incoherent optical system have been reported by O'Neill and Goodman [4.1, 4.2]. They used the concept of system transfer function, as a criterion, to evaluate the image quality of a coherent and an incoherent optical system. However, a strictly coherent or incoherent optical wave field that occurs in practice consists of a very limited degree of coherence, because the electromagnetic radiation from a real physical source is never strictly monochromatic. In reality, a physical source cannot be a point, but rather a finite extension that consists of many elementary radiators.

4.1. INTRODUCTION

The optical system under the partially coherent regime has been studied by Becherer and Parrent [4.3], and by Swing and Clay [4.4]. They have shown that there are difficulties in applying the linear system theory to the evaluation of imagery at high spatial frequencies. These difficulties are primarily due to the inapplicability of the linear system theory under a partial coherence regime. Nevertheless, using a sinusoidal analysis, an apparent transfer function for a white-light or partially coherent optical system can be obtained. The result is appreciably different from those that would have been obtained from a linear system concept, either in intensity or in complex amplitude. In a more recent paper, Dutta and Goodman [4.5] described a procedure for sampling the mutual intensity function so that the image of a partially coherent object can be reconstructed.

In this chapter, we show that an apparent transfer function for a white-light optical signal processing system can be derived from the partial coherence theory of Wolf [4.6, 4.7]. We show that the apparent transfer function of a white-light or partially coherent optical processor is dependent upon the temporal and spatial coherence of the processor. We also show that the concept of transfer function is a valuable one that can be used as a criterion for selecting an appropriate incoherent light source for a specific information processing operation.

4.2. SYSTEM RESPONSE UNDER STRICTLY COHERENT AND INCOHERENT REGIMES

We have, in Section 2.2, discussed signal processing under strictly coherent and incoherent regimes. We have shown that under strictly coherent illumination (i.e., spatially and temporally coherent), the optical system operates in a complex wave field, such that the output complex light field is described by a complex amplitude convolution integral, such as

$$g(\alpha, \beta) = \iint_{-\infty}^{\infty} f(x, y)h(\alpha - x, \beta - y)dx\, dy, \tag{4.1}$$

which yields a spatially invariant property of the complex amplitude transformation, where $f(x, y)$ is a coherent complex wave field that impinges at the input end of an optical system, and $h(x, y)$ describes the spatial impulse response of the optical system. We have further shown that Eq. (4.1) can be written in the following Fourier transform forms:

$$G(p, q) = F(p, q)H(p, q), \tag{4.2}$$

where $G(p, q)$, $H(p, q)$, and $F(p, q)$ are the Fourier transforms of $g(x, y)$, $h(x, y)$, and $f(x, y)$, respectively. Thus the effect of Eq. (4.2) describes the optical system operation in the spatial frequency domain, where $H(p, q)$ can be referred to as the *coherent* or *complex amplitude transfer function* of the optical system.

On the other hand, if the optical system is under strictly incoherent illumination, we have shown that the optical system operates in intensity, rather than in complex wave field, such that the output irradiance is described by the following intensity convolution integral:

$$I(\alpha, \beta) = \iint_{-\infty}^{\infty} I_i(x, y)h_i(\alpha - x, \beta - y)dx\, dy, \tag{4.3}$$

which yields a spatially invariant form of the intensity operation, where $I_i(x, y) = |f(x, y)|^2$ is the input irradiance and $h_i(x, y) = |h(x, y)|^2$ is the intensity spatial impulse response of the optical system. Similarly, Eq. (4.3) can be written in Fourier transform forms, that is,

$$I(p, q) = F_i(p, q)H_i(p, q), \tag{4.4}$$

where $I(p, q)$, $F_i(p, q)$, and $H_i(p, q)$ are the Fourier transforms of $I(x, y)$, $f_i(x, y)$, and $h_i(x, y)$, respectively, and $H_i(p, q)$ can be referred to as the *incoherent* or *intensity transfer function* of the optical system.

Let us now define the following normalized quantities:

$$\tilde{I}(p, q) = \frac{\iint\limits_{-\infty}^{\infty} I(\alpha, \beta)\exp[-i(p\alpha + q\beta)]d\alpha\, d\beta}{\iint\limits_{-\infty}^{\infty} I(\alpha, \beta)d\alpha\, d\beta}, \tag{4.5}$$

$$\tilde{F}_i(p, q) = \frac{\iint\limits_{-\infty}^{\infty} f_i(x, y)\exp[-i(px + qy)]dx\, dy}{\iint\limits_{-\infty}^{\infty} f_i(x, y)dx\, dy}, \tag{4.6}$$

and

$$\tilde{H}_i(p, q) = \frac{\iint\limits_{-\infty}^{\infty} h_i(x, y)\exp[-i(px + qy)]dx\, dy}{\iint\limits_{-\infty}^{\infty} h_i(x, y)dx\, dy}. \tag{4.7}$$

These normalized quantities would give a set of convenient mathematical forms and also provide a concept of image contrast interpretation. Since the quality of a visual image is to a large extent dependent on the contrast, or the relative irradiance, of the information, a normalized intensity distribution would certainly enhance its information bearing capacity.

With the application of the Fourier convolution theorem to Eq. (4.3), we can write the following relationship:

$$\tilde{I}_i(p, q) = \tilde{F}_i(p, q)\tilde{H}_i(p, q), \tag{4.8}$$

where $\tilde{H}_i(p, q)$ of Eq. (4.7) is commonly referred to as the *optical transfer function* (OTF) of the optical system, and the modulus of $\tilde{H}_i(p, q)$ is known as the *modulation transfer function* (MTF) of the optical system.

We further note that Eq. (4.7) can also be written as [see Eq. (2.20)]

$$\tilde{H}_i(p, q) = \frac{\iint_{-\infty}^{\infty} H(p', q')H^*(p' - p, q' - q)dp'\, dq'}{\iint_{-\infty}^{\infty} |H(p', q')|^2\, dp'\, dq'}. \tag{4.9}$$

By changing the variables

$$p'' = p' + \frac{p}{2}, \qquad q'' = q' + \frac{q}{2},$$

Eq. (4.9) results in the following symmetrical form:

$$\tilde{H}_i(p, q) = \frac{\iint_{-\infty}^{\infty} H(p'' + p/2, q'' + q/2)H^*(p'' - p/2, q'' - q/2)dp''\, dq''}{\iint_{-\infty}^{\infty} |H(p'', q'')|dp''\, dq''}. \tag{4.10}$$

We note that the definition of OTF is valid for any linear spatially invariant optical system regardless of whether the system is with or without aberrations. We emphasize that Eq. (4.10) would serve as the primary link between the strictly coherent and strictly incoherent systems. A number of basic properties of the OTF can be easily summarized in the following:

1. $\tilde{H}_i(p, q)$ is positive and real.
2. $\tilde{H}_i(0, 0) = 1$.
3. $\tilde{H}_i(-p, -q) = \tilde{H}_i^*(p, q)$.
4. $|\tilde{H}_i(p, q)| \leqq |\tilde{H}_i(0, 0)|$.

4.3. BASIC FORMULATION

In Chapter 3, we have evaluated the coherence requirement for a partially coherent optical signal processing system. We have shown that the temporal and spatial coherence requirements are, respectively, dependent upon the

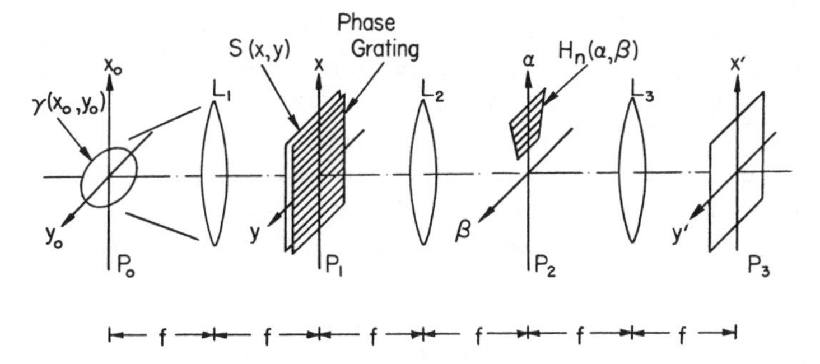

Figure 4.1. A partially coherent optical signal processor. L, achromatic lens; P_0, source plane; P_1, input plane; P_2, Fourier plane; P_3, output plane.

spectral bandwidth of a set of narrow spectral band filters $H_n(\alpha, \beta)$, and dependent upon the size of the light source. In the analysis of the temporal coherence requirement, we have assumed an infinitely small light source, while for the evaluation of the spatial coherence requirement, we have assumed a monochromatic source with finite extension. We use these same basic approaches in the evaluation of the apparent transfer for a white-light, or partially coherent, processing system.

We now evaluate the relationship between the apparent transfer function of an optical system and the spectral bandwidth of a set of narrow spectral band filters. Let us now refer to the partially coherent optical information processing system of Fig. 4.1. We note that a diffraction grating is used at the input signal plane (x, y). The purpose of using a diffraction grating at the input plane is to disperse the input signal in the spatial frequency plane (α, β) so that a high degree of temporal coherence can be achieved in the Fourier domain [4.8]. Thus the input signal can be processed in a complex amplitude, rather than in intensity, with a broad-band white-light source. By using the partial coherence theory, we have computed a general output intensity distribution for the partially coherent processor, as shown in the following [see Eq. (3.10)]:

$$I(x', y') = \sum_{n=1}^{N} \int_{\lambda_n - \Delta\lambda_n/2}^{\lambda_n + \Delta\lambda_n/2} \iint \gamma(x_0, y_0; \lambda) S(\lambda) C(\lambda)$$

$$\cdot \left| \iint S(x_0 + \alpha - \lambda f \nu_0, y_0 + \beta) \right.$$

$$\left. \cdot H_n(\alpha, \beta) \exp\left[-i \frac{2\pi}{\lambda f} (x'\alpha + y'\beta) \right] d\alpha \, d\beta \right|^2 dx_0 \, dy_0 \, d\lambda, \quad (4.11)$$

where: N is the total number of the spectral band filters; λ_n and $\Delta\lambda_n$ are the central wavelength and wavelength spread of the nth narrow spectral band filter $H_n(\alpha, \beta)$; $\gamma(x_0, y_0)$ is the source intensity distribution; $S(\lambda)$ and $C(\lambda)$ are the source spectral distribution and the spectral sensitivity of the recording material, respectively; $H_n(\alpha, \beta)$ is the nth narrow spectral band filter; $S(\alpha, \beta)$ is the Fourier spectrum of the input signal; and ν_0 is the spatial frequency of the diffraction grating. For simplicity, we assume that both $S(\lambda)$ and $C(\lambda)$ are constants that can be ignored in Eq. (4.11).

We now use a sinusoidal input object transparency to evaluate the apparent transfer function of the optical signal processing system. The overall input signal transmittance, which includes the diffraction grating, can be written as

$$s(x) = (1 + c \cos 2\pi\nu x)(1 + \cos 2\pi\nu_0 x), \tag{4.12}$$

where ν is the spatial frequency of the sinusoidal signal and c is the contrast. For simplicity, we use a one-dimensional notation. The output intensity distribution of Eq. (4.11) can be reduced to

$$I(x'; \Delta s) = \sum_{n=1}^{N} \iint_{\lambda_n - \Delta\lambda/2}^{\lambda_n + \Delta\lambda_n/2} \gamma(x_0) |A_n(x'; x_0; \lambda)|^2 \, dx_0 \, d\lambda, \tag{4.13}$$

where Δs is the size of the light source, N is total number of the narrow spectral band filters, and

$$A(x'; x_0; \lambda) = \int_{-\infty}^{\infty} S(\alpha; x_0; \lambda) H_n(\alpha) \exp\left(-i \frac{2\pi}{\lambda f} \alpha x'\right) d\alpha, \tag{4.14}$$

is a complex amplitude function.

In order to achieve a high degree of temporal coherence for the complex filtering process, we limit the width of the spatial filter $H_n(\alpha)$, which is placed at the first diffraction order in the Fourier plane, to a finite extension in the α-direction, that is,

$$H_n(\alpha) = \operatorname{rect}\left(\frac{\alpha - \lambda_n f \nu_0}{\Delta\alpha_n}\right), \tag{4.15}$$

where $\Delta\alpha_n$ is the spatial width of the nth spatial filter $H_n(\alpha)$, λ_n is the center wavelength of the filter, and the filter is centered at $\alpha = \lambda_n f v_0$.

Since the Fourier spectrum of the input signal at the spatial frequency plane P_2 is

$$S(\alpha; x_0; \lambda) = \int_{-\infty}^{\infty} (1 + c\cos 2\pi\nu x)(1 + \cos 2\pi\nu_0 x) \cdot \exp\left[-i\frac{2\pi}{\lambda f}(x_0 + \alpha)x\right]dx, \tag{4.16}$$

we obtain

$$\begin{aligned} S(\alpha; x_0; \lambda) = {} & \delta\left(\frac{\alpha + x_0}{\lambda f}\right) + \frac{c}{2}\delta\left(\frac{\alpha + x_0}{\lambda f} - \nu\right) + \frac{c}{2}\delta\left(\frac{\alpha + x_0}{\lambda f} + \nu\right) \\ & + \frac{1}{2}\delta\left(\frac{\alpha + x_0}{\lambda f} - \nu_0\right) + \frac{1}{2}\delta\left(\frac{\alpha + x_0}{\lambda f} + \nu_0\right) \\ & + \frac{c}{4}\delta\left[\frac{\alpha + x_0}{\lambda f} - (\nu_0 + \nu)\right] + \frac{c}{4}\delta\left[\frac{\alpha + x_0}{\lambda f} + (\nu_0 - \nu)\right] \\ & + \frac{c}{4}\delta\left[\frac{\alpha + x_0}{\lambda f} - (\nu_0 - \nu)\right] + \frac{c}{4}\delta\left[\frac{\alpha + x_0}{\lambda f} + (\nu_0 + \nu)\right], \end{aligned} \tag{4.17}$$

where we assume that the spatial frequency of the diffraction grating is much higher than that of the input object, that is, $\nu_0 \gg \nu$. This implies that there will be no overlapping among the various diffraction orders. In optical signal processing we focus only on the first-order term of Eq. (4.17), that is,

$$\begin{aligned} S(\alpha; x_0; \lambda) = {} & \frac{1}{2}\delta\left(\frac{\alpha + x_0}{\lambda f} - \nu_0\right) \\ & + \frac{c}{4}\delta\left[\frac{\alpha + x_0}{\lambda f} - (\nu_0 - \nu)\right] + \frac{c}{4}\delta\left[\frac{\alpha + x_0}{\lambda f} - (\nu_0 + \nu)\right]. \end{aligned} \tag{4.18}$$

By substituting Eqs. (4.15) and (4.18) into Eq. (4..14), and by using Parseval's theorem, we have

$$A_n(x'; x_0; \lambda) = \int_{-\infty}^{\infty} \operatorname{rect}\left(\frac{\alpha - \lambda_n f \nu_0}{\Delta\alpha_n}\right) \exp\left(-i\frac{2\pi}{\lambda f} a x'\right) d\alpha$$
$$* \int_{-\infty}^{\infty} \frac{1}{2} \delta\left(\frac{\alpha + x_0}{\lambda f} + \nu_0\right) \exp\left(-i\frac{2\pi}{\lambda f} \alpha x'\right) d\alpha$$
$$+ \int_{-\infty}^{\infty} \operatorname{rect}\left(\frac{\alpha - \lambda_n f \nu_0}{\Delta\alpha_n}\right) \exp\left(-i\frac{2\pi}{\lambda f} \alpha x'\right) d\alpha * \int_{-\infty}^{\infty} \frac{c}{4}$$
$$\cdot \delta\left(\frac{\alpha + x_0}{\lambda f} + \nu_0 - \nu\right) \exp\left(-i\frac{2\pi}{\lambda f} \alpha x'\right) d\alpha$$
$$+ \int_{-\infty}^{\infty} \operatorname{rect}\left(\frac{\alpha - \lambda_n \nu_0}{\Delta\alpha_n}\right) \exp\left(-i\frac{2\pi}{\lambda f} d x'\right) d\alpha$$
$$* \int_{-\infty}^{\infty} \frac{c}{4} \delta\left(\frac{\alpha + x_0}{\lambda f} + \nu_0 + \nu\right) \exp\left(-i\frac{2\pi}{\lambda f} \alpha x'\right) d\alpha$$
$$= \left\{\frac{1}{2} \exp\left(i\frac{2\pi}{\lambda f} x_0 x'\right) \operatorname{rect}\left(\frac{x_0}{\Delta\alpha_n}\right)\right.$$
$$+ \frac{c}{4} \exp\left[i2\pi\left(\frac{x_0}{\lambda f} + \nu\right) x'\right] \operatorname{rect}\left(\frac{x_0 + \lambda f \nu}{\Delta\alpha_n}\right)$$
$$\left. + \frac{c}{4} \exp\left[i2\pi\left(\frac{x_0}{\lambda f} - \nu\right) x'\right] \operatorname{rect}\left(\frac{x_0 - \lambda f \nu}{\Delta\alpha_n}\right)\right\}$$
$$\cdot \exp(-i2\pi\nu_0 x'), \tag{4.19}$$

where $*$ denotes the convolution operation. In the following sections we utilize the result of Eq. (4.19) to evaluate the apparent transfer functions under the temporal and spatial coherence regimes.

4.4. APPARENT TRANSFER FUNCTION UNDER TEMPORAL COHERENCE ILLUMINATION

Let us now consider the effect of the temporal coherence illumination. We assume that the incoherent source is a point source [i.e., $\gamma(x_0) = \delta(x_0)$]; thus Eq. (4.13) becomes

$$I(x') = \sum_{n=1}^{N} \int_{\lambda_n - \Delta\lambda_n/2}^{\lambda_n + \Delta\lambda_n/2} |A_n(x'; \lambda)|^2 \, d\lambda, \tag{4.20}$$

where

$$|A_n(x';\lambda)|^2 = \frac{1}{4} + \frac{c^2}{8}\operatorname{rect}\left(\frac{\lambda f \nu}{\Delta\alpha_n}\right) + \frac{c}{2}\cos(2\pi\nu x')\operatorname{rect}\left(\frac{\lambda f \nu}{\Delta\alpha_n}\right)$$
$$+ \frac{c^2}{8}\cos(4\pi\nu x')\operatorname{rect}\left(\frac{\lambda f \nu}{\Delta\alpha_n}\right). \qquad (4.21)$$

By letting $\lambda' = \lambda - \lambda_n$, Eq. (4.20) can be written in the following form:

$$I(x') = \sum_{n=1}^{N} \int_{-\infty}^{\infty} \operatorname{rect}\left(\frac{x'}{\Delta\lambda_n}\right)\left[\frac{1}{4} + \frac{c^2}{8}\operatorname{rect}\left(\frac{\lambda_n f\nu + \lambda' f\nu}{\Delta\alpha_n}\right) + \frac{c}{2}\cos(2\pi\nu x')\right.$$
$$\left.\cdot\operatorname{rect}\left(\frac{\lambda_n f\nu + \lambda' f\nu}{\Delta\alpha_n}\right) + \frac{c^2}{8}\cos(4\pi\nu x')\operatorname{rect}\left(\frac{\lambda_n f\nu + \lambda' f\nu}{\Delta\alpha_n}\right)\right]d\lambda'. \qquad (4.22)$$

We now attempt to evaluate the output intensity distribution for the following separate cases:

1. For $\nu < \Delta\alpha_n/(2\lambda_n + \Delta\lambda_n)f$. Let us denote the cross product rectangular function of Eq. (4.22) as

$$g_n(\lambda') \triangleq \operatorname{rect}\left(\frac{\lambda'}{\Delta\lambda_n}\right)\operatorname{rect}\left(\frac{\lambda_n f\nu + \lambda' f\nu}{\Delta\alpha_n}\right), \qquad (4.23)$$

where $g_n(\lambda')$ is always equal to unity within the region $-\Delta\lambda_n/2 < \lambda' < \Delta\lambda_n/2$, as illustrated in Fig. 4.2. The shaded area is the overlapping region of the two

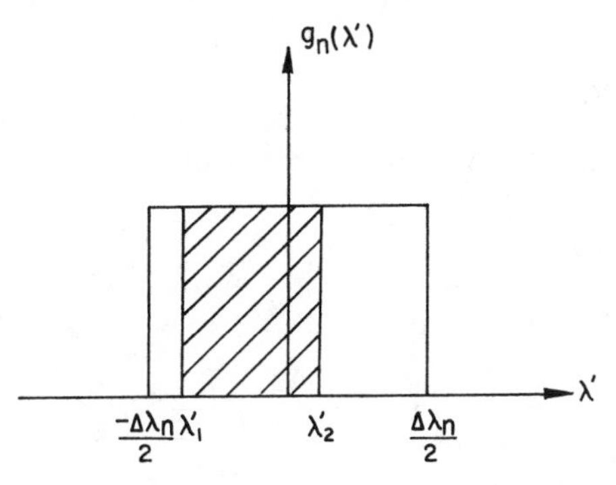

Figure 4.2. Plot of $g_n(\lambda')$ for $\nu < \dfrac{\Delta\alpha_n}{(2\lambda_n + \Delta\lambda_n)f}$.

rectangular functions, and λ_1' and λ_2' denote the extensions of $g_n(\lambda')$ (i.e., the shaded area), which can be written as

$$\lambda_1' = -\frac{\Delta\alpha_n}{2fv} - \lambda_n,$$

and

$$\lambda_2' = \frac{\Delta\alpha_n}{2fv} - \lambda_n.$$

Thus the output intensity distribution of Eq. (4.22), for this case, is

$$I^{(1)}(x') = \sum_{n=1}^{N} \int_{-\Delta\lambda_n/2}^{\Delta\lambda_n/2} \left[\frac{1}{4} + \frac{c^2}{8} + \frac{c}{2}\cos(2\pi v x') + \frac{c^2}{8}\cos(4\pi v x')\right] d\lambda'$$

$$\cong N\left[\frac{1}{4} + \frac{c^2}{8} + \frac{c}{2}\cos(2\pi v x') + \frac{c^2}{8}\cos(4\pi v x')\right]\Delta\lambda_n.$$

From this equation, we see that the output signal contrasts for the fundamental and the second harmonic frequency are

$$\gamma^{(1)}(v) = \frac{4c}{2 + c^2}, \tag{4.24}$$

and

$$\gamma^{(1)}(2v) = \frac{c^2}{2 + c^2}. \tag{4.25}$$

2. For $\Delta\alpha_n/(2\lambda_n + \Delta\lambda_n)f < v < \Delta\alpha_n/(2\lambda_n - \Delta\lambda_n)f$. With reference to the $g_n(\lambda')$ shown in Fig. 4.3, the output intensity distribution would be

$$I^{(2)}(x') = \sum_{n=1}^{N} \int_{-\Delta\lambda_n/2}^{\Delta\lambda_n/2} \frac{1}{4}\, d\lambda' + \int_{-\Delta\lambda_n/2}^{b} \left[\frac{c^2}{8} + \cos(2\pi v x') + \frac{c^2}{8}\cos(4\pi v x')\right] d\lambda'$$

$$= N\left[\frac{1}{4}\Delta\lambda_n + \frac{c^2}{8}\left(\frac{\Delta\alpha_n}{2fv} - \lambda_n + \frac{\Delta\lambda_n}{2}\right) + \frac{c}{2}\cos(2\pi v x')\left(\frac{\Delta\alpha_n}{2fv} - \lambda_n + \frac{\Delta\lambda_n}{2}\right)\right.$$

$$\left. + \frac{c^2}{8}\cos(4\pi v x')\left(\frac{\Delta\alpha_n}{\lambda f v} - \lambda_n + \frac{\Delta\lambda_n}{2}\right)\right], \tag{4.26}$$

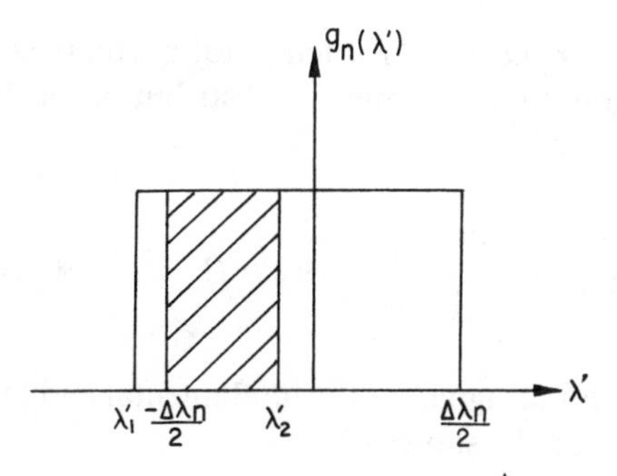

Figure 4.3. Plot of $g_n(\lambda')$ for $v > \dfrac{\Delta\alpha_n}{(2\lambda_n + \Delta\lambda_n)f}$.

where the fundamental and second harmonic image contrasts would be

$$\gamma^{(2)}(v) = \frac{4c(\Delta\alpha_n - 2fv\lambda_n + fv\Delta\lambda_n)}{4fv\Delta\lambda_n + c^2(\Delta\alpha_n - 2fv\lambda_n + fv\Delta\lambda_n)}, \tag{4.27}$$

and

$$\gamma^{(2)}(2v) = \frac{c^2(\Delta\alpha_n - 2fv\lambda_n + fv\Delta\lambda_n)}{4fv\Delta\lambda_n + c^2(\Delta\alpha_n - 2fv\lambda_n + fv\Delta\lambda_n)}. \tag{4.28}$$

3. For $v > \Delta\alpha_n/(2\lambda_n - \Delta_n)f$. Since there is no overlapping between the two rectangular functions, $g_n(\lambda')$ is always zero. Thus the output irradiance would be uniform, that is,

$$I^{(3)}(x') = \sum_{n=1}^{N} \int_{-\Delta\lambda_n/2}^{\Delta\lambda_n/2} \frac{1}{4}\, d\lambda' = \frac{N}{4}\, \Delta\lambda_n. \tag{4.29}$$

The contrast of the output image would be zero, that is,

$$\gamma^{(3)}(v) = \gamma^{(3)}(2v) = 0. \tag{4.30}$$

It is now apparent that we can use the maximum allowable spatial frequency, that is,

$$v_c \triangleq \frac{\Delta\alpha_n}{(2\lambda_n - \Delta\lambda_n)f}, \tag{4.31}$$

to define the *cut-off frequency* for a temporal partially coherent optical system. We further note that the intensity distribution of the input sinusoidal object is [see Eq. (4.12)]

$$I(x) = 1 + \frac{c^2}{2} + 2c\cos(2\pi\nu x) + \frac{c^2}{2}\cos(4\pi\nu x). \tag{4.32}$$

The contrast of the input signal for the fundamental and the second harmonic frequency can therefore be shown as

$$\gamma^{(0)}(\nu) = \frac{4c}{2 + c^2}, \tag{4.33}$$

and

$$\gamma^{(0)}(2\nu) = \frac{c^2}{2 + c^2}. \tag{4.34}$$

These results are identical to the output signal contrast of Eqs. (4.24) and (4.25), respectively.

Since the apparent modulation transfer function (MTF) can be defined as the ratio of the contrast of the output fundamental frequency signal to the contrast of the input signal, from Eqs. (4.24) to (4.34) we have

$$\mathrm{MTF}(\nu) = \begin{cases} 1, & \text{for } \nu < \dfrac{\Delta\alpha_n}{(2\lambda_n + \Delta\lambda_n)f}, \\[2ex] \dfrac{(2 + c^2)(\Delta\alpha_n - 2f\nu\lambda_n + f\nu\Delta\lambda_n)}{4f\nu\Delta\lambda_n + c^2(\Delta\alpha_n - 2f\nu\lambda_n + f\nu\Delta\lambda_n)}, & \text{for } \dfrac{\Delta\alpha_n}{(2\lambda_n + \Delta\lambda_n)f} < \nu < \dfrac{\Delta\alpha_n}{(2\lambda_n - \Delta\lambda_n)f}, \\[2ex] 0, & \text{for } \nu > \dfrac{\Delta\alpha_n}{(2\lambda_n - \Delta\lambda_n)f}. \end{cases} \tag{4.35}$$

Thus we see that the MTF depends on the spatial frequency of the diffraction grating ν_0, the spatial width of the filter $\Delta\alpha_n$, and the spectral bandwidth $\Delta\lambda_n$ of $H_n(\alpha)$. We note that the width of the filter $\Delta\alpha_n$ should be chosen equal to the product of the spatial frequency of the diffraction grating, the focal length of the achromatic transform lens, and the spectral bandwidth of the filter [see Eq. (3.14)], that is,

$$\Delta\alpha_n = f\nu_0\Delta\lambda_n. \tag{4.36}$$

By substituting Eq. (4.36) into Eq. (4.35), we have

$$\mathrm{MTF}(\nu) = \begin{cases} 1, & \text{for } \nu < \dfrac{\nu_0 + \Delta\lambda_n}{2\lambda_n + \Delta\lambda_n}, \\[2ex] \dfrac{(2 + c^2)(\nu_0\Delta\lambda - 2\nu\lambda_n + \nu\Delta\lambda_n)}{4\nu\Delta\lambda + c^2(\nu_0\Delta\lambda - 2\nu\lambda_n + \nu\Delta\lambda_n)}, & \text{for } \dfrac{\nu_0\Delta\lambda_n}{2\lambda_n + \Delta\lambda_n} < \nu < \dfrac{\nu_0\Delta\lambda_n}{2\lambda_n - \Delta\lambda_n}, \\[2ex] 0, & \text{for } \nu < \dfrac{\nu_0\Delta\lambda_n}{2\lambda_n - \Delta\lambda_n}, \end{cases} \tag{4.37}$$

in which the apparent transfer is a function of ν_0 and $\Delta\lambda_n$.

We now illustrate the dependence of the apparent transfer function on the spectral width $\Delta\lambda_n$ of $H_n(\alpha)$, the spatial width $\Delta\alpha_n$ of the filter, and the spatial frequency ν_0 of the grating. Figure 4.4 shows plots of the apparent transfer

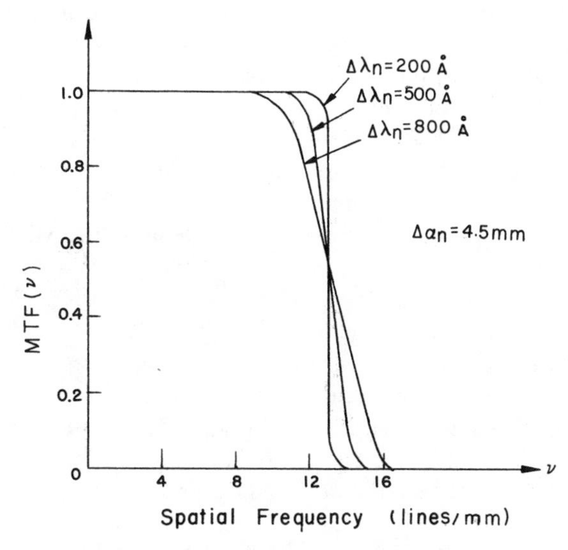

Figure 4.4. Temporal coherence apparent transfer function as a function of input signal frequency ν, for various values of spectral width $\Delta\lambda_n$.

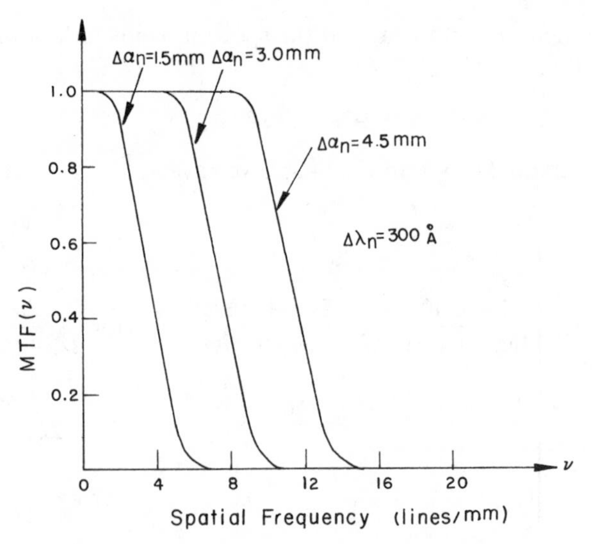

Figure 4.5. Temporal coherence apparent transfer function as a function of ν, for values of spatial width of filter $\Delta\alpha_n$.

function (MTF) as a function of the input signal frequency ν for various values of spectral bandwidth $\Delta\lambda_n$. From this figure we note that the MTF is not appreciably affected by the spectral bandwidth $\Delta\lambda_n$ of the spatial filter $H_n(\alpha)$, except for some slight changes in frequency response. For example, as $\Delta\lambda_n$ becomes broader a slight reduction in frequency response is expected. However, the system bandwidth is somewhat broader. Figure 4.5 shows the variation of the MTF as a function of ν for various values of spatial width $\Delta\alpha_n$ of the filter $H_n(\alpha)$, for a given wavelength spread $\Delta\lambda_n$. From this figure we see that the system bandwidth is linearly related to the spatial width of the filter. In other words, an increase in the spatial width of the filter results in a wider system bandwidth. Figure 4.6 shows the dependence of the MTF on the spatial frequency of the grating ν_0. Again, we see that the system bandwidth is linearly proportional to ν_0, as expected from the relationship of Eq. (4.37). In other words, a higher frequency grating has the advantage of achieving finer image resolution. However, this advantage is somewhat reduced with the use of larger achromatic transform lenses, which are generally more expensive.

Furthermore, the cut-off frequency ν_c of Eq. (4.31) can also be expressed as a function of the spatial frequency of the diffraction grating ν_0 and the spectral bandwidth $\Delta\lambda_n$ of the spatial filter $H_n(\alpha)$, that is,

$$\nu_c = \frac{\Delta\lambda_n \nu_0}{2\lambda_n - \Delta\lambda_n}. \tag{4.38}$$

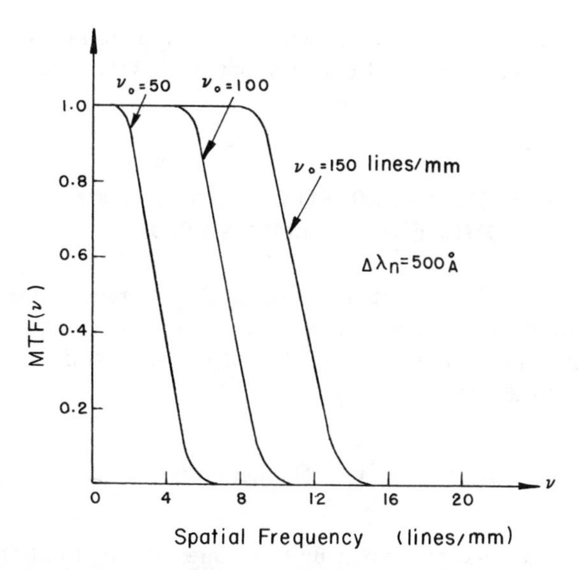

Figure 4.6. Temporal coherence apparent transfer function as a function of ν, for various values of grating frequency ν_0.

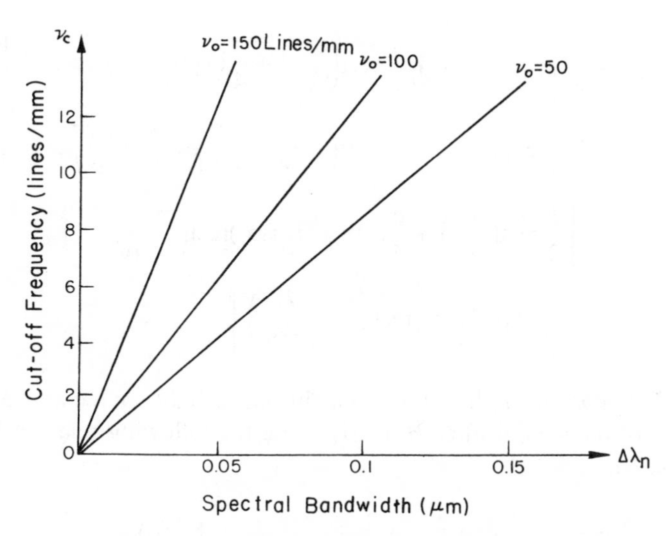

Figure 4.7. Cut-off frequency ν_c as a function of the spectral bandwidth $\Delta\lambda_n$ for various grating frequencies ν_0.

Figure 4.7 shows the cut-off frequency ν_c as a function of $\Delta\lambda_n$ for various ν_0. It is evident that the cut-off frequency increases as the spectral bandwidth increases.

4.5. APPARENT TRANSFER FUNCTION UNDER SPATIAL COHERENCE ILLUMINATION

Now we determine the dependence of the apparent transfer function on the source size Δs. To simplify our analysis, we assume that the light source is monochromatic, but of finite extent. The output intensity distribution of Eq. (4.13) can be written as

$$I(x'; \Delta s) = \int_{-\infty}^{\infty} \gamma(x_0)|A(x'; x_0)|^2 \, dx_0. \tag{4.39}$$

For simplicity, we assume that a uniform intensity distribution of the extended light source exists, that is,

$$\gamma(x_0)\mathrm{rect}\left(\frac{x_0}{\Delta s}\right), \tag{4.40}$$

where Δs is the width of the source. With reference to Eq. (4.19), we see that

$$\begin{aligned} |A(x'; x_0)|^2 = {} & \exp\left(i\frac{2\pi}{\lambda f}x_0 x'\right)\left[\frac{1}{2}\mathrm{rect}\left(\frac{x_0}{\Delta\alpha}\right) + \frac{c}{4}\exp(i2\pi\nu x')\mathrm{rect}\left(\frac{x_0 + \lambda f\nu}{\Delta\alpha}\right)\right. \\ & \left. + \frac{c}{4}\exp(-i2\pi\nu x')\mathrm{rect}\left(\frac{x_0 - \lambda f\nu}{\Delta\alpha}\right)\right]\exp\left(-i\frac{2\pi}{\lambda f}x_0 x'\right) \\ & \cdot\left[\frac{1}{2}\mathrm{rect}\left(\frac{x_0}{\Delta\alpha}\right) + \frac{c}{4}\exp(-i2\pi\nu x')\mathrm{rect}\left(\frac{x_0 + \lambda f\nu}{\Delta\alpha}\right)\right. \\ & \left. + \frac{c}{4}\exp(i2\pi\nu x')\mathrm{rect}\left(\frac{x_0 - \lambda f\nu}{\Delta\alpha}\right)\right], \end{aligned} \tag{4.41}$$

where λ is the wavelength of the monochromatic light source and $\Delta\alpha$ is the extension of the spatial filter $H(\alpha)$. By using the following relationship (see Appendix B):

$$\mathrm{rect}\left(\frac{x}{a}\right)\mathrm{rect}\left(\frac{x + b}{a}\right) = \mathrm{rect}\left(\frac{x + b/2}{a - |b|}\right), \tag{4.42}$$

Eq. (4.41) can be expressed as

$$|A(x'; x_0)|^2 = \frac{1}{4}\operatorname{rect}\left(\frac{x_0}{\Delta\alpha}\right) + \frac{c^2}{16}\left[\operatorname{rect}\left(\frac{x_0 + \lambda f v}{\Delta\alpha}\right) + \operatorname{rect}\left(\frac{x_0 - \lambda f v}{\Delta\alpha}\right)\right]$$
$$+ \frac{c}{4}\cos(2\pi v x')\left\{\operatorname{rect}\left[\frac{x_0 + 1/2(\lambda f v)}{\Delta\alpha - \lambda f v}\right] + \operatorname{rect}\left[\frac{x_0 - 1/2(\lambda f v)}{\Delta\alpha - \lambda f v}\right]\right\}$$
$$+ \frac{c^2}{8}\cos(4\pi v x')\operatorname{rect}\left(\frac{x_0}{\Delta\alpha - 2\lambda f v}\right). \tag{4.43}$$

By substituting Eqs. (4.40) and (4.43) into Eq. (4.39), the output irradiance becomes

$$I(x'; \Delta s) = \int_{-\infty}^{\infty} \frac{1}{4}\operatorname{rect}\left(\frac{x_0}{\Delta\alpha}\right) + \frac{c^2}{16}\left[\operatorname{rect}\left(\frac{x_0}{\Delta\alpha + 2\lambda f v}\right)\right.$$
$$\left. + \operatorname{rect}\left(\frac{x_0}{\Delta\alpha - 2\lambda f v}\right)\operatorname{rect}\left(\frac{x_0}{\Delta s}\right) dx_0\right.$$
$$+ \frac{c}{4}\cos(2\pi v x')\int_{-\infty}^{\infty}\left[\operatorname{rect}\left(\frac{x_0}{\Delta\alpha}\right) + \operatorname{rect}\left(\frac{x_0}{\Delta\alpha - 2\lambda f v}\right)\right]$$
$$\cdot \operatorname{rect}\left(\frac{x_0}{\Delta s}\right) dx_0 + \frac{c^2}{8}\cos(4\pi v x')\int_{-\infty}^{\infty}\operatorname{rect}\left(\frac{x_0}{\Delta\alpha - 2\lambda f v}\right)\operatorname{rect}\left(\frac{x_0}{\Delta s}\right) dx_0. \tag{4.44}$$

In order to reduce the complexity of our notation, let us define the following relationships:

$$g_1(x_0) = \operatorname{rect}\left(\frac{x_0}{\Delta\alpha}\right)\operatorname{rect}\left(\frac{x_0}{\Delta s}\right),$$

$$g_2(x_0) = \operatorname{rect}\left(\frac{x_0}{\Delta\alpha - 2\lambda f v}\right)\operatorname{rect}\left(\frac{x_0}{\Delta s}\right),$$

$$g_3(x_0) = \operatorname{rect}\left(\frac{x_0}{\Delta\alpha + 2\lambda f v}\right)\operatorname{rect}\left(\frac{x_0}{\Delta s}\right).$$

Thus Eq. (4.44) becomes

$$I(x';\Delta s) = \int_{-\infty}^{\infty}\left[\frac{1}{4}g_1(x_0) + \frac{c^2}{16}g_3(x_0) + g_2(x_0)\right]dx_0 + \frac{c}{4}\cos(2\pi\nu x') \cdot \int_{-\infty}^{\infty}[g_1(x_0) + g_2(x_0)]dx_0 + \frac{c^2}{8}\cos(4\pi\nu x')\int_{-\infty}^{\infty}g_2(x_0)dx_0. \tag{4.45}$$

Equation (4.45) can be separately evaluated for the following cases:

1. *For* $0 < \Delta s < \Delta\alpha - 2\lambda f\nu$. We have

$$g_1(x_0) = g_2(x_0) = g_3(x_0) = \operatorname{rect}\left(\frac{x_0}{\Delta s}\right).$$

The output irradiance of Eq. (4.45) becomes

$$\begin{aligned} I^{(1)}(x';\Delta s) &= \int_{-\Delta s/2}^{\Delta s/2}\left[\left(\frac{1}{4} + \frac{c^2}{8}\right) + \frac{c}{2}\cos(2\pi\nu x') + \frac{c^2}{8}\cos(4\pi\nu x')\right]dx_0 \\ &= \left[\left(\frac{1}{4} + \frac{c^2}{8}\right) + \frac{c}{2}\cos(2\pi\nu x') + \frac{c^2}{8}\cos(4\pi\nu x')\right]\Delta s. \end{aligned} \tag{4.46}$$

Thus the output signal contrasts for the fundamental and the second harmonic frequency are

$$\gamma^{(1)}(\nu) = \frac{4c}{2 + c^2}, \tag{4.47}$$

$$\gamma^{(1)}(2\nu) = \frac{c^2}{2 + c^2}. \tag{4.48}$$

2. *For* $\Delta\alpha - 2\lambda f < \Delta s < \Delta\alpha$. We have

$$g_1(x_0) = g_3(x_0) = \operatorname{rect}\left(\frac{x_0}{\Delta s}\right),$$

and

$$g_2(x_0) = \operatorname{rect}\left(\frac{x_0}{\Delta\alpha - 2\lambda_0 f\nu}\right).$$

The output irradiance can be written as

$$I^{(2)}(x'; \Delta s) = \left[\left(\frac{1}{4} + \frac{c^2}{16}\right)\Delta s + \frac{c^2}{16}(\Delta\alpha - 2\lambda f v)\right] + \frac{c}{4}(\Delta s + \Delta\alpha - 2\lambda f v)$$
$$\cdot \cos(2\pi v x') + \frac{c^2}{8}(\Delta\alpha - 2\lambda f v)\cos(4\pi v x'). \tag{4.49}$$

And the corresponding output fundamental and the second harmonic signal contrasts are

$$\gamma^{(2)}(v) = \frac{4c(\Delta s + \Delta\alpha - 2\lambda f v)}{4\Delta s + c^2(\Delta s + \Delta\alpha - 2\lambda f v)}, \tag{4.50}$$

$$\gamma^{(2)}(2v) = \frac{2c^2(\Delta\alpha - 2\lambda f v)}{4\Delta s + c^2(\Delta s + \Delta\alpha - 2\lambda f v)}, \tag{4.51}$$

3. *For* $\Delta\alpha < \Delta S < \Delta\alpha + 2\lambda f v$. We have

$$g_1(x_0) = \text{rect}\left(\frac{x_0}{\Delta\alpha}\right),$$

$$g_2(x_0) = \text{rect}\left(\frac{x_0}{\Delta\alpha - 2\lambda_0 f v}\right),$$

and

$$g_3(x_0) = \text{rect}\left(\frac{x_0}{\Delta s}\right).$$

The output irradiance becomes

$$I^{(3)}(x'; \Delta s) = \left(\frac{1}{4} + \frac{c^2}{16}\right)\Delta\alpha + \frac{c^2}{16}(\Delta s - 2\lambda f v) + \frac{c}{2}(\Delta\alpha - \lambda f v)\cos(2\pi v x')$$
$$+ \frac{c^2}{8}(\Delta\alpha - 2\lambda f v)\cos(2\pi v x'). \tag{4.52}$$

The fundamental and second harmonic output signal contrasts are therefore

$$\gamma^{(3)}(v) = \frac{8c(\Delta\alpha - \lambda f v)}{4\Delta\lambda + c^2(\Delta s + \Delta\alpha - 2\lambda f v)}, \tag{4.53}$$

$$\gamma^{(3)}(2v) = \frac{2c^2(\Delta\alpha - 2\lambda f v)}{4\Delta\alpha + c^2(\Delta\alpha + \Delta s - 2\lambda f v)}. \tag{4.54}$$

4. *For* $\Delta s > \Delta\alpha + 2\lambda v f$. We have

$$g_1(x_0) = \mathrm{rect}\left(\frac{x_0}{\Delta\alpha}\right),$$

$$g_2(x_0) = \mathrm{rect}\left(\frac{x_0}{\Delta\alpha - 2\lambda f v}\right),$$

and

$$g_3(x_0) = \mathrm{rect}\left(\frac{x_0}{\Delta\alpha + 2\lambda f v}\right).$$

The output irradiance is therefore

$$I^{(4)}(x'; \Delta s) = \left(\frac{1}{4} + \frac{c^2}{8}\right)\Delta\alpha + \frac{c}{2}(\Delta\alpha - \lambda f v)\cos(2\pi v x') + \frac{c^2}{8}(\Delta\alpha - 2\lambda f v)\cos(2\pi 2 v x'). \tag{4.55}$$

And the fundamental and second harmonic signal contrast are

$$\gamma^{(4)}(v) = \frac{4c(\Delta\alpha - \lambda f v)}{(2 + c^2)}, \tag{4.56}$$

$$\gamma^{(4)}(2v) = \frac{c^2(\Delta\alpha - 2\lambda f v)}{(2 + c^2)\Delta\alpha}. \tag{4.57}$$

5. *For* $\Delta\alpha < \lambda_0 f v$. We obtain

$$I^{(5)}(x'; \Delta s) = \text{constant}, \tag{4.58}$$

and

$$\gamma^{(5)}(v) = \gamma^{(5)}(2v) = 0. \tag{4.59}$$

Now, with reference to these results, the cut-off frequency of the optical system,

$$\nu_c = \frac{\Delta\alpha}{\lambda_0 f}, \tag{4.60}$$

and the input signal contrast of Eq. (4.32), the apparent transfer function (MTF) for the fundamental and the second harmonic frequency can be shown in the following:

$$\text{MTF}(\nu) = \begin{cases} 0, & \text{for } 0 < \Delta s < \Delta\alpha - 2\lambda f\nu, \\ \dfrac{2(2 + c^2)(\Delta\alpha - 2\lambda f\nu)}{4\Delta s + c^2(\Delta s + \Delta\alpha - 2\lambda f\nu)}, & \text{for } \Delta\alpha - 2\lambda f\nu < \Delta s < \Delta\alpha, \\ \dfrac{2(2 + c^2)(\Delta\alpha - \lambda f\nu)}{4\Delta\alpha + c^2(\Delta s + \Delta\alpha - 2\lambda f\nu)}, & \text{for } \Delta\alpha < \Delta s < \Delta\alpha + 2\lambda f\nu, \\ \dfrac{\Delta\alpha - 2\lambda f\nu}{\Delta\alpha}, & \text{for } \Delta\alpha + 2\lambda f\nu < \Delta s, \\ 0, & \text{for } \Delta\alpha < \lambda f\nu, \end{cases} \tag{4.61}$$

and

$$\text{MTF}(2\nu) = \begin{cases} 1, & \text{for } 0 < s < \Delta\alpha - 2\lambda f\nu, \\ \dfrac{2(2 + c^2)(\Delta\alpha - 2\lambda f\nu)}{4\Delta\alpha + c^2(\Delta s + \Delta\alpha - 2\lambda f\nu)}, & \text{for } \Delta\alpha - 2\lambda f\nu < \Delta s < \Delta\alpha, \\ \dfrac{2(2 + c^2)(\Delta\alpha - 2\lambda f\nu)}{4\Delta\alpha + c^2(\Delta s + \Delta\alpha - 2\lambda f\nu)}, & \text{for } \Delta\alpha < \Delta s < \Delta\alpha + 2\lambda f\nu, \\ \dfrac{\Delta\alpha - 2\lambda f\nu}{\Delta\alpha}, & \text{for } \Delta\alpha + 2\lambda f\nu < \Delta s, \\ 0, & \text{for } \Delta\alpha < \lambda f\nu. \end{cases} \tag{4.62}$$

From the above equations, one may see that the MTF, either for the fundamental or for the second harmonic frequency, decreases rather rapidly as the source size Δs increases. Figure 4.8*a* shows the MTF(ν) as a function of the input spatial frequency for various source sizes Δs. From this figure, we

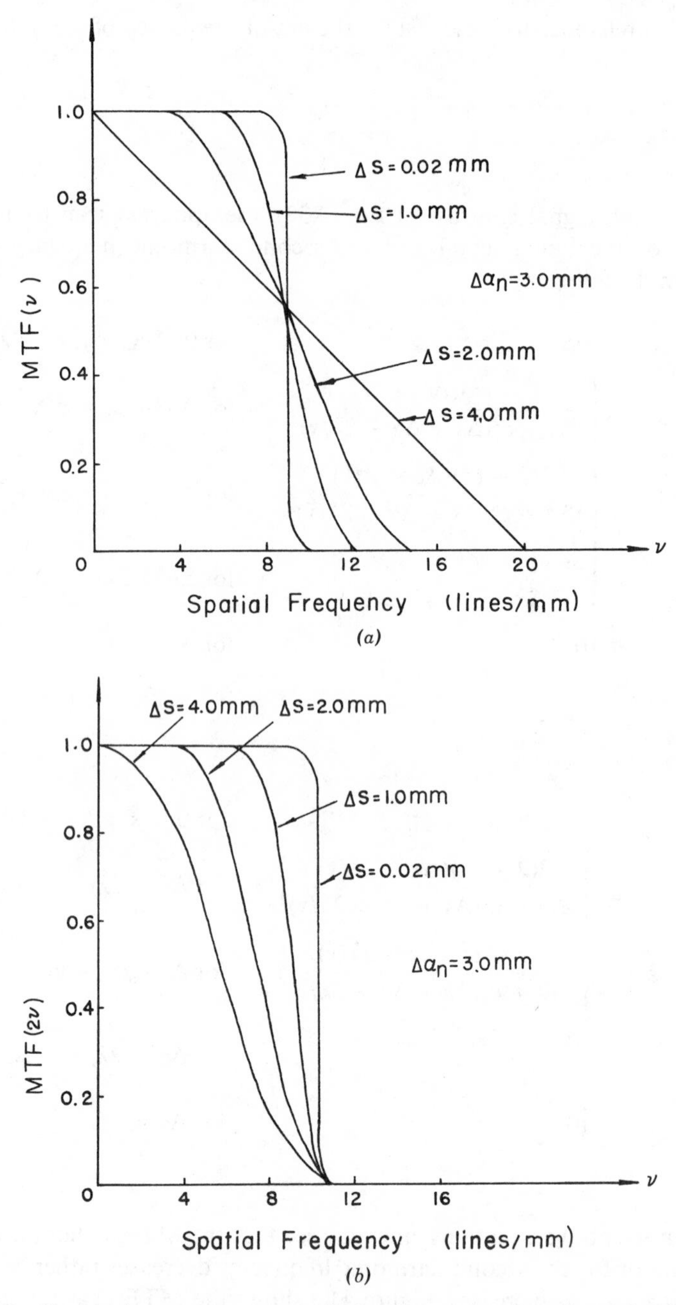

Figure 4.8. Spatial coherence apparent transfer function as a function of input signal frequency v, for various values of source size Δs. (*a*) Fundamental. (*b*) Second harmonic MTF.

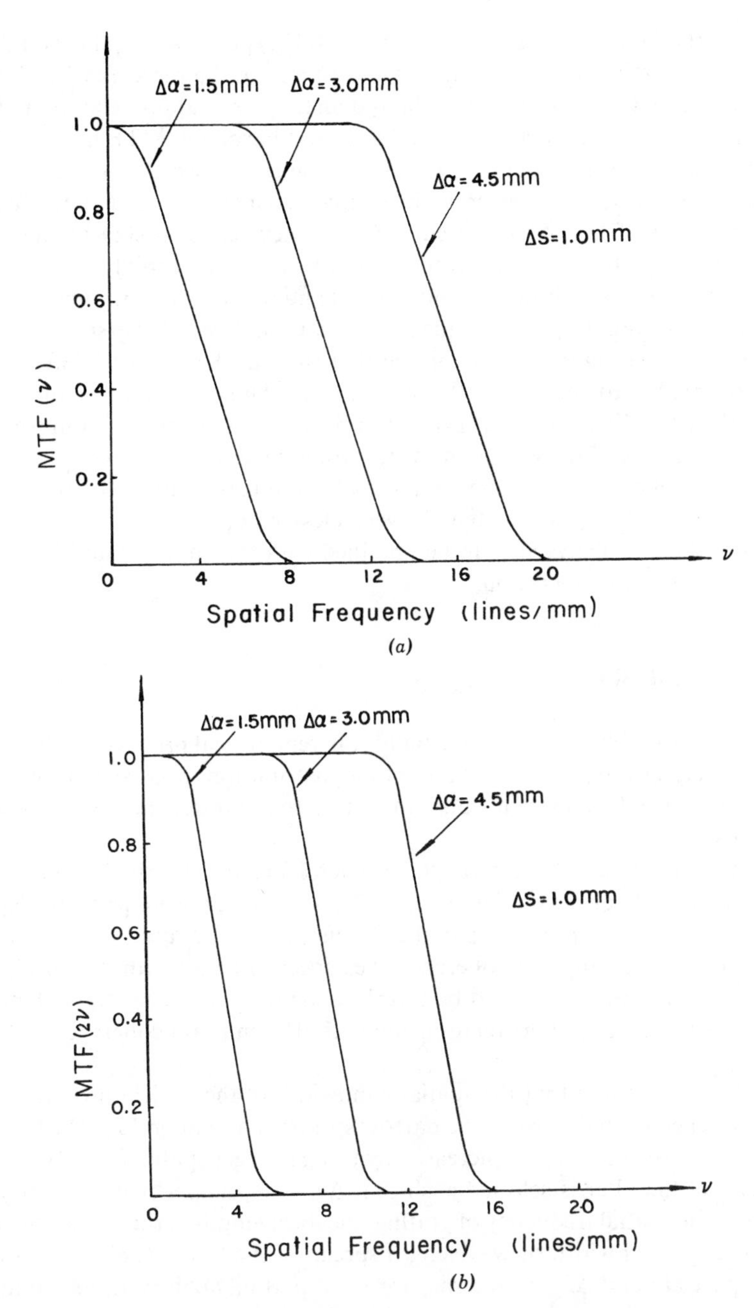

Figure 4.9. Spatial coherence apparent transfer function as a function of input signal frequency ν, for various values of filter width $\Delta\alpha$, (*a*) Fundamental. (*b*) Second harmonic MTF.

see that the frequency response [i.e., the MTF(ν)] decreases quite rapidly as the source size increases. In other words, for a fixed filter size, the smaller the source size that is used, the better the system frequency approaches that of the incoherent case. Figure 4.8*b* shows the second harmonic MTF as a function of the input spatial frequency. From this figure, we see that the frequency response decreases even faster as the source size increases. Unlike the fundamental MTF of Fig. 4.8*a*, the cut-off frequency tends to stay at the same value, although the frequency response decreases rather rapidly.

Figure 4.9*a* and *b* show the fundamental and the second harmonic MTF as a function of input spatial frequency for various sizes of the spatial filters. From these two figures, we again see that the transfer system bandwidth is linearly related to the size of the spatial filter. Thus, the MTF is obviously limited by the filter bandwidth of the optical processor. In addition we note that the increase of the filter bandwidth would also cause a reduction in temporal coherence, as seen in Section 4.4, which in turn reduces the processing capabilities of the optical system. Nevertheless, an optimum processing capability of the optical system can be obtained with the appropriate MTF for a specific processing operation.

4.6. SUMMARY

The nonlinear behavior of the partially coherent optical processor, when considering either the intensity or amplitude distribution input signals, necessitates the use of the apparent transfer function to accurately predict the system response.

In this chapter, we have derived the general formulas for the MTF in terms of the theory of partially coherent light. These derivations indicate the dependence of the MTF upon the spatial coherence (i.e., the source size Δs) as well as the degree of temporal coherence [i.e., spectral bandwidth $\Delta\lambda_n$ of $H_n(\alpha)$]. The MTF has been shown to be less dependent upon the spatial coherence requirement than upon its relationship with the temporal coherence requirement.

It has been noted that the spatial bandwidth of the optical processor primarily depends on the size of the narrow spectral band filter $\Delta\alpha_n$. The transfer systems bandwidth may be increased by using a larger spatial filter. However, the size of the filter is selected such that $\Delta\alpha_n = \nu_0 f \Delta\lambda_n$, which is linearly related to the spatial frequency of grating, the focal length of the transform lens, and the spectral width or wavelength spread of the filter $H_n(\alpha)$. Since a narrow spectral band $\Delta\lambda_n$ is necessary for most partially coherent optical information processing operations, to achieve the required $\Delta\lambda_n$ for a wide strip of spatial filter in the spatial frequency plane, a diffraction sampling grating of sufficiently high frequency at the input plane is needed. As apparent for partially coherent processing with the white-light source shown in Fig. 4.1, a set of narrow spectral band filters, each with a narrow spectral bandwidth $\Delta\lambda_n$, can be used in the spatial frequency plane. Finally, we stress that the apparent transfer functions that we derived are rather general, and may be applied to most of the partially coherent optical processing systems.

REFERENCES

4.1. E. L. O'Neill, *Introduction to Statistical Optics*, Addison-Wesley, Reading, MA, 1963.

4.2. J. W. Goodman, *Introduction to Fourier Optics*, McGraw-Hill, New York, 1968.

4.3. R. J. Becherer and G. B. Parrent, "Nonlinearity in optical image system," *J. Opt. Soc. Am.* **57**, 1479 (1967).

4.4. R. E. Swing and J. R. Clay, "Ambiguity of the transfer function with partially coherent illumination," *J. Opt. Soc. Am.* **57**, 1180 (1967).

4.5. K. Dutta and J. W. Goodman, "Reconstruction of image of partially coherent objects from samples of mutual intensity," *J. Opt. Soc. Am.* **67**, 796 (1977).

4.6. M. Born and E. Wolf, *Principle of Optics*, 4th ed., Pergamon Press, New York, 1970.

4.7. S. L. Zhuang and F. T. S. Yu, "Apparent transfer function for partially coherent optical information processing," *Appl. Phys.* **B28**, 366 (1982).

4.8. S. L. Zhuang and F. T. S. Yu, "Coherence requirement for partially coherent optical processing, *Appl. Opt.* **21**, 2587 (1982).

5

Noise Performance

White-light optical processors perform better under noisy environments than their coherent counterparts. Here we give an analysis of their performance under partially coherent illumination and obtain quantitative results to support this claim. The output signal-to-noise ratio (SNR) for the proposed white-light optical signal processor is evaluated as a measure of its performance. We study the effects of temporal partial coherence on the noise performance of the white-light optical processor separately from those of spatial partial coherence.

Important noise sources in an optical signal processor are usually the random defects in the input to the system at the object plane or in the filters to be used at the Fourier plane; both cases are considered in this chapter. Two important types of noise analyzed are the amplitude noise and the phase noise.

We show that, except for the case when the amplitude noise is present at the input plane, the resulting output SNR improves considerably when a wider spectral band of the light source is utilized. A similar set of results for the noise performance of the optical system is obtained when the requirements on spatial coherence are relaxed. It is shown that the noise at the Fourier plane can be efficiently suppressed, since the number of filter channels in a typical white-light optical processor is large.

5.1. INTRODUCTION

Coherent optical signal processing offers a powerful alternative to its electronic and digital counterparts. Their ability to perform a large number of important operations at extremely high speeds is well demonstrated [5.1–5.3]. However, the lack of versatility, the high cost of coherent sources, and the unavoidable coherent artifact noise have impeded the development of dependable practical optical signal processing systems. To alleviate some of these problems, we have recently proposed a white-light processing technique

[5.4–5.6] with the essential capability of processing the optical signal in complex amplitude.

Chavel and Lowenthal [5.7] and Chavel [5.8] have studied the relationship between the noise and coherence of a conventional optical imaging system. They show that the noise fluctuations caused by the pupil plane can be reduced considerably if broad-band illumination is employed. They have also shown that the noise at the object plane due to defects other than the phase deviations cannot be suppressed by partially coherent illumination. Leith and Roth [5.9] have also studied the noise performance of an achromatic coherent optical system. They analyzed the problem by introducing the concept of a three-dimensional transfer function to describe the noise suppression properties of the system. They demonstrated that such a system shows considerable noise immunity if a broad spectral light source is employed.

In this chapter, we study the performance of a white-light optical signal processor under a noisy environment [5.10, 5.11]. We compute the output SNR as the criterion for the noise performance of the system.

5.2. NOISE PERFORMANCE UNDER THE TEMPORAL COHERENCE REGIME

Since the problem of coherent artifact noise in an optical processing system is intimately related to the coherence properties of the source, we propose to study the cases under temporal and spatial coherence illumination independently. In this section, we study the noise performance of a white-light optical signal processor under temporal coherence illumination.

The important noise sources in an optical system are usually the random defects in the input transparencies at the object plane or those in the spatial filters at the Fourier plane; both these cases are analyzed here. There are, however, two types of noise to be considered; namely the amplitude and the phase noise. In the following sections we develop the models for phase as well as for amplitude noise. We indicate the conditions under which the phase noise can be considered weak and use these facts in our calculations. Also, when necessary, the example of a Gaussian process is used to model the noise in the system to bring out the qualitative performance of the optical processor. However, we restrict our analysis to the effects of noise at the input and at the Fourier plane, since these are the most common sources in an optical processing system.

5.2.1. Problem Formulation

The white-light optical signal processor to be studied here is described in Fig. 5.1. The output light intensity distribution due to the nth narrow spectral

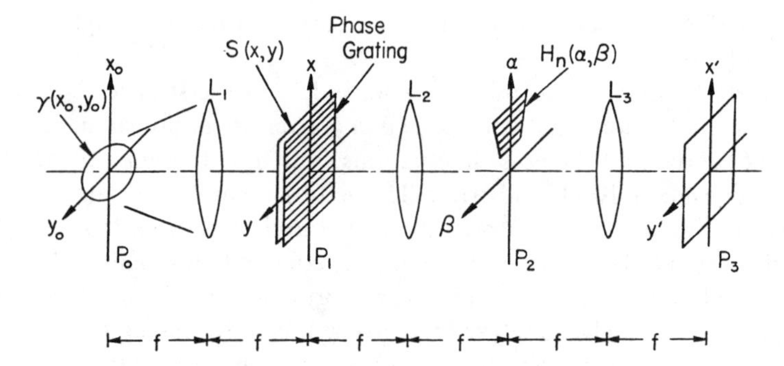

Figure 5.1. A white-light optical processor. $\gamma(x_0, y_0)$, source intensity distribution; $s(x, y)$, input transmittance; $H(\alpha, \beta)$, spectral band filter; P_0, source plane; P_1, input plane; P_2, Fourier plane; P_3 output plane.

band spatial filter is given by Eq. (3.9), such as

$$I_n(x', y') = \int_{\lambda_n - \Delta\lambda_n/2}^{\lambda_n + \Delta\lambda_n/2} \iint_{-\infty}^{\infty} \gamma(x_0, y_0) S(\lambda) C(\lambda) \cdot \left| \iint_{-\infty}^{\infty} S(x_0 + \alpha - \lambda f \nu_0, y_0 + \beta) H_n(\alpha, \beta) \exp\left[-i \frac{2\pi}{\lambda f} (x'\alpha + y'\beta) \right] \cdot d\alpha\, d\beta \right|^2 dx_0\, dy_0\, dy, \tag{5.1}$$

where: $\gamma(x_0, y_0)$ is the spatial intensity distribution of the white-light source; $S(\lambda)$ and $C(\lambda)$ are the relative spectral intensity of the light source and the relative spectral response sensitivity of the detector, respectively; λ_n and $\Delta\lambda_n$ are the mean wavelength and the bandwidth, respectively, of the nth spectral band filter; $S(x_0 + \alpha - \lambda f \nu_0, y_0 + \beta)$ represents the Fourier spectrum of the input transmittance $s(x, y)$ due to wavelength λ; ν_0 is the spatial frequency of the phase sampling grating; $H_n(\alpha, \beta)$ is the nth spectral band spatial filter corresponding to the nth channel in the system, and f is the focal length of the achromatic Fourier transform lens.

We now rewrite Eq. (5.1) for the case of temporal partial coherence (TPC). For a spectrally broad-band point source the spatial intensity distribution of the light can be described by a δ-function [i.e., $\gamma(x_0, y_0) = \delta(x_0, y_0)$]. We also make the assumption that $C(\lambda)$ and $S(\lambda)$ are constant over the spectral width of the light source. Thus Eq. (5.1) can be written as

$$I_n(x', y') = \int_{\lambda_n - \Delta\lambda_n/2}^{\lambda_n + \Delta\lambda_n/2} \left| \iint_{-\infty}^{\infty} S(\alpha - \lambda f \nu_0, \beta) H_n(\alpha, \beta) \cdot \exp\left[-i \frac{2\pi}{\lambda f} (x'\alpha + y'\beta) \right] d\alpha \, d\beta \right|^2 d\lambda, \tag{5.2}$$

where the unimportant proportionality constant is omitted for convenience. Let us further assume that the input signal is a one-dimensional object and is placed along the y-direction, that is, $s(x, y) = s(y)$. The phase grating at the input plane is inserted with the signal and can be represented by the expression $\exp[i2\pi\nu_0 x]$. The corresponding fan-shaped Fourier spectrum is also one-dimensional along the β-axis; however, it is smeared in rainbow colors due to the phase grating in the α-direction. Let the width of the nth narrow spectral band filter be $\Delta\alpha_n$. If the filter is placed in the appropriate spectral band of the smeared Fourier spectra, the corresponding spectral bandwith $\Delta\lambda_n$ can be approximated by [see Eq. (3.14)]

$$\Delta\lambda_n = \frac{\Delta\alpha_n}{\nu_0 f}. \tag{5.3}$$

The redundancy in the filtered signal is proportional to the spectral bandwidth $\Delta\lambda_n$ of the nth filter and also to the number of spectral band filters employed in the Fourier plane. From Eq. (5.3), it is evident that for a fixed spatial width of the filter, either a low spatial frequency of phase grating or a shorter focal length of the transform lens should be used to improve the SNR at the output plane. However, a decrease in the grating spatial frequency also reduces the number of spectral band filters that can be placed in the Fourier plane, limiting the processing capability of the system. An appropriate relationship between the two parameters is given by Yu [5.12]:

$$N \triangleq \frac{\Delta\lambda}{\Delta\lambda_n} \simeq \frac{\Delta\lambda\nu_0}{4\bar{\lambda}\Delta\nu_\alpha}; \qquad \text{for } \nu_0 \gg \Delta\nu_\alpha, \tag{5.4}$$

where: $\Delta\lambda$ and $\bar{\lambda}$ are the spectral bandwidth and the mean wavelength of the light source, respectively; and $\Delta\nu_\alpha$ is the spatial frequency bandwidth of input object in the α-direction.

Although the output SNR from each spectral band filter would decrease as ν_0 increases, the problem can be offset by the use of a larger number of spectral band filters in the Fourier plane. Thus the overall output SNR would improve considerably with an appropriate choice of the above-mentioned parameters.

We define the output SNR for the nth spectral band in a standard manner:

$$\mathrm{SNR}_n(y') \triangleq \frac{E[I_n(y')]}{\sigma_n(y')}, \tag{5.5}$$

where $E[\cdot]$ represents the ensemble average, and

$$\sigma_n^2(y') \triangleq E[I_n^2(y')] - \{E[I_n(y')]\}^2. \tag{5.6}$$

Since we are interested in the comparisons of a partially coherent to a coherent system, let us define the normalized SNR for the nth spectral band as

$$\overline{\mathrm{SNR}} \triangleq \frac{\mathrm{SNR}_n \text{ for partially coherent illumination}}{\mathrm{SNR(coherent)}}. \tag{5.7}$$

This definition conveniently sets the output SNR for strictly coherent illumination to unity. Also, to simplify the analysis, it is useful to define a relative spatial frequency. If ν_c is the cut-off frequency of the system [see Eq. (4.38)], then the relative spatial frequency for some real frequency ν can be defined, such as

$$\Omega \triangleq \frac{\nu}{\nu_c}. \tag{5.8}$$

We note that $\Omega = 1$ for $\nu = \nu_c$.

5.2.2. Noise at the Input Plane

We now evaluate the noise performance of the proposed white-light optical signal processor when the noise is present at the object plane. Consider the output intensity distribution due to the nth spectral band filter given by Eq. (5.2). The output intensity distribution in a one-dimensional form may be written as

$$I_n(y') = \int_{\lambda_n - \Delta\lambda_n/2}^{\lambda_n + \Delta\lambda_n/2} \left| \int_{-\infty}^{\infty} s(y)\,\mathrm{sinc}[2\pi\nu_c(y' - y)]dy \right|^2 d\lambda, \tag{5.9}$$

where the subscript n denotes the nth spectral band filter, and $s(y)$ represents the one-dimensional input signal with noise. We have also assumed that the nth spectral band filter is simply given by

$$H_n(\beta) = \operatorname{rect}\left(\frac{\beta}{\lambda f v_c}\right), \tag{5.10}$$

where

$$\operatorname{rect}\left(\frac{\beta}{B}\right) \triangleq \begin{cases} 1, & |\beta| \leqq B, \\ 0, & \text{otherwise}, \end{cases} \tag{5.11}$$

and

$$v_c \triangleq \frac{\Delta\lambda_n v_0}{2\lambda_n - \Delta\lambda_n}, \tag{5.12}$$

is defined as the cut-off frequency of the optical system [see Eq. (4.38)].

The phase sampling grating at the input plane of the optical processor is employed to disperse the signal spectrum into rainbow colors in the Fourier plane, which results in separation of filter channels, as illustrated in Fig. 5.1. Its use is critical when we consider the noise defects at the Fourier plane. However, in the present case we assume, without loss of generality, that its spatial frequency is zero. Hence, the overall noisy transmittance at the object plane can be written as

$$s(y) = a(y)\exp[ik\phi(y)] + n(y), \tag{5.13}$$

where $k = 2\pi/\lambda$, and $a(y)$ is a one-dimensional signal. The multiplicative exponent represents the noise due to a random thickness fluctuation of phase distribution $\phi(y)$, and $n(y)$ includes other additive amplitude noise sources at the input plane.

5.2.2.1. Weak Phase Noise. To study the phase noise alone we let $n(y) = 0$. Also, we assume that $\phi(y)$ is wide sense stationary with zero mean and variance σ_p^2 and that its probability density is identical throughout the recording medium [5.13, 5.14]. The expressions for the moments are

$$E[\phi(y)] = 0, \tag{5.14}$$

and

$$E[\phi(y_1)\phi(y_2)] = \sigma_p^2 \exp\left(-\frac{|y_1 - y_2|}{d}\right), \tag{5.15}$$

where we have chosen an exponential form for the correlation in which quantity d is called *scale size* of the irregularities, or more commonly the *correlation distance.* If we further assume that the phase noise is weak with a very small variance [i.e., $k\phi(y) \ll 1$], the third and higher order moments can then be neglected in a Taylor series of the exponential. This makes the problem tractable by reducing it to additive noise; however, the terms are not independent of the spectral bandwidth, and as we show, the output SNR improves slowly with an expanding spectral bandwidth. Experimental evidence suggests that the improvement of SNR would be more significant if higher order terms could also be included in the calculations. Under these assumptions the relevant moments for the phase noise may be written as

$$E[\exp\{ik[\phi(y_1) - \phi(y_2)]\}] = 1 - k^2\sigma_p^2 + k^2\sigma_p^2\left(-\frac{|y_1 - y_2|}{d}\right), \tag{5.16}$$

and

$$\begin{aligned} E[\exp\{ik_1[\phi(y_1) - \phi(y_2)] + ik_2[\phi(\bar{y}_1) - \phi(\bar{y}_2)]\}] \\ = 1 - (k_1^2 + k_2^2)\sigma_p^2 + k_1^2 E[\phi(y_1)\phi(y_2)] + k_2^2 E[\phi(\bar{y}_1)\phi(\bar{y}_2)] \\ + k_1 k_2 E[\phi(y_1)\phi(\bar{y}_1) - \phi(y_1)\phi(\bar{y}_2) - \phi(y_2)\phi(\bar{y}_1) + \phi(y_2)\phi(\bar{y}_2)]. \end{aligned} \tag{5.17}$$

Let us assume that the object transmittance is $a(y) = 1 + \sin(2\pi\Omega y)$, where Ω is the relative frequency of the input grating.

After substituting Eqs. (5.13)–(5.17) into Eq. (5.9), and performing straight-forward calculation, we have

$$E[I_n(y')] = \pi^2\left[1 + (F_1 - \pi^2)\frac{4\sigma_p^2}{\lambda_n^2 - (\Delta\lambda/2)^2}\right]\Delta\lambda_n, \tag{5.18}$$

where

$$F_1 = \iint_{-\infty}^{\infty} a(y)a^*(\bar{y})\operatorname{sinc}[2\pi(y' - y)]\operatorname{sinc}[2\pi(y' - \bar{y})]\exp\left(-\frac{|y - \bar{y}|}{d}\right)dy\, d\bar{y}, \tag{5.19}$$

and

$$E[I_n^2(y')] = \pi^4\left[1 + (F_1 - \pi^2)\frac{8\sigma_p^2}{\lambda_n^2 - (\Delta\lambda_n/2)^2}\right]\Delta\lambda_n^2. \tag{5.20}$$

From the definition of Eq. (5.5), the output SNR due to the nth spectral band filter would be

$$\mathrm{SNR}_n = \frac{1 + (F_1 - \pi^2)\dfrac{4\sigma_p^2}{\lambda_n^2 - (\Delta\lambda_n/2)^2}}{\left|(F_1 - \pi^2)\dfrac{4\sigma_p^2}{\lambda_n^2 - (\Delta\lambda_n/2)^2}\right|}, \tag{5.21}$$

where $\Delta\lambda_n$ is the spectral bandwidth of $H_n(\beta)$. Note that, for a white-light source spanning the visible spectrum, the magnitude of the overall spectral bandwidth of the spectral band filters, that is $\Delta\lambda = \Sigma\Delta\lambda_n$, would be as large as the mean wavelength $\bar{\lambda}$ of the light source. The SNR for the strictly coherent case can be obtained readily by assuming the light source characteristic to be in the form of a delta function, that is $S(\lambda) = \delta(\lambda - \bar{\lambda})$, and carrying out the necessary calculations. Here, the mean wavelength λ_n of the nth spectral filter is assumed to be the *characteristic wavelength* of the coherent counterpart. Thus we conclude, using Eq. (5.7), that the normalized SNR in the case of weak input phase noise for the nth spectral band filter is

$$\overline{\mathrm{SNR}} = 1 + \frac{(\Delta\lambda_n/\lambda_n)^2}{|4[1 + (F_1 - \pi^2)(4\sigma_p^2/\lambda_n^2)]|}. \tag{5.22}$$

It is evident that the normalized SNR would improve with the spectral bandwidth of the processor. Since the channels are not physically separated, the observed improvement in SNR is primarily due to the optical path difference for the wavelengths in the source spectrum. The effect is quite noticeable for broad spectral band filters [5.15, 5.16]. Numerical results based on the above equation are summarised in Fig. 5.2. The output SNR improves with the spectral bandwidth $\Delta\lambda_n$; also, it is somewhat higher for shorter mean wavelengths λ_n. We emphasize that Eq. (5.22) obtained under simplified conditions to make the problem tractable, is an approximation. It only shows a trend of improvement in the SNR. The SNR of the white-light optical processor improves with increasing spectral bandwidth $\Delta\lambda_n$ of the spatial filter $H_n(\alpha, \beta)$; also it is somewhat higher for shorter wavelengths. We expect that

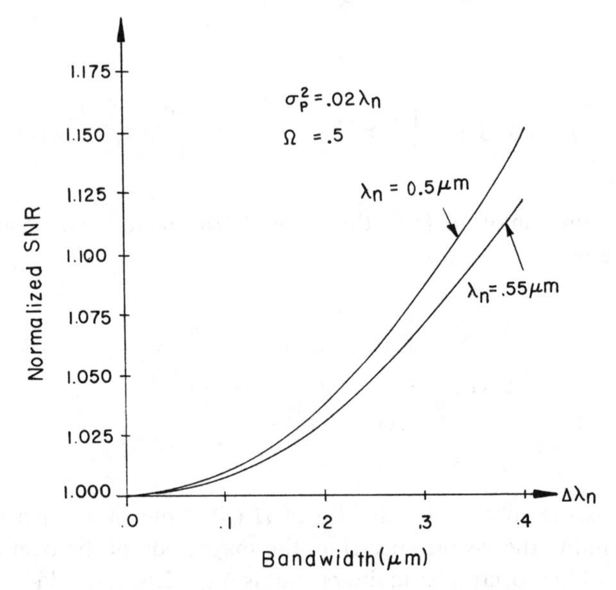

Figure 5.2. Normalized SNR for the phase noise at the input plane as a function of the spectral bandwidth for two different base wavelengths.

the result would be more realistic if higher terms for the phase noise could be included in the calculation.

5.2.2.2 Amplitude Noise. It is experimentally known that amplitude noise, for example, noise due to film granularity [5.17], at the object plane behaves as a part of the signal and that its size and shape does not change with the wavelength of the system. Since we are unable to distinguish the signal from the amplitude noise at the object plane by varying the wavelength of the system, we expect to find no significant advantage in using a broad-band source. With $\phi(y) = 0$, the output intensity for a coherent source can be written as

$$I(y') = \left| \int_{-\infty}^{\infty} [a(y) + n(y)]\text{sinc}[2\pi(y' - y)]dy \right|^2. \tag{5.23}$$

Since the above expression is independent of wavelength, the SNR for both a temporal partially coherent and a monochromatic source would have the same value. This implies, from the definition of Eq. (5.7), that the normalized SNR will be unity for any spectral bandwidth of the system.

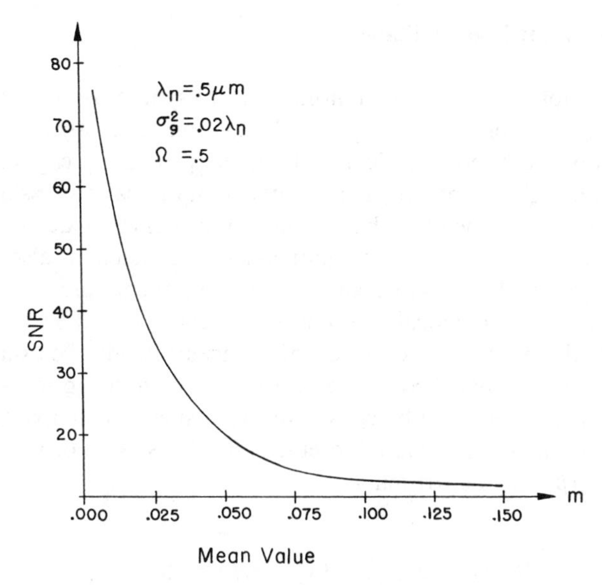

Figure 5.3. A plot of SNR for additive Gaussian noise at the input plane as a function of its mean value.

If we now assume that the additive noise is a stationary Gaussian process with mean m and a correlation function of the form

$$E[n(y_1)n(y_2)] = \sigma_g^2 \exp\left(-\frac{|y_1 - y_2|}{b}\right) + m^2, \tag{5.24}$$

where b is the *correlation distance*, then the third- and fourth-order moments may be expressed in terms of the second-order moments [5.18]. If we also choose some suitable functional form for the input signal, the expression for the output SNR can be simplified to a larger extent, and then can be numerically evaluated. The results we have obtained with $a(y) = 1 + \sin(2\pi\Omega y)$ show that the output SNR depends strongly on the mean value m of the input additive noise, as shown in Fig. 5.3.

Furthermore, for the case of amplitude noise, the separation of filter channels, for example, using a phase sampling grating at a later stage, will not by itself help improve the noise performance of the optical system. However, when spectral band filters specifically designed to reduce the input noise are inserted at the Fourier plane, considerable improvement in the output SNR may be expected. The analysis here agrees with the results obtained by Chavel and Lowenthal [5.7, 5.8] for a simple optical imaging system.

5.2.3. Noise at the Fourier Plane

Since photographic plates are commonly used to synthesize the spatial filters for optical signal processing, the phase fluctuations and other additive random defects of the filters significantly limit the processing capability of the optical system. This is especially true for coherent optical processing systems. The *setup noise*, as defined by Chavel and Lowenthal [5.6], due to scattering and reflections from the optical components of the system, is also an important consideration. However, if all optical components are sufficiently far away as compared to the focal depth of the system, it would be reasonable to assume that all the setup noise sources are concentrated in the Fourier plane. Thus if high quality optical components are utilized in the optical system, the setup noise would generally be considered very small as compared with the inherent noise in the photographic emulsion. Thus the noisy nth spectral band filter $H_n(\beta)$ may be written as

$$H_n(\beta) = \{\exp[ik\phi(\beta)] + n(\beta)\}\text{rect}\left(\frac{\beta}{v_c \lambda f}\right). \tag{5.25}$$

The rectangular function represents the size of the filter in the β-direction. Again, for simplicity in calculation, a one-dimensional signal is considered.

Let the transmittance at the input plane be

$$s(x, y) = a(y)\exp(i2\pi v_0 x). \tag{5.26}$$

Notice that we now have the phase sampling grating in place. It is used to separate the filter channels by dispersion. Apart from providing flexibility in the design of spectral band filters, it also accounts for the decorrelation in the noise encountered at the Fourier plane for various information carrying channels. As long as a suitable spatial frequency is chosen for the phase sampling grating, the processor will achieve a desirable separation of channels along the α-direction. Under these conditions, Eq. (5.2) for the output intensity can now be written as

$$I_n(y') = \int_{\lambda_n - \Delta\lambda_n/2}^{\lambda_n + \Delta\lambda_n/2} \left| \int_{-\infty}^{\infty} S(\beta)H_n(\beta)\exp\left(-i\frac{2\pi}{\lambda f}\beta y'\right)d\beta \right|^2 d\lambda, \tag{5.27}$$

where $S(\beta)$ is the Fourier transform of the input signal, and the trivial integration in the α-direction is omitted. We now consider the effects due to the phase and the amplitude noise independently in the following.

5.2.3.1. Weak Phase Noise. Under the weak phase noise approximation, the output SNR can be calculated from Eqs. (5.25)–(5.27) and the definition of Eq. (5.5). Assuming $a(y) = 1 + \cos(2\pi\Omega y)$, we have

$$\mathrm{SNR}_n = \Bigg\{ \left(\frac{1}{\pi^2\sigma_p^2} - \frac{4}{\lambda_n^2 - (\Delta\lambda_n/2)^2} \right) [i + \cos(2\pi\Omega y')]^2 + \frac{6}{\lambda_n^2 - (\Delta\lambda_n/2)^2} + 4g_1 \cos(2\pi\Omega y') + g_2 \cos(4\pi\Omega y') \Bigg\} \Bigg/ \Bigg| \Bigg\{ \frac{6}{\lambda_n^2 - (\Delta\lambda_n/2)^2} + 4g_1 \cos(2\pi\Omega y') + g_2 \cos(4\pi\Omega y') + \frac{4}{\lambda_n^2 - (\Delta\lambda_n/2)^2} [1 + \cos(2\pi\Omega y')]^2 \Bigg\}^2 - \frac{16}{\lambda_n^2 - (\Delta\lambda_n/2)^2} [1 + \cos(2\pi\Omega y')]^2 \Bigg|^{1/2}, \tag{5.28}$$

where

$$g_1 \triangleq \frac{1}{\Delta\lambda_n} \int_{\lambda_n - \Delta\lambda_n/2}^{\lambda_n + \Delta\lambda_n/2} \frac{1}{\lambda^2} \exp\left(\frac{-\lambda f \Omega}{d} \right) d\lambda, \tag{5.29}$$

and

$$g_2 \triangleq \frac{1}{\Delta\lambda_n} \int_{\lambda_n - \Delta\lambda_n/2}^{\lambda_n + \Delta\lambda_n/2} \frac{1}{\lambda^2} \exp\left(\frac{-2\lambda f \Omega}{d} \right) d\lambda. \tag{5.30}$$

Also, the output SNR for the coherent case with $S(\lambda) = \delta(\lambda - \lambda_n)$ can also be written as

SNR (coherent)

$$= \frac{(1 - 4\pi^2\sigma_p^2/\lambda_n^2)[1 + \cos(2\pi\Omega y')]^2 + (4\pi^2\sigma_p^2/\lambda_n^2) g_3}{\left| \dfrac{4\pi^2\sigma_p^2}{\lambda_n^2} \left\{ \dfrac{4\pi^2\sigma_p^2}{\lambda_n^2} [1 + \cos(2\pi\Omega y')^2 - g_3] + [1 + \cos(2\pi\Omega y')]^2 g_3 \right\} \right|^{1/2}}, \tag{5.31}$$

where

$$g_3 \triangleq \frac{3}{2} + \exp\left(-\frac{\lambda_n f \Omega}{d} \right) \cos(2\pi\Omega y') + \frac{1}{4} \exp\left(-\frac{2\lambda_n f \Omega}{d} \right) \cos(4\pi\Omega y'). \tag{5.32}$$

Finally, the normalized output SNR of Eq. (5.28), due to the nth spectral band filter $H_n(\beta)$, can be calculated from the definition of Eq. (5.7): $\overline{\text{SNR}} = \text{SNR}_n/\text{SNR(coherent)}$.

Qualitative effects pertain to the noise performance of the optical processing system are rather difficult to deduce from the foregoing equations. Several numerical computations were made to develop a feeling for the above result. We find that the normalized SNR improves with the spectral bandwidth of the processor and that it is higher for shorter wavelengths, as shown in Fig. 5.4. We note that the comments in the previous section about weak phase noise at the input plane can also be applied to the observations that we have obtained in this figure.

Furthermore, contributions from separate spectral bands to the output intensity would have a smoothing effect on the phase noise and the overall SNR would show considerable improvement. Since the channels are physically separated by the action of the phase sampling grating at the input plane and encounter independent realizations of noise, approximately $\sqrt{N}$-fold improvement in the output SNR would result, where N is the total number of narrow spectral band filters employed in the Fourier plane. We note that this result is quite consistent with the experimental results for broad-band deblurring with a white-light source [5.14].

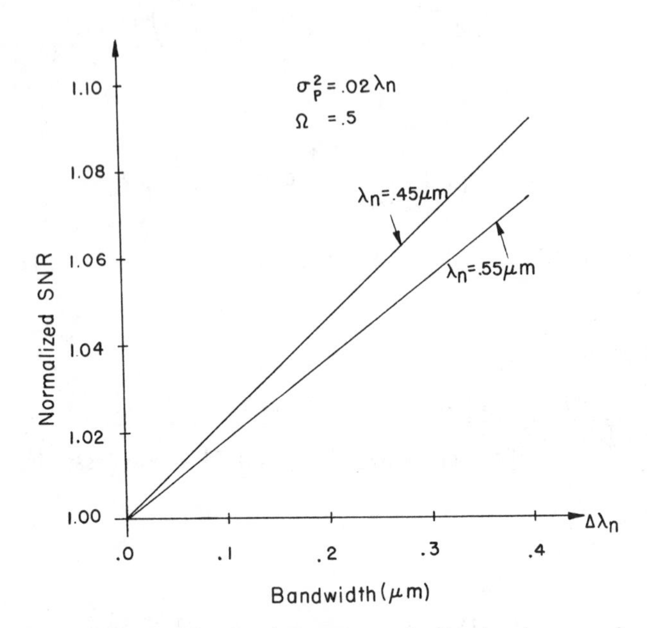

Figure 5.4. Normalized SNR for the phase noise at the Fourier plane as a function of the spectral bandwidth for two different wavelengths.

5.2.3.2. Amplitude Noise. Another source of noise in photographic filters is granularity. But available models for the grain noise are inadequate and are mathematically intractable. As before, we consider an example of additive Gaussian noise with mean m and correlation function described by Eq. (5.24).

The desired size of the spectral spread $\Delta\lambda_n$ over the nth spatial filter $H_n(\beta)$ along the α-direction can be obtained by a suitable choice of the spatial frequency of the input phase sampling grating. The physical size of the narrow spectral band filters can be chosen to be of the order of several correlation distances, b. The corresponding spectral bandwidth of a channel can then be easily evaluated from Eq. (5.3).

Under these conditions, for a single filter channel, the noise is highly correlated at the Fourier plane and, for reasons discussed in Section 5.2.2.2, there is little improvement in the output SNR. But the channels in this case are corrupted by independent additive noise as they have been sufficiently separated from each other by the use of phase sampling grating at the input plane. Since all channels are carrying the same information with independent Gaussian noise, their superposition at the image plane results in an $\sqrt{N}$-fold improvement for the white-light processing system. Figure 5.5 shows the improvement in the output SNR as a function of the number N of filter channels in

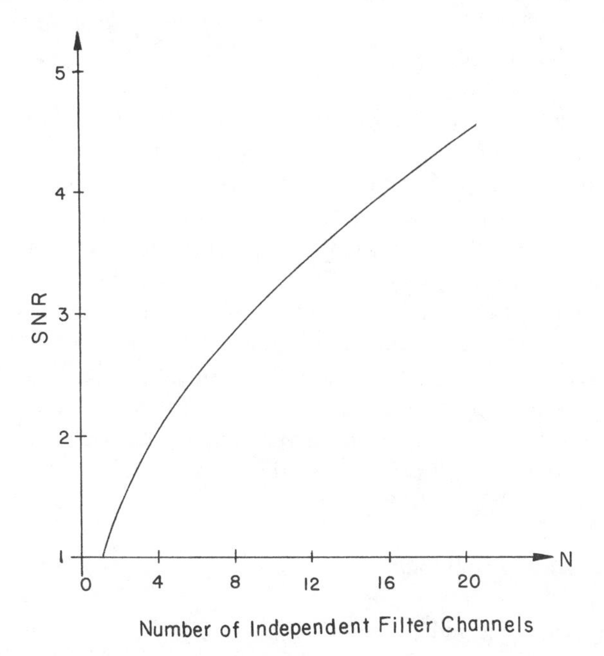

Figure 5.5. The output SNR for the additive Gaussian noise at the Fourier plane as a function of the number of filter channels in the system.

the system. Thus we see that the improvement in the output SNR depends on the extent of spectral decomposition at the Fourier plane. However, we note that the number of channels or the optical system is restricted by Eq. (5.4) and also by practical considerations in generating the narrow spectral band filters.

5.3 NOISE PERFORMANCE UNDER THE SPATIAL COHERENCE REGIME

We have shown in the previous section that a broad-band illumination with strict spatial coherence (point source) can be used advantageously to suppress coherent artifact noise. This was possible due to the redundancy in information at the output plane resulting from a large spectral bandwidth of the light source. Alternatively, a monochromatic light source of a finite size can also be employed to attenuate noise in an optical system. Such an illumination is called spatially partially coherent (SPC). Each point on the extended light source, called a channel, produces an object image at the output plane. While a single channel corresponds to a strict spatially coherent source, a large source with infinitely many channels signifies the other extreme, namely, *spatial incoherence.* To quantify these ideas we shortly give a definition of the degree of coherence and delineate an approximate region where the source can be termed spatially partially coherent.

In this section we study mainly the noise performance of optical systems employing SPC illumination. A typical SPC light source corresponds to many channels, each projecting an object image at the output plane. The information about the object carried by each of these channels is superimposed incoherently at the image plane to give the net observable output irradiance. If the channels are physically separated when they hit the noise plane and if the noise is assumed to be spatially independent and wide sense stationary, the noise effects would generally cancel each other, providing considerable improvement in the output SNR of the optical system. The effects of the phase and the amplitude noise are considered in some detail, again at the object plane and at the Fourier plane. When necessary, we again assume that the noise is Gaussianly distributed, so that qualitative features of noise performance can be calculated.

5.3.1. Problem Formulation

The source plane P_0 and the Fourier plane P_2 shown in Fig. 5.1, form a conjugate pair in the sense that the light source is imaged in the Fourier plane by lenses L_1 and L_2. The ratio of the source image size to the spatial filter size may be defined as a parameter of spatial coherence. If lenses L_1 and L_2 have the same focal length, the *spatial coherence parameter* ρ for the optical system can be written as [5.19]

$$\rho \triangleq \frac{\Delta s}{\Delta H}, \tag{5.33}$$

where Δs and ΔH are the source and the filter size, respectively. For $\rho \ll 1$ we have strong spatial coherence. As it increases to infinity an extremely incoherent illumination results. A typical range for an SPC source is $0.2 < \rho < 0.7$. We now study the noise performance of the proposed optical processor shown in Fig. 5.1 in detail for the indicated region.

The number of channels, or the space bandwidth product, can be simply related to the spatial coherence parameter ρ. For convenience, if all focal lengths f for the achromatic transform lenses are chosen to be equal, then Eq. (5.33) can be written as

$$\rho = \frac{N_s}{N_f} = \frac{\Delta a \Delta s/\lambda^2 f^2}{\Delta H \Delta a/\lambda^2 f^2}, \tag{5.34}$$

where Δa is the input object area, and N_s and N_f define the space bandwidth product, or the number of channels, for the source and the Fourier plane, respectively.

The one-dimensional expression for the output intensity distribution for the SPC optical system of Fig. 5.1 can be written as [see Eq. (3.66)]

$$I_n(x') = \int_{\Delta s} \gamma(x_0) \left| \int_{-\infty}^{\infty} S(x_0 + \alpha) H(\alpha) \exp\left(-i \frac{2\pi}{\lambda f} \alpha x'\right) d\alpha \right|^2 dx_0. \tag{5.35}$$

where: $\gamma(x_0)$ is the spatial intensity distribution of the monochromatic source; $S(\alpha)$ and $H(\alpha)$ are the spectrum of the input signal transmittance and the spatial filtering function, respectively; and Δs is the size of the light source. Also, without loss of generality, we have assumed that the spatial frequency of the phase sampling grating at the object plane is zero, since we would be concerned only with the monochromatic illumination.

Equation (5.35) can be put in a more appropriate form if we define the following transformed coordinates:

$$u_0 = \left(\frac{1}{\Delta s}\right)x_0, \tag{5.36}$$

$$u = \left(\frac{\Delta H}{\lambda f}\right)x, \tag{5.37}$$

$$\tau = \left(\frac{1}{\Delta H}\right)\alpha, \tag{5.38}$$

and

$$u' = \left(\frac{\Delta H}{\lambda f}\right)x'. \tag{5.39}$$

Using the transformed coordinates above and also assuming that

$$\gamma(x_0) = \text{rect}\left(\frac{x_0}{\Delta s}\right), \tag{5.40}$$

where

$$\text{rect}\left(\frac{x_0}{\Delta s}\right) = \begin{cases} 1, & \text{for } |x_0| \leqq \Delta s/2, \\ 0, & \text{otherwise,} \end{cases}$$

Eq. (5.35) can be written as

$$I(u') = C \iint \text{sinc}[2\pi\rho(u_1 - u_2)]S(u_1)S^*(u_2)K(u' - u_1)K^*(u' - u_2)du_1\, du_2, \tag{5.41}$$

where C is an appropriate constant and is ignored in further discussion, and

$$K(u) \triangleq \int_{-\infty}^{\infty} H(\tau)\exp(-i2\pi\tau u)d\tau. \tag{5.42}$$

Equation (5.41) is the basic formula that may be used to calculate the output SNR of an SPC optical processor for various cases of interest.

We obtain the output SNR of the optical system as a measure of its performance, which is defined as [see Eq. (5.5)]

$$\mathrm{SNR} \triangleq \frac{E[I(u')]}{\sigma_I(u')}, \tag{5.43}$$

where $E[\ ]$ denotes the ensemble average and σ_I^2 is the variance in intensity at the output image plane. To make the numerical results more meaningful for the noise performance in using partially coherent illumination, again a normalized SNR can be defined as [see Eq. (5.7)]

$$\overline{\mathrm{SNR}} = \frac{\mathrm{SNR(SPC)}}{\mathrm{SNR(SC)}}, \tag{5.44}$$

where SPC and SC stand for spatially partially coherent and strictly coherent illumination, respectively.

5.3.2. Noise at the Input Plane

Consider the input transmittance in the form

$$s(u) = a(u)\exp[ik\phi(u)] + n(u) \tag{5.45}$$

where $\phi(u)$ and $n(u)$ are the random phase fluctuations and the amplitude noise, respectively; $a(u)$ is the input signal; $k = 2\pi/\lambda$; and $u \triangleq [\Delta H/(\lambda f)]x$.

The additive noise at the input plane behaves as if it were a part of the signal to be processed. All channels see the same "realization" of the amplitude noise process. Since we are dealing with a particular realization superimposed on the signal, no averaging effect for the additive noise is observed at the output image plane. We expect its behavior to show little dependence on the degree of spatial coherence as was the case with TPC illumination (see Section 5.2). A detailed analysis of this type of noise is given by Chavel [5.8]. It is therefore sufficient to say that SPC illumination does not help improve the SNR of the optical system in this case.

Experiments show that the phase defects at the input plane, a severe limitation of coherent systems, can be handled effectively when spatially incoherent light is employed. Neglecting $n(u)$ in Eq. (5.45) and assuming that $a(u) = 1 + \sin(2\pi\Omega u)$, we have

$$s(u) = [1 + \sin(2\pi\Omega u)]\exp[ik\phi(u)], \tag{5.46}$$

where

$$\Omega \triangleq \frac{\lambda f}{\Delta H}, \tag{5.47}$$

is defined as the input relative spatial frequency, v is the real spatial frequency, and $\Delta H/\lambda f$ is the cut-off spatial frequency of the optical processing system. We further assume that the spatial filter is a rectangular function, that is, $H(\tau) = \text{rect}(\tau)$. Therefore

$$K(u) = \int_{-1}^{1} \exp(-i2\pi\tau u)d\tau = C_0 \operatorname{sinc}(2\pi u), \tag{5.48}$$

where C_0 is an appropriate constant and is ignored in the following discussion. Using the expression for the output intensity given in Eq. (5.41) and substituting from Eqs. (5.46) and (5.48), we have

$$I(u') = \iint F(u', u_1, u_2, \rho)a(u_1)a^*(u_2)\exp\{ik[\phi(u_1) - \phi(u_2)]\}du_1\, du_2, \tag{5.49}$$

where

$$F(u', u_1, u_2, \rho) = \operatorname{sinc}[2\pi\rho(u_1 - u_2)]\operatorname{sinc}[2\pi(u' - u_1)]\operatorname{sinc}[2\pi(u' - u_2)]. \tag{5.50}$$

Assuming that the phase noise is weak $[k\phi(u) \ll 1]$ with zero mean, the exponential in Eq. (5.49) can be expanded into a Taylor series and a solution for the above equation can be obtained by neglecting higher order terms. The relevant moments for this case are

$$E[\phi(u)] = 0, \tag{5.51}$$

and

$$E[\phi(u_1)\phi(u_2)] = k^2\sigma_p^2 \exp\left(-\frac{|u_1 u_2|}{d}\right), \tag{5.52}$$

where σ_p^2 is the variance and d is the correlation distance for the phase noise. The SNR for the SPC illumination can now be easily calculated from Eqs. (5.43), (5.44), and (5.49) to give

$$\text{SNR} = \frac{G_1(u') + k^2\sigma_p^2[G_1(u') - G_2(u')]}{|k^2\sigma_p^2[G_1(u') - G_2(u')]|}, \tag{5.53}$$

where

$$G_1(u') = \iint F(u', u_1, u_2, \rho)a(u_1)a^*(u_2)du_1\, du_2, \tag{5.54}$$

and

$$G_2(u') = \iint F(u', u_1, u_2, \rho)a(u_1)a^*(u_2)\exp\left(-\frac{|u_1 - u_2|}{d}\right)du_1\, du_2. \tag{5.55}$$

When $\rho \ll 1$ we approach the coherent limit. Here the spatial intensity distribution of the source can be replaced by a delta function. Assuming a point source at the origin, we have $\gamma(x_0) = \delta(x_0)$. Equation (5.35) can now be approximately modified to obtain the SNR for the coherent case. It can be shown that

$$\text{SNR(SC)} = \frac{G_3(u') + k^2\sigma_p^2[G_3(u') - G_4(u')]}{|k^2\sigma_p^2[G_3(u') - G_4(u')]|}, \tag{5.56}$$

where

$$G_3(u') = \iint \text{sinc}[2\pi(u' - u_1)]\text{sinc}[2\pi(u' - u_2)]a(u_1)a^*(u_2)du_1\, du_2, \tag{5.57}$$

and

$$G_4(u') = \iint \text{sinc}[2\pi(u' - u_1)]\text{sinc}[2\pi(u' - u_2)]\exp\left(-\frac{|u_1 - u_2|}{d}\right)du_1\, du_2. \tag{5.58}$$

The ratio of the expressions in Eqs. (5.53) and (5.56) then gives the normalized SNR for weak phase noise at the output plane. Although the equations

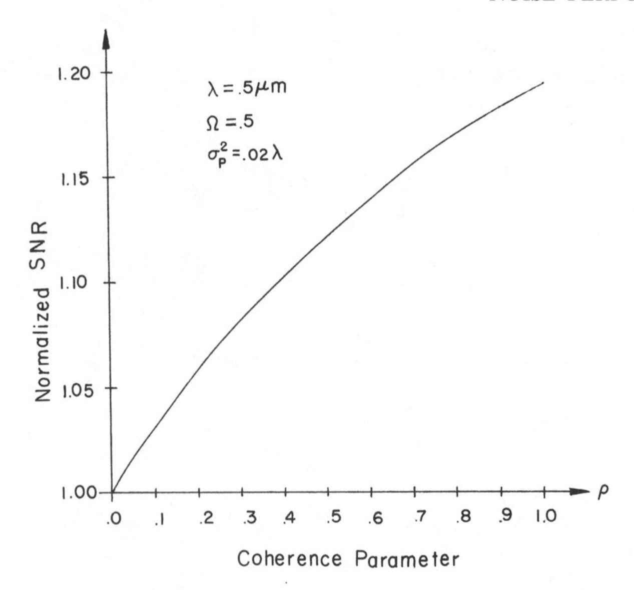

Figure 5.6. Normalized SNR for the phase noise at the input plane as a function of the spatial coherence parameter.

are complex, it is easy to see that the SNR would increase to infinity as the variance in noise is decreased to zero (i.e., the noise is eliminated from the system). Also, as necessary, Eq. (5.53) does reduce to Eq. (5.56) in the limit when ρ goes to zero. A numerical evaluation of the results given above is summarized in Fig. 5.6. It is seen that the normalized SNR increases as the requirements on spatial coherence are relaxed, that is, as ρ is increased. Note that we have considered the problem in one dimension for calculational simplicity, whereas the phenomenon is truely two-dimensional. We expect, as indicated by experimental work, that the normalized SNR would be much higher if the problem could be worked in two dimensions and also if higher terms in the Taylor expansion of phase noise could be included. We also note that the problem for strong phase can be worked out following Parry [5.20–5.21] since a Gaussian character for the noise can be assumed.

5.3.3. Noise at the Fourier Plane

The results obtained for the input noise show that while a reduced degree of spatial coherence considerably attenuates the phase fluctuations, it is ineffective in reducing amplitude noise. However, we see that the situation is somewhat different for the noise at the Fourier plane.

We assume that the noise at the Fourier plane is wide sense stationary Gaussian. Although this assumption is only approximate, it has been shown [5.7] that the output SNR obtained under more realistic non-Gaussian form is in general very close or larger than the Gaussian case. We also assume that the correlation distance for the complex noise is much smaller than the diffraction limited image of the source point at the Fourier plane.

Under the foregoing assumptions it can be easily seen that various channels in the system would be independent. They would see different "realizations" of the complex noise at the Fourier plane. Each channel would be diffracted at the noisy Fourier plane, resulting in high contrast noise fringes at the output image plane. Since the overall output irradiance is the superposition of images from each channel, the visibility of noise fringes decreases with an increasing size of the source. More precisely, the output SNR for independent Gaussian channels increases as $\sqrt{N}$, where N is the total number of channels in the SPC system and is proportional to the area of the source. Except for some intensity attenuation, the amplitude noise at the Fourier plane can be eliminated by using an extended light source with $\rho = 1$ (a typical incoherent source corresponds to $\sim 10^6$ channels).

The number of channels that eventually contribute to the output image is constrained by the limiting aperture of the system. Usually this critical aperture for the system is determined by the filter size. Figure 5.7 shows a plot of

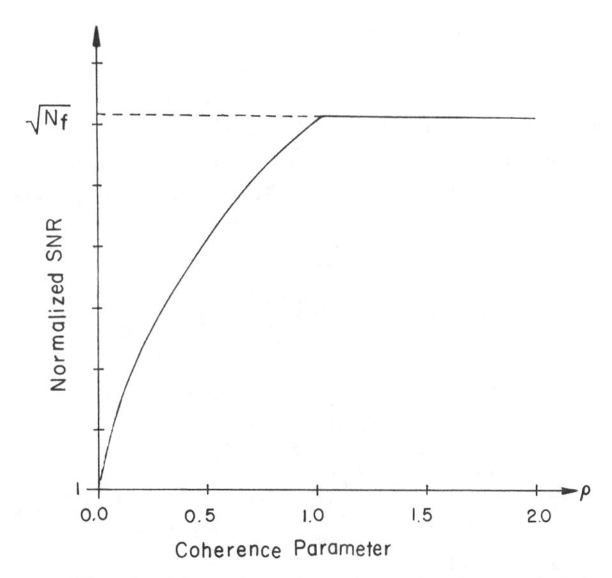

Figure 5.7. A plot of SNR for the complex noise at the Fourier plane as a function of the spatial coherence parameter.

the output SNR as a function of the spatial coherence parameter ρ. For $\rho < 1$ when the source size is smaller than the filter size, the output normalized SNR improves as $\sqrt{N_s}$, where N_s denotes the number of channels from the source. However, when the source size becomes larger than the filter size, $\rho > 1$, the system no longer supports the additional channels and the output normalized SNR stays at a constant level, $\sqrt{N_f}$, where N_f is the total number of the independent filter channels.

5.4. MEASUREMENT OF NOISE PERFORMANCE

In the previous sections, we have analyzed the performance of a white-light optical processor under noisy environment. We have shown that, except for the case when the amplitude noise is present at the input plane, the resulting output SNR improves considerably under partially coherent illumination.

In this section, however, we experimentally measure the noise performance of the proposed white-light optical processor [5.22], We show that the measured results conform with the analytical results that we have calculated. Aside from the noise effects at the input and the Fourier planes, we have also measured the noise performance due to thin noise effect along the longitudinal direction (i.e., the Z-axis) of the proposed white-light optical system.

5.4.1. Noise Measurement

A schematic diagram for the noise measurement of a white-light optical signal processor is depicted in Fig. 5.8. We first investigate the noise performance under the spatially coherent illumination, that is, the effects due to source size. For convenience, we assume that the source irradiance is uniform over a square aperture at source plane P_0, which can be written as

$$\gamma(x_0, y_0) = \text{rect}\left(\frac{x_0}{a}\right)\text{rect}\left(\frac{y_0}{a}\right), \tag{5.59}$$

where

$$\text{rect}\left(\frac{x_0}{a}\right) \triangleq \begin{cases} 1, & |x_0| \leqq \dfrac{a}{2}, \\ 0, & |x_0| > \dfrac{a}{2}. \end{cases}$$

For simplicity, we assume that the input signal is a one-dimensional object independent of the y-axis. The Fourier spectrum would also be one-dimen-

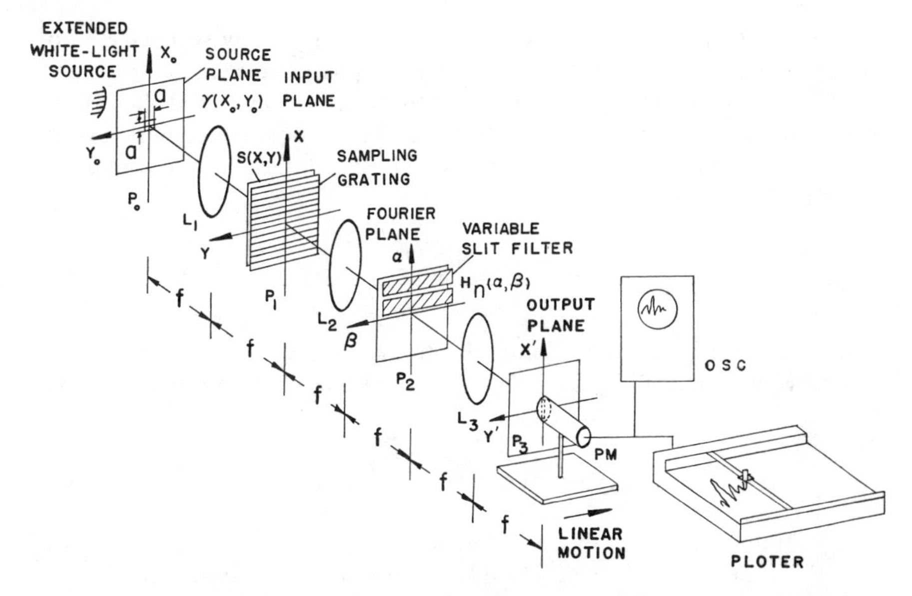

Figure 5.8. Grating-based white-light optical processor. $\gamma(x_0, y_0)$, source intensity distribution; P_0, source plane; P_1, input plane; P_2, Fourier plane; P_3, output plane; $s(x, y)$, object transparency; $H_n(\alpha, \beta)$, slit filter; *PM*, photometer; *OSC*, oscilloscope; *L*, achromatic transform lenses.

sional in the β-axis, but smeared into rainbow colors along the α-direction. Let the width of the nth narrow spectral band filter be $\Delta\alpha_n$. If the filter is placed in the smeared Fourier spectra, the spectral bandwidth of the filter can be written as [see Eq. (5.3)]

$$\Delta\lambda_n = \frac{\Delta\alpha_n}{\nu_0 f}. \tag{5.60}$$

The total number of filter channels can therefore be determined as

$$N \triangleq \frac{\Delta\lambda}{\Delta\lambda_n} \simeq \frac{\Delta\lambda \nu_0 f}{\Delta\alpha_n}, \tag{5.61}$$

where $\Delta\lambda$ is the spectral bandwidth of the white-light source. Thus we see that the degree of temporal coherence, at the Fourier plane, increases as the spatial frequency of the sampling grating ν_0 increases.

For noise measurement under the temporally coherent regime, we would use a variable slit representing a broad spectral filter in the Fourier plane. The output noise fluctuation can be traced out with a linearly scanned photometer, as illustrated in Fig. 5.8. It is therefore apparent that the output noise fluctuation due to the spectral bandwidth of the slit filter and due to the source size can then be separately determined. We adopt the definition of an output SNR proposed in Eq. (5.5) of Section 5.2.1 for the measurement of the noise performance, that is,

$$\mathrm{SNR}_n(y') \triangleq \frac{E\,[I_n(y')]}{\sigma_n(y')}, \tag{5.62}$$

where $I_n(y')$ is the output irradiance due to the nth channel, $E[\ ]$ denotes the ensemble average, and $\sigma_n^2(y')$ is the variance of the output noise fluctuation, that is,

$$\sigma_n^2(y') \triangleq E[I_n^2(y')] - \{E[I_n(y')]\}^2. \tag{5.63}$$

Evidently, the output intensity fluctuation $I_n(y')$ can be traced out by a linearly scanned photometer. The dc component of the output traces is obviously the out signal irradiance (i.e., $E[I_n]$), and the mean square fluctuation of the traces is the variance of the output noise (i.e., σ_n^2). Thus we see that the effect of the output SNR due to spectral bandwidth (i.e., temporal coherence) and source size (i.e., spatial coherence) can readily be obtained with the proposed measurement technique.

5.4.2. Experimental Results

In our experiments, a 75-W xenon arc lamp is used as a white-light source. The spatial frequency of the sampling phase grating used is about 50 lines/mm, the focal length of the achromatic transform lenses is about 380 mm. Both the amplitude and the phase noise plates used in the experiments are generated by photographing a laser speckle pattern, and the phase plate is obtained with a surface relieving technique through an R-10 bleaching process [5.23]. Shower glass, for strong phase perturbation, is also utilized in the experiments. We first demonstrate the noise performance due to perturbation at the input plane. For the amplitude noise at the input plane, the experiments have shown that there is no apparent improvement in noise performance under the partially coherent illumination. The result is quite consistent with the prediction by Chavel and Lowenthal [5.7, 5.8]. However, for the phase noise at the input plane, the noise performance of the system is largely improved with a partially coherent illumination. We first utilize a weak phase model (obtained by laser speckle and a photographic bleaching process described earlier) as an input noise. Now we consider the situation of

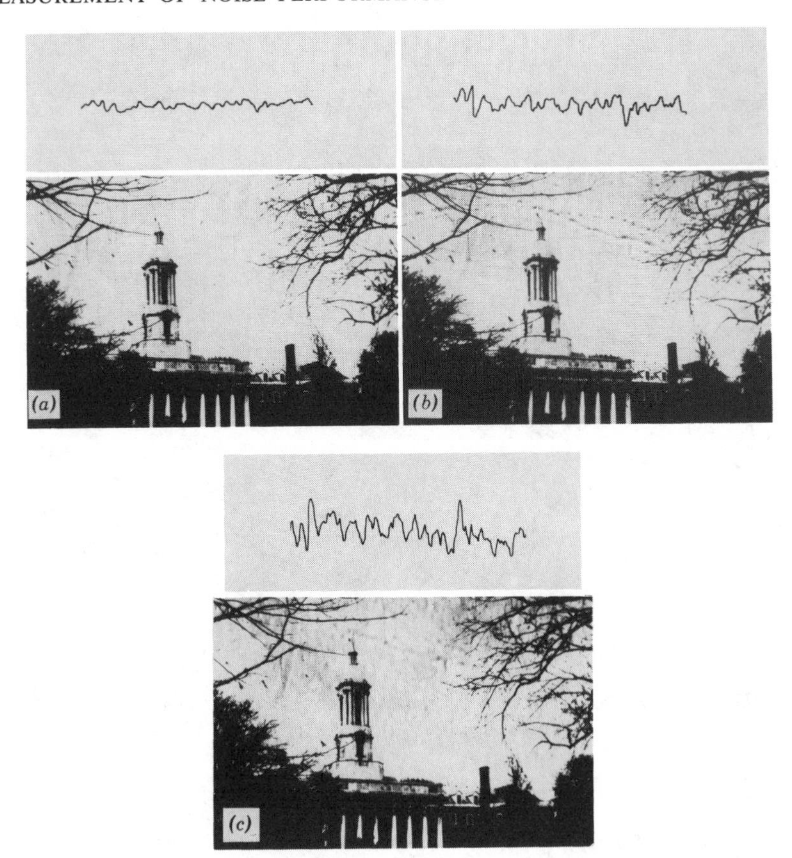

Figure 5.9. Effect on output image (with a section of photometer traces) due to phase noise at input plane for different spectral bandwidths. The source size used is 0.7 mm^2 (i.e., $a = 0.7$ mm). (a) For $\Delta\lambda_n = 1500$ Å. (*b*) For $\Delta\lambda_n = 1000$ Å. (*c*) For $\Delta\lambda_n = 500$ Å.

an input object transparency superimposing with the phase noise at the input plane. The effect of the noise performance of the optical system can then be obtained by varying the source size and the spectral bandwidth of the slit filter, as described in Fig. 5.8. Figure 5.9 shows a set of output photographic images with sections of photometer traces to illustrate the output noise due to spectral bandwidth of the slit filter. From these pictures, we see that the noise performance (i.e., SNR) improves as the spectral bandwidth (i.e., temporal coherence) of the slit filter increases.

Quantitative mesurements of the noise performance due to phase noise at the input plane are plotted in Figs. 5.10 and 5.11. From the figures, we see that the output SNR increases monotonically as the spectral bandwidth $\Delta\lambda_n$

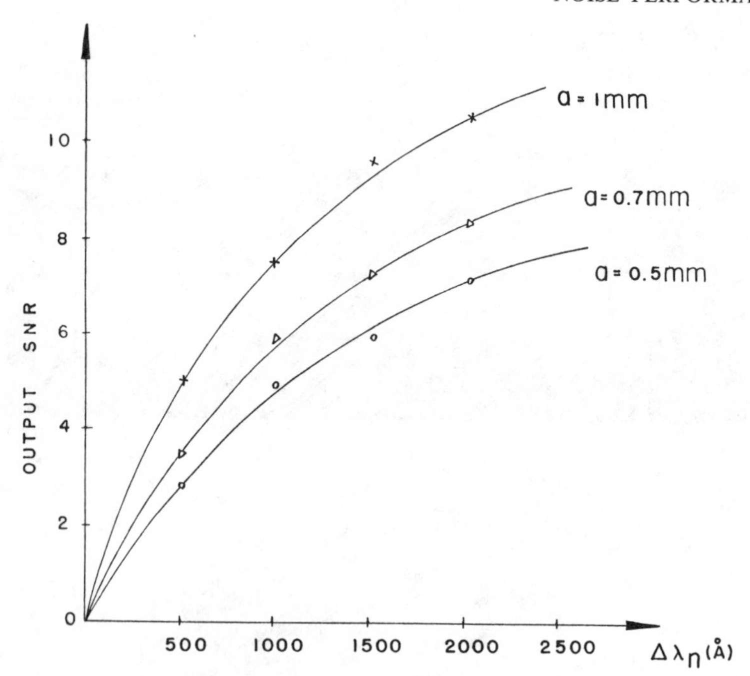

Figure 5.10. Output SNR for phase noise at input plane as a function of spectral bandwidth of the slit filter for various values of source sizes.

of the slit filter increases and that it increases linearly as the source size enlarges. Thus the noise performance for a partially coherent optical system improves as the degree of coherence (i.e., temporal and spatial coherence) relaxes. In other words, to improve the output SNR of the white-light (i.e., partially coherent) processor, one can relax the spatial coherence (i.e., the source size), or the temporal coherence (i.e., the spectral bandwidth of the filter), or both in the optical processing system. We stress that the experimental results are quite compatible with the previous analysis (i.e., Fig. 5.5).

Let us now demonstrate the effect of the noise performance due to strong phase noise. In the experiments, a conventional shower glass is used as an input noise. Figure 5.12 shows a set of results that we have obtained under various spectral bandwidth illuminations. Figure 5.12*a* shows an output result obtained under entirely broad-band white-light illumination. Although this image is somewhat aberrated due to the thick phase perturbation, the image is relatively immune from random noise fluctuation. Comparing the results obtained in Fig. 5.12*a* to those of 5.12*d*, we see that the output SNR decreases rather rapidly as the spectral bandwidth of the slit filter decreases.

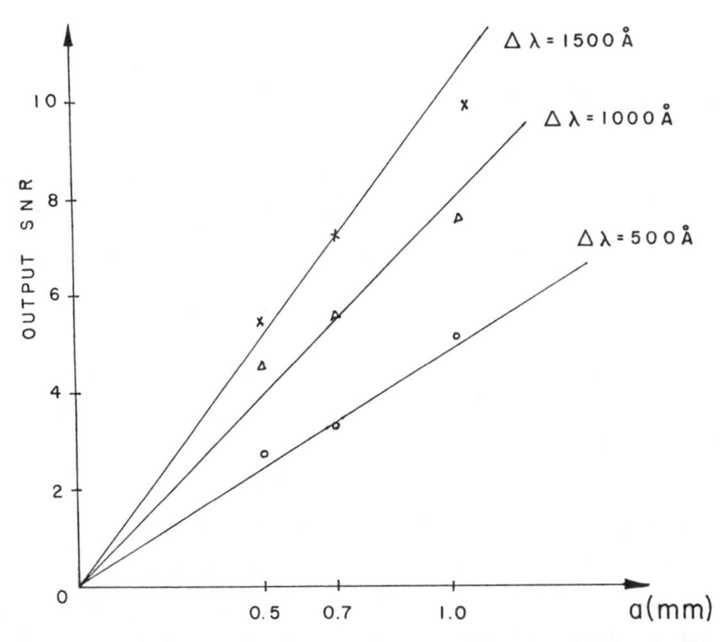

Figure 5.11. Output SNR for phase noise at input plane as a function of source size for various values of spectral bandwidths.

Figure 5.12. Effect on output image due to strong phase perturbation at input plane. $a = 0.7$ mm. (*a*) For $\Delta\lambda_n = 3000$ Å. (*b*) For $\Delta\lambda_n = 1500$ Å. (*c*) For $\Delta\lambda_n = 500$ Å. (*d*) Obtained with a HeNe laser.

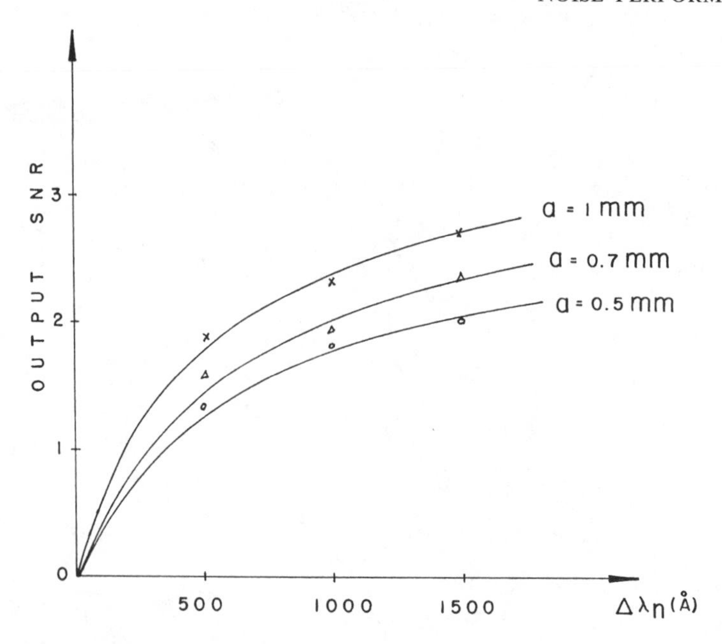

Figure 5.13. Output SNR for amplitude noise at Fourier plane as a function of spectral bandwidth for various source sizes.

Furthermore, Fig. 5.12*d* shows a result obtained with a narrow-band coherent source (i.e., a HeNe laser). Aside from the poor noise performance, we have noted that the output image is also severely corrupted by coherent artifact noise.

We now demonstrate the noise performance of a white-light optical processing system due to noise at the Fourier plane. With reference to the same measurement technique as proposed in Fig. 5.8, the effects of amplitude noise at the Fourier plane are plotted in Fig. 5.13. In contrast with the amplitude noise at the input plane, we see that the output SNR increases monotonically as the spectral bandwidth of the slit filter increases. The output SNR also improves as the source size enlarges. Thus, for amplitude noise at the Fourier plane, the noise performance of a white-light optical processor improves as the degree of temporal and spatial coherence decreases.

Figure 5.14 shows the noise performance of a white-light processor for phase noise at the Fourier plane. From this figure, once again we see that the output SNR is a monotonic increasing function of spectral bandwidth $\Delta\lambda_n$. The SNR also increases as the source size increases. However, as compared with the case of phase noise at the input plane of Fig. 5.9, the improvement of the noise performance is somewhat less effective. Once again we see that the

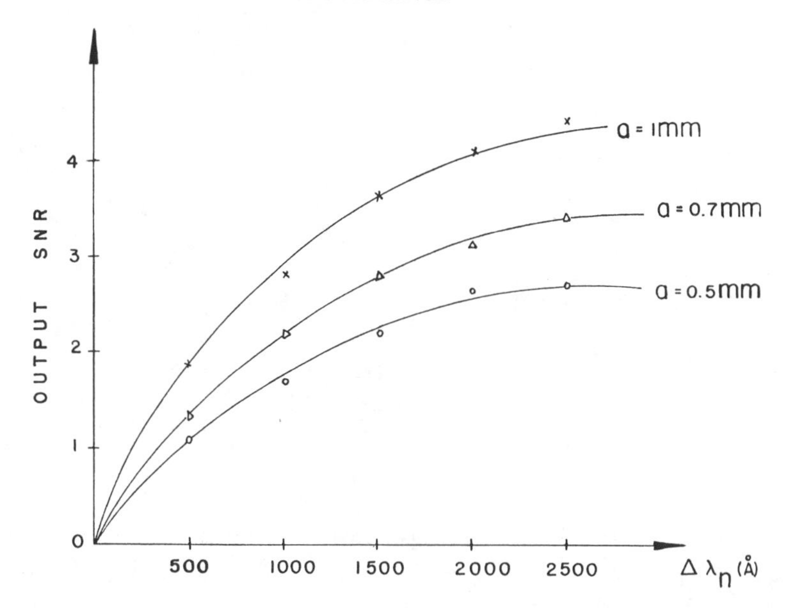

Figure 5.14. Output SNR for phase noise at Fourier plane as a function of spectral bandwidth for various source sizes.

experimental results are compatible with the analytical results that we have obtained previously.

We now provide the result of noise performance due to thin phase noise along the optical axis (i.e., the Z-axis) of the optical system. Figure 5.15 shows the variation of output SNR due to phase noise inserted in various planes of the optical system. From this figure, we see that the output SNR improves drastically for phase noise inserted at the input and at the output plane under TPC and SPC illumination. The noise performance is somewhat less effective for the phase noise at the Fourier plane, even though under partially coherent illumination. Nonetheless, the phase noise at the Fourier plane can, in principle, be totally eliminated under very broad-band illumination, if each of the noise channels is uncorrelated. We have also noted that the output SNR is somewhat lower for higher spatially coherent illumination (i.e., smaller source size). In other words, the output SNR can also be improved with extended source illumination.

The noise performance due to amplitude noise along the Z-axis of the proposed optical processor is plotted in Fig. 5.16. From this figure, we see that the output SNR improves when the noise perturbation moves away from the input and the output plane, and when the optimum SNR occurs at the Fourier plane. Again, we see that the output SNR is somewhat higher for

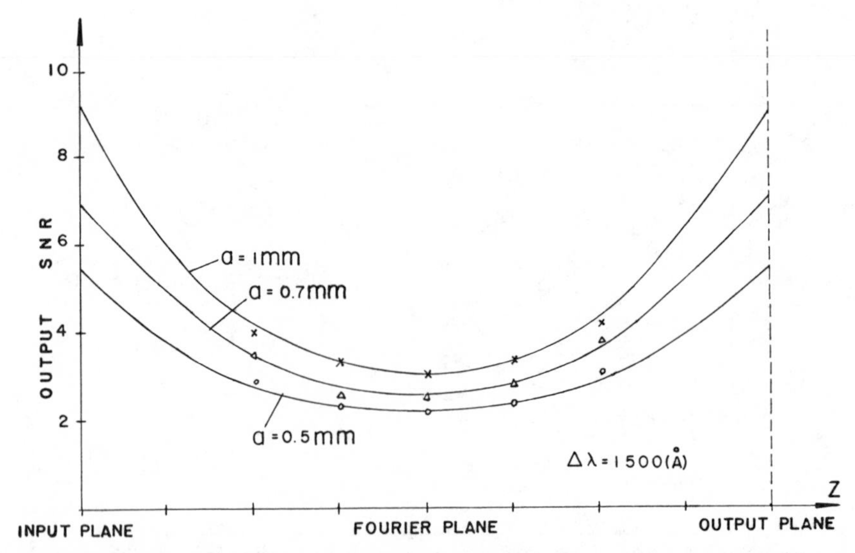

Figure 5.15. The variation of output SNR due to thin phase noise as a function of the Z-direction for various source sizes.

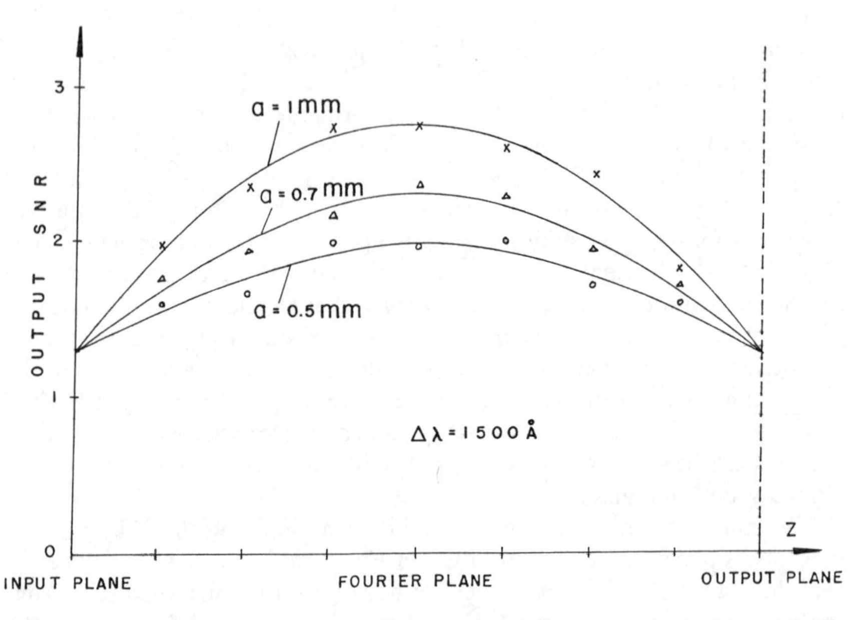

Figure 5.16. The variation of output SNR due to thin amplitude noise as a function of the Z-direction for various source sizes.

larger source size (i.e., lower degree of spatial coherence). However, we stress that the noise performance cannot be improved with partially coherent illumination if the amplitude noise is placed at the input or at the output plane.

5.5. SUMMARY AND REMARK

We have studied the effects of partial temporal coherence on a white-light optical signal processor under a noisy environment and have evaluated expressions for the output SNR of the optical system. Results have been obtained for noise both at the object and the Fourier plane and it is found that a broad-band optical system has superior noise suppression capabilities compared to coherent optical systems. We think the results would have been more realistic for the phase noise if higher order terms in its Taylor series expansion could have been included in the calculation. Also, we have not considered the case of strong phase noise. For this extreme it is possible to assume that the phase noise is Gaussian, which results in a considerable simplification of the problem [5.20, 5.21].

Significant output noise reduction is possible for the white-light optical processor when broad-band illumination is employed. However, the results for the amplitude noise at the input plane show no improvement over the coherent case. Except for the amplitude noise at the object plane, the following observations are true:

1. The output normalized SNR is an increasing function of the spectral bandwidth $\Delta\lambda_n$ of the nth spatial filter.
2. The SNR is found to be higher when shorter wavelengths are employed in the processing.
3. The major advantage of the white-light optical processor results from its ability to process signals in a multichannel mode. The overall SNR approximately improves as the square-root of the number of spectral band filters in the Fourier plane.

We have also studied the case of SPC illumination. A set of results very similar to the case of temporal coherence is obtained; that is, with the exception of amplitude noise at the object plane, the output normalized SNR improves as the requirements on spatial coherence are relaxed.

We have considered the noise problem at the object as well as at the Fourier plane. We have found that the output normalized SNR at the Fourier plane is much higher than it is at the object plane. The noise at the Fourier plane can be effectively suppressed, if a spatially incoherent source is employed. But this is to be expected: The system channels become increasingly decorrelated as we move toward the Fourier plane. Though we have not analyzed the noise at an arbitrary plane, a problem relatively difficult to formulate, its behavior is expected to lie somewhere between the two aforementioned planes. Here we should also mention that we have not dealt with detector noise; it is an important problem and we need to consider its effect on performance while designing an optical system.

We have also devised a technique of measuring the noise performance to confirm our claims. The technique utilizes a scanning photometer to trace out the output noise intensity fluctuation of the optical system. The effect on performance of noise perturbation at the object and at the Fourier plane are measured. The experimental results, except for the amplitude noise at the input plane, show better noise immunity if the optical system is operating in the partially coherent regime. We have also measured the noise performance due to perturbation along the optical axis of the system. The experimental results show that the resulting output SNR improves considerably with increasing bandwidth and source size of the illuminator. The optimum noise immunity occurs for phase noise at the input and at the output plane. For amplitude noise, the optimum SNR occurs at the Fourier plane. In brief, the experimental results confirm the claims of our analytical study.

Finally, we note that it has been shown here and elsewhere that the partially coherent optical systems can be used advantageously when noise is an important consideration. Since most partially coherent optical systems simultaneously relax the requirements on temporal and spatial coherence, a creative design may result in a sizable improvement in output SNR. We hope that this work will provide further insight into the workings of partially coherent, or white-light, optical signal processing systems.

REFERENCES

5.1. L. J. Cutrona et al., "Optical data processing and filtering systems," *IRE Trans. Inform. Theory* **IT-6**, 386 (1960).

5.2. A. Vander Lugt, "Signal detection by complex spatial filtering," *IEEE Trans. Inform. Theory* **IT-10**, 139 (1964).

5.3. A. Vander Lugt, "Coherent optical processing," *Proc. IEEE*, **62**, 1300 (1974).

5.4. F. T. S. Yu, "A new technique of incoherent complex detection," *Opt. Commun.* **27**, 23 (1978).

5.5. F. T. S. Yu, "Restoration of smeared photographic image by incoherent optical processing," *Appl. Opt.* **17**, 3571 (1978).

5.6. F. T. S. Yu, "A technique of white-light optical processing with diffraction grating method," *SPIE* **232**, 9 (1980)

5.7. P. Chavel and S. Lowenthal, "Noise and coherence in optical image processing. II. Noise fluctuations," *J. Opt. Soc. Am.* **68**, 721 (1978)

5.8. P. Chavel, "Optical noise and temporal coherence," *J. Opt. Soc. Am.* **70**, 935 (1980)

5.9. E. N. Leith and J. Roth, "Noise performance of an achromatic optical system," *Appl. Opt.* **16**, 2803 (1979).

5.10. F. T. S. Yu, K. S. Shaik, and S. L. Zhuang, "Noise performance of a white-light optical signal processor: Part I. Temporally partially coherent illumination," *J. Opt. Soc. Am.* A, **1**, 489 (1984).

5.11. F. T. S. Yu, K. S. Shaik, and S. L. Zhuang, "Noise performance of a white-light optical signal processor: Part II. Spatially partially coherent illumination," *Appl. Phys.*, **B36**, 11(1985).

5.12. F. T. S. Yu, *Optical Information Processing*, Wiley-Interscience, New York, 1983, p. 266.

5.13. E. L. O'Neill, *Introduction to Statistical Optics*, Pergamon Press, Reading, MA, 1963.

5.14. B. J. Uscinski, *The Elements of Wave Propagation in Random Media*, McGraw-Hill, New York, 1977.

5.15. S. L. Zhuang, T. H. Chao, and F. T. S. Yu, "Smeared-photographic image deblurring utilizing white-light processing technique," *Opt. Lett.* **6**, 102 (1981).

5.16. T. H. Chao, S. L. Zhuang, S. Z. Mao, and F. T. S. Yu, "Broad spectral band color image deblurring," *Appl. Opt.* **22**, 1439 (1983).

5.17. F. T. S. Yu, "Markov photographic noise," *J. Opt. Soc. Am.* **59**, 342 (1969).

5.18. J. B. Thomas, *Statistical Communication Theory*, John Wiley & Sons, New York, 1969.

5.19. M. Born and E. Wolf, *Principles of Optics*, 2nd rev. ed., Pergamon Press, New York, 1964, p. 524.

5.20. G. Parry, "Speckle patterns in partially coherent light," in *Laser Speckle and Related Phenomena*, Topics in Applied Physics, Vol. 9, J. C. Dainty, Ed., Springer-Verlag, Berlin, 1975, pp. 78–122.

5.21. G. Parry, "Some effects of temporal coherence on the first order statistics of speckle," *Opt. Acta* **21**, 763 (1974).

5.22. F. T. S. Yu, L. N. Zheng, and F. K. Hsu, "Measurement of noise performance for a white-light optical processor," *Appl. Opt.*, **24**, 173 (1985).

5.23. B. J. Chang and K. Winick, "Silver-halide gelatin holograms," *SPIE Proc.* **215**, 172 (1980).

6

Techniques and Applications

In the preceding chapters we have described in detail the coherence requirement, apparent transfer function, and noise performance of a white-light optical signal processor. We have, on various occasions, shown that the proposed white-light optical signal processor is capable of processing the signal, not in irradiance, but in complex wave field, as a coherent system would, and at the same time the white-light system is capable of suppressing the coherent artifact noise, as an incoherent system would. We have also mentioned several times that the white-light processor is rather suitable for color image processing, since the light source emanates all the visible wavelengths. In this chapter we discuss some basic techniques and applications of the proposed white-light optical signal processor. We see that the proposed white-light system is indeed easy to operate and economical to maintain for some specific processing operations. The white-light system, if it is assembled, would cost much less than its digital counterpart, and the resolution is limited, in principle, only by the input image provided.

Because of the wide variety of applications of white-light optical processing, in this and the next chapter we illustrate some examples that should be interesting to general readers. It is, however, not our intention to cover the vast domain of white-light applications.

6.1. SOURCE ENCODING, SIGNAL SAMPLING, AND PARTIALLY COHERENT FILTERING

In this section, we describe a linear transformation relationship between the spatial coherence function and the source encoding intensity transmittance function. Since the spatial coherence requirement depends on the signal processing operation, a more relaxed spatial coherence function may be used for specific signal processing operations. By Fourier transforming this spatial coherence function, a source encoding intensity transmittance function may be found.

The purpose of source encoding is to relax the stringent spatial coherence requirement so that an extended white-light source can be used for the optical signal processing. In other words, the source encoding technique is capable of generating an appropriate spatial coherence function for a specific optical signal processing application such that the available light power from the source can be utilized more efficiently. We illustrate by examples that complex optical signal processing can be carried out by an encoded extended white-light source [6.1].

6.1.1. Source Encoding

We begin our discussion with Young's experiment under an extended incoherent source illumination, as shown in Fig. 6.1. First, we assume that a narrow slit is placed in the source plane P_0 behind an extended monochromatic source. To maintain a high degree of spatial coherence between the slits Q_1 and Q_2 at plane P_1, the source size should be very narrow. If the separation between Q_1 and Q_2 is large, then a narrower slit size S_1 is required. Thus to maintain a high degree of spatial coherence between Q_1 and Q_2, the slit width should be

$$w \leqq \frac{\lambda R}{2h_0}, \tag{6.1}$$

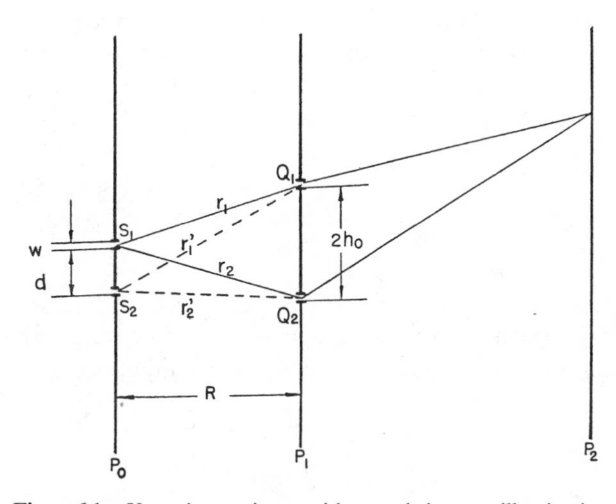

Figure 6.1. Young's experiment with extended source illumination.

where R is the distance between planes P_0 and P_1, and $2h_0$ is the separation between Q_1 and Q_2. Let us now consider two narrow slits of S_1 and S_2 located in source plane P_0. We assume that the separation between S_1 and S_2 satisfies the following path length relation:

$$r_1' - r_2' = (r_1 - r_2) + m\lambda, \tag{6.2}$$

where the r's are the respective distances from S_1 and S_2 to Q_1 and Q_2 as shown in the figure. m is an arbitrary integer, and λ is the wavelength of the extended source. Then the interference fringes due to each of the two source slits S_1 and S_2 should be in phase. A brighter fringe pattern can be seen at plane P_2. To further increase the intensity of the fringe pattern, one would simply increase the number of source slits in appropriate locations in the source plane P_0 such that every separation between slits satisfied the coherence or fringe condition of Eq. (6.2). If separation R is large, that is, if $R \gg d$ and $R \gg 2h_0$, then the spacing d between the source slits becomes

$$d = m\frac{\lambda R}{2h_0}. \tag{6.3}$$

From the above illustration, we see that, by properly encoding an extended source, it is possible to maintain the spatial coherence between Q_1 and Q_2, and at the same time, to increase the intensity of illumination. Thus with a specific source encoding technique for a given optical signal processing operation, an efficient utilization of an extended source may result.

To encode an extended source, we would first search for a spatial coherence function for a specific signal processing operation. With reference to the extended source optical signal processor shown in Fig. 6.2, the spatial coherence function at input plane P_1 can be written as [6.2]

$$\Gamma(\mathbf{x}_1, \mathbf{x}_1') = \iint \gamma(\mathbf{x}_0)K(\mathbf{x}_0, \mathbf{x}_1)K^*(\mathbf{x}_0, \mathbf{x}_1)d\mathbf{x}_0 \tag{6.4}$$

where the integration is over the source plane P_0, $\gamma(\mathbf{x}_0)$ is the intensity transmittance function of a source encoding mask, and $K(\mathbf{x}_0, \mathbf{x}_1)$ is the transmittance function between the source plane P_0 and the input plane P_1, which can be written

$$K(\mathbf{x}_0, \mathbf{x}_1) = \exp\left(i2\pi\frac{\mathbf{x}_0\mathbf{x}_1}{\lambda f}\right). \tag{6.5}$$

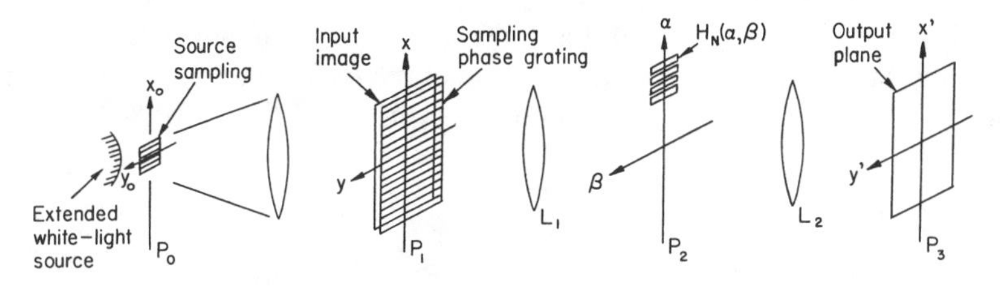

Figure 6.2. A white-light optical signal processor.

By substituting $K(\mathbf{x}_0, \mathbf{x}_1)$ into Eq. (6.4), we have

$$\Gamma(\mathbf{x}_1 - \mathbf{x}_1') = \iint \gamma(\mathbf{x}_0)\exp\left[i2\pi \frac{\mathbf{x}_0}{\lambda f}(\mathbf{x}_1 - \mathbf{x}_1')\right]d\mathbf{x}_0. \tag{6.6}$$

From the above equation we see that the spatial coherence function and source encoding intensity transmittance function form a Fourier transform pair, that is,

$$\gamma(\mathbf{x}_0) = \mathscr{F}[\Gamma(\mathbf{x}_1 - \mathbf{x}_1')], \tag{6.7}$$

and

$$\Gamma(\mathbf{x}_1 - \mathbf{x}_1') = \mathscr{F}^{-1}[\gamma(\mathbf{x}_0)], \tag{6.8}$$

where $\mathscr{F}$ denotes the Fourier transformation operation. We note that the Fourier transform relationship of Eqs. (6.7) and (6.8) is also known as the Van Cittert-Zernike theorem [6.3, 6.4]. Thus we see that if a reduced spatial coherence function for optical signal processing is calculated, then a source encoding intensity transmittance function can be found through the Fourier transformation. We further note that in practice the source encoding function should be a positive real function that satisfies the following physical realizable condition:

$$0 \leqq \gamma(\mathbf{x}_0) \leqq 1. \tag{6.9}$$

6.1.2. Signal Sampling and Partially Coherent Filtering

There is, however a temporal coherence requirement for white-light optical signal processing. In optical signal processing, the scale of the signal spectrum varies with the wavelength of the light source. Therefore, a temporal coherence requirement should be imposed on every processing operation, as has been discussed in detail in Chapter 3. If we restrict the signal spectra, due to wavelength spread, within a small fraction of the fringe spacing d of a narrow spectral band filter $H_n(\alpha, \beta)$ (e.g., deblurring filter), then we have

$$\frac{P_m f \Delta\lambda_n}{2\pi} \ll d, \tag{6.10}$$

where $1/d$ is the highest spatial frequency of the filter, p_m is the angular spatial frequency limit of the input signal transparency, f is the focal length of the achromatic transform lens, and $\Delta\lambda_n$ is the spectral bandwidth of the narrow spectral band filter $H_n(\alpha, \beta)$. The spectral width or the temporal coherence requirement of the spatial filter is, therefore,

$$\frac{\Delta\lambda_n}{\lambda_n} \ll \frac{\pi}{h_0 p_m}, \tag{6.11}$$

where λ_n is the center wavelength of the nth narrow spectral band filter, $2h_0$ is the main separation of the input signal transparencies, and $2h_0 = \lambda_n f/d$.

In order to gain some feeling of magnitude, we provide a numerical example. Let us assume that the size of the input image is $2h_0 = 5$ mm, that the center wavelength of the filter $H_n(\alpha, \beta)$ is $\lambda_n = 5461$ Å, and that we can take a factor 10 for Eq. (6.11) for consideration, that is,

$$\Delta\lambda_n = \frac{10\pi\lambda_n}{h_0 p_m}. \tag{6.12}$$

Several values of the spectral width requirement $\Delta\lambda_n$ for various spatial frequencies p_m are tabulated in Table 6.1.

From Table 6.1, we see that, if the spatial frequency of the input signal transparency is low, a broader spectral width of the narrow spectral band filters can be used. In other words, if higher spatial frequency is required for a specific processing operation, then narrower spectral width spatial filters are needed. Evidently, a narrower spectral spread $\Delta\lambda_n$ corresponds to a higher temporal coherence requirement, which can be obtained by increasing the signal sampling frequency p_0. However, the higher image sampling frequency

Table 6.1. Temporal Coherence Requirement for $H_n(\alpha, \beta)$

$p_m/2\pi$(lines/mm)	0.5	1	5	20	100
$\Delta\lambda_n$(Å)	218.4	109.2	21.8	5.46	1.09

used may require larger apertures for the transform lenses in the optical system, which tend to be more expensive. Nevertheless, in practice, high quality images have been obtained with relatively low-cost transform lenses.

6.1.3. An Illustrative Application

We now illustrate an application of the source encoding, signal sampling, and filtering for a white-light optical signal processor. Let us consider a polychromatic image subtraction problem. The image subtraction of Lee [6.5] that we consider is essentially a one-dimensional processing operation, in which a one-dimensional fan-shaped diffraction grating should be utilized, as illustrated in Fig. 6.3. We note that the fan-shaped grating (i.e., filter) is imposed by the temporal coherence condition of Eq. (6.11). Since the image subtraction is a point-pair processing operation, a strictly broad spatial coherence function at the input plane is not required. In other words, if one maintains the spatial coherence between the corresponding image points to be subtracted at the input plane, then the subtraction operation can be carried out

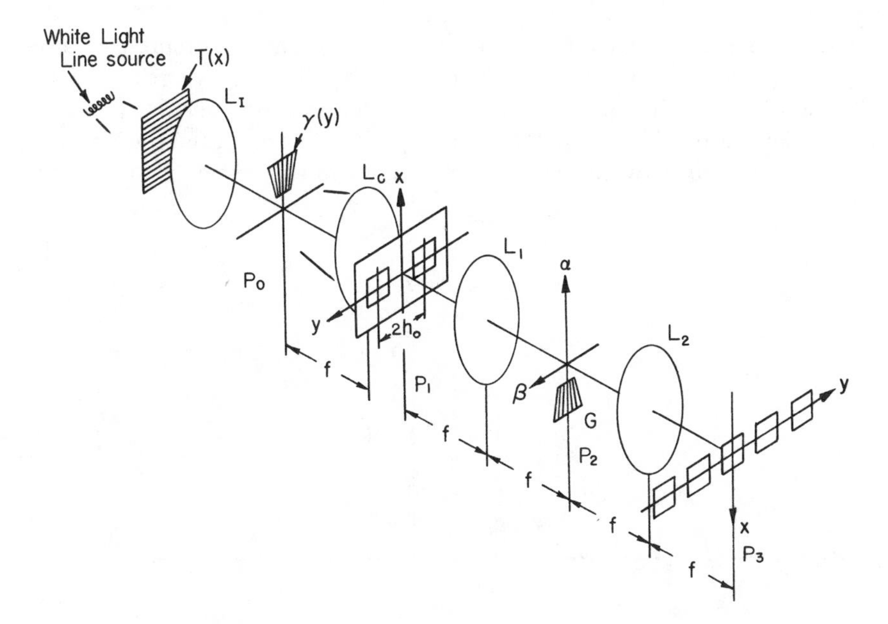

Figure 6.3. A white-light image subtraction processor. $T(x)$, phase grating; L_I, imaging lens; L_C, collimated lens; L_1 and L_2, achromatic transform lenses; $\gamma(y)$, fan-shaped source encoding mask: G, fan-shaped diffraction grating.

at the output image plane. Thus instead of using a strictly broad spatial coherence function, a reduced spatial coherence function may be utilized,

$$\Gamma(y - y') = \delta(y - y' - h_0) + \delta(y - y' + h_0), \tag{6.13}$$

where $2h_0$ is the main separation between the two input color image transparencies. The source encoding function can therefore be evaluated through the Fourier transform of Eq. (6.7):

$$\gamma(y_0) = 2 \cos\left(\frac{2\pi h_0}{\lambda f} y_0\right). \tag{6.14}$$

Unfortunately Eq. (6.14) is a bipolar function that is not physically realizable. To ensure a physically realizable source encoding function, we let a reduced spatial coherence function with the point-pair coherence requirement be [6.2]

$$\Gamma(|y - y'|) = \frac{\sin[(N\pi/h_0)|y - y'|]}{N \sin[(\pi/h_0)|y - y'|]} \operatorname{sinc}\left(\frac{\pi w}{h_0 d}|y - y'|\right), \tag{6.15}$$

where $N \gg 1$ a positive integer, and $w \ll d$. Equation (6.15) represents a sequence of narrow pulses that occur at every $|y - y'| = nh_0$, where n is a positive integer, and their peak values are weighted by a broader sinc factor, as shown in Fig. 6.4. Thus, a high degree of spatial coherence can be achieved at every point-pair between the two input image transparencies. By taking the

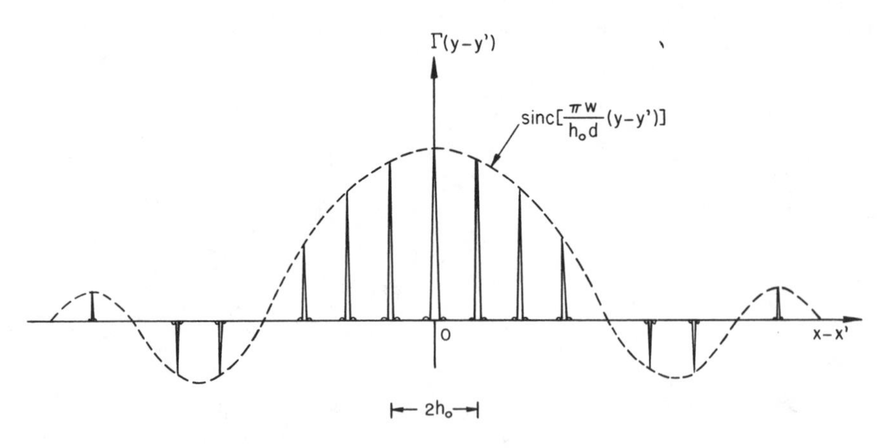

Figure 6.4. A spatial coherence function.

Fourier transformation of the reduced spatial coherence function of Eq. (6.15), the corresponding source encoding function is

$$\gamma(|y|) = \sum_{n=1}^{N} \operatorname{rect} \frac{|y - nd|}{w}, \tag{6.16}$$

where w is the slit width, $d = (\lambda f/h_0)$ is the separation between the slits, and N is the number of the slits. Since $\gamma(|y|)$ is a positive real function that satisfies the constraint of Eq. (6.9), the proposed source encoding function of Eq. (6.16) is physically realizable.

In view of Eq. (6.16), we note that the separation of slit d is linearly proportional to λ. The source encoding is a fan-shaped function, as shown in Fig. 6.5. To obtain lines of rainbow color spectral light sources for the signal processing, we would utilize a linear extended white-light source with a dispersive phase grating, as illustrated in Fig. 6.3. Thus with the described broad-band source encoding mask, image sampling grating, and fan-shaped sinusoidal grating, a substrated image can be seen at the output plane with a broad-band white-light source.

It would occupy exhaustive pages to describe all the recent advances in white-light optical signal processing. However, we must restrict our discussion to a few recent results that may be interesting to the readers. Since the

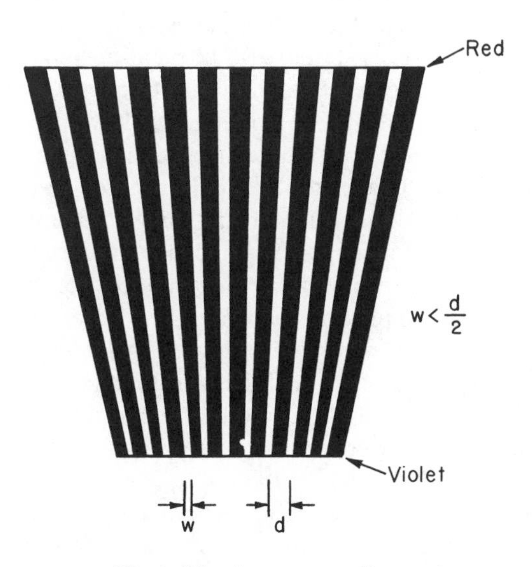

Figure 6.5. A source encoding mask.

white-light optical signal processor is particularly suitable for color signal processing, in the following sections, most of our attention is on some applications in polychromatic processings.

6.2. COLOR IMAGE DEBLURRING

One of the interesting applications of white-light optical signal processing may be the restoration of blurred color photographic images. We have recently presented a broad-band white-light optical signal processing technique for deblurring smeared color photographic images [6.6]. Since linear smeared deblurring is a one-dimensional problem, its deblurring filter operates on a point-by-point basis. Thus a *fan-shaped* deblurring filter can be utilized.

We note that monochrome photographic image deblurring with a coherent image processor was illustrated by Tsujiuchi [6.7] in 1963. The inverse spatial filter was synthesized by the combination of an amplitude and a pure phase filter. This can also be accomplished by a holographic synthesis technique. The preparation of such a filter by holographic techniques had been studied by Stroke and Zech [6.8], and by Lohmann and Paris [6.9]. Nevertheless, the holographic synthesis technique also suffers one disadvantage, namely a low diffraction efficiency. Mention must also be made of image deblurring with a computer generated phase filter obtained by Tsujiuchi, Honda, and Fukaya [6.10]. They have shown various effects due to amplitude, phase, and amplitude-phase filtering. Another interesting result obtained by Horner [6.11, 6.12] should also be mentioned. He has shown how optimum image deblurring can be obtained with a least-mean-square error Wiener filter. Nevertheless, as those works evolved, two problems are still of intense interest; namely, the coherent artifact reduction and the color image deblurring. Recently, Yang and Leith [6.13] proposed a spatial domain deconvolution technique for image deblurring. They used an extended incoherent line source for deblurring, and the coherent noise was remarkably reduced. However, their technique is only suitable for processing monochrome blurred images. In previous papers [6.14, 6.15] we presented a white-light processing technique for linearly smeared image deblurring. We have shown that the white-light image deblurring technique is capable of eliminating the coherent artifact noise and is suitable for color image deblurring. However, the results that we obtained were primarily restricted to the concept of narrow spectral band deblurring.

Here we extend the image deblurring technique to the entire spectral band of the white-light source. To obtain this broad-band deblurring effect, a fan-shaped spatial filter to compensate the scale of the Fourier spectra is utilized at the Fourier plane. In addition, we briefly discuss a technique of synthesizing a fan-shaped broad-band deblurring filter, and experimental demonstrations are given.

6.2.1. Broad-Band Image Deblurring

Let us now discuss a broad-band image deblurring technique utilizing a fan-shaped deblurring filter for color image deblurring. Let a linear smeared color image be described as

$$\hat{s}(x, y) = s(x, y) * \operatorname{rect}\left(\frac{y}{W}\right), \tag{6.17}$$

where $\hat{s}(x, y)$ and $s(x, y)$ are the smeared and unsmeared images,

$$\operatorname{rect}\left(\frac{y}{w}\right) \triangleq \begin{cases} 1, & y \leqq \dfrac{w}{2}, \\ 0, & \text{otherwise}, \end{cases} \tag{6.18}$$

and w is the smeared length.

Let us insert the smeared image transparency of Eq. (6.17) into the input plane P_1 of a white-light optical processor as shown in Fig. 6.6. The complex light distribution for every wavelength λ at the back focal length of the achromatic transform lens L_2 would be

$$E(\alpha, \beta; \lambda) = C \iint \hat{s}(x, y; \lambda) \exp(ip_0 x) \exp\left[-i \frac{2\pi}{\lambda f}(x\alpha + y\beta)\right] dx\, dy, \tag{6.19}$$

where p_0 is the angular spatial frequency of the phase grating, $\alpha = (\lambda f/2\pi)p$ and $\beta = (\lambda f/2\pi)q$ represent the spatial coordinate system of Fourier plane P_2, (p, q) is the corresponding angular spatial frequency coordinate system, f is the focal length of the achromatic transform lens, and C is an appropriate complex constant. Thus, Eq. (6.19) can be written as

$$E(\alpha, \beta; \lambda) = C\hat{S}\left(\alpha - \frac{\lambda f}{2\pi} p_0, \beta\right), \tag{6.20}$$

where

$$\hat{S}\left(\alpha - \frac{\lambda f}{2\pi} p_0, \beta\right) = S\left(\alpha - \frac{\lambda f}{2\pi} p_0, \beta\right) \operatorname{sinc}\left(\frac{\pi w}{\lambda f} \beta\right) \tag{6.21}$$

is the linear smeared image spectrum.

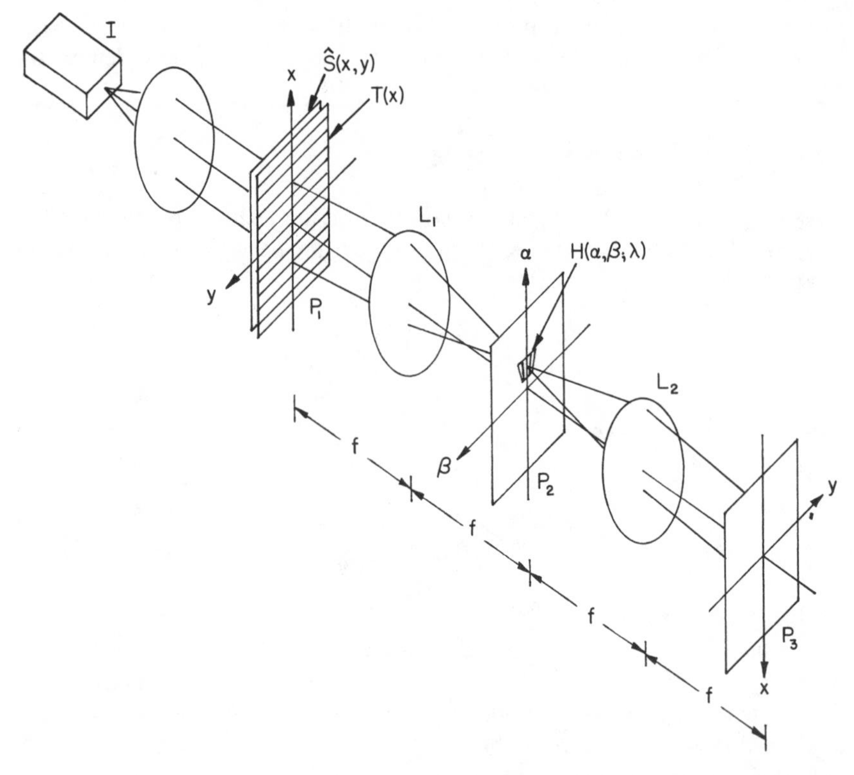

Figure 6.6. White-light processor for smeared color image deblurring: I, white-light point source; $\hat{s}(x, y)$, smeared color image transparency; $T(x)$, diffraction grating; L_1 and L_2, achromatic transform lenses; $H(\alpha, \beta; \lambda)$, broad spectral band deblurring filter.

Since scale of the image spectrum is proportional to the wavelength of the light source, the corresponding image spectra would smear into a fan-shaped rainbow color, as can be seen from Eq. (6.21). In other words, the top of the smeared spectra is in red and the bottom is in violet.

Let us assume that a fan-shaped broad spectral band deblurring filter (i.e., a broad-band inverse filter) to accommodate the variation of the scale of the signal spectra is available. This fan-shaped filter is described in the following equation:

$$H(\alpha, \beta; \lambda) = \delta\left(\alpha - \frac{\lambda f}{2\pi} p_0, \beta\right)\left\{\int \left[\operatorname{rect}\left(\frac{y}{w}\right)\exp\left(-i\frac{2\pi}{\lambda f}\beta y\right)\right]dy\right\}^{-1},$$

$$= \delta\left(\alpha - \frac{\lambda f}{2\pi} p_0, \beta\right)\left[\operatorname{sinc}\left(\frac{\pi w}{\lambda f}\beta\right)\right]^{-1}. \tag{6.22}$$

In image deblurring we would insert this deblurring filter of Eq. (6.21) in the spatial frequency plane of P_2. The complex light distribution for every λ at the output image plane P_3 can be written as

$$g(x, y; \lambda) = \mathscr{F}^{-1}\left[\hat{S}\left(\alpha - \frac{\lambda f}{2\pi} p_0, \beta\right) H(\alpha, \beta; \lambda)\right], \tag{6.23}$$

where $\mathscr{F}^{-1}$ denotes the inverse Fourier transform; by substituting Eqs. (6.21) and (6.22) into Eq. (6.23), we have

$$g(x, y; \lambda) = s(x, y)\exp(ip_0 x), \tag{6.24}$$

which is independent of the wavelength of the light source. The resultant output intensity distribution can be shown as

$$I(x, y) = \int_{\Delta\lambda} |g(x, y; \lambda)|^2 \, d\lambda \cong \Delta\lambda |s(x, y)|^2, \tag{6.25}$$

which is proportional to the entire spectral bandwidth $\Delta\lambda$ of the white-light source. Thus we see that this proposed white-light deblurring technique is capable of processing the image with the entire visible spectral band, and it is very suitable for the application to color-image deblurring. Since the integration of Eq. (6.25) is taken from the entire spectral band of the white-light source, the coherent artifact noise can be eliminated.

6.2.2. Deblurring Filter Synthesis

We now briefly describe the synthesis of a fan-shaped (i.e., broad spectral band) deblurring filter. The synthesis is a combination of an absorptive-amplitude filter and a phase filter. A fan-shaped phase filter is composed of several slanted bar-type phase objects, as illustrated in Fig. 6.7. Each phase bar would give rise to specific π phase retardation for a predescribed dispersion of a rainbow color wavelength. We note that the height of deblurring filter is, of course, dependent on the grating frequency p_0 at the input plane. The periodicity of the deblurred filter is certainly determined by the smeared length of the blurred object, and the width of the filter defines the degree of deblurring [6.16]. In constructing a broad spectral band phase deblurring filter, we utilize a vacuum deposition technique. It can be accomplished by depositing the magnesium fluoride (MgF_2) on the surface of an optical flat glass substrate. In this technique, a blocking mask of a fan-shaped bar pattern, as shown in Fig. 6.7, is used for the vacuum deposition. The MgF_2 vapor is deposited

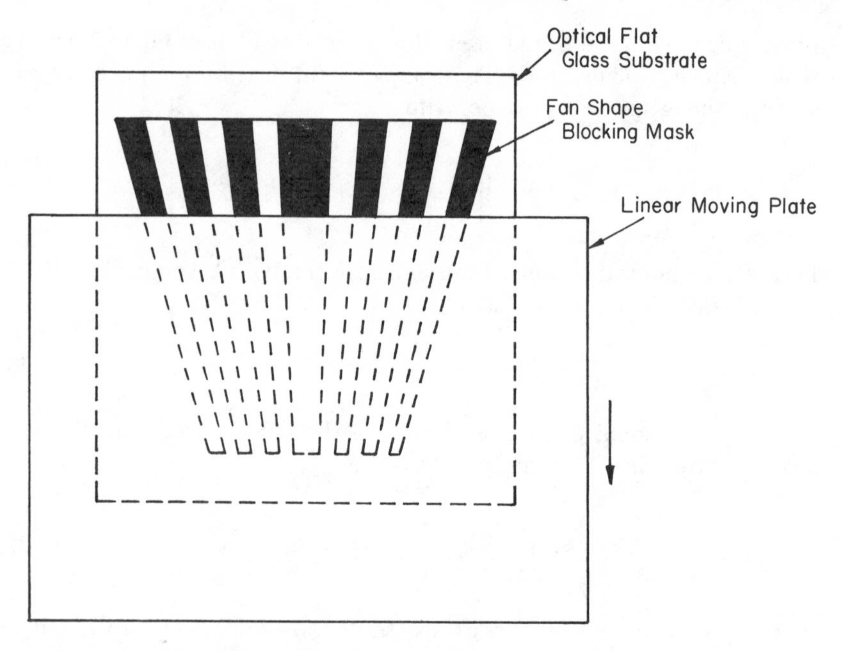

Figure 6.7. Phase filter mask for MgF_2 vapor deposition.

through this blocking mask, together with a linear moving covering plate, from top to bottom as illustrated in the figure.

The thickness of the deposited coating can be determined by the following equation:

$$t = \frac{\lambda}{2(\eta - 1)}, \tag{6.26}$$

where η is the refractive index of the coating material. This coating thickness is linearly proportional to the dispersion of the illuminating wavelength. We note that a strictly linear control of coating thicknesses is very essential. The advantage of this phase-type filter is to improve the transmission efficiency, since the overall deblurring filter is, in general, highly absorptive.

In principle, it is a straightforward method to synthesize a fan-shaped amplitude filter. The synthesis can be accomplished by inserting a slit aperture with a width equal to the smeared length of the blurred image at the input plane P_1 of the white-light optical processor shown in Fig. 6.6. The size of the white-light source should be adequately small under the spatial coherence regime to obtain a smeared sinc factor (i.e., smeared Fourier spectra of the slit

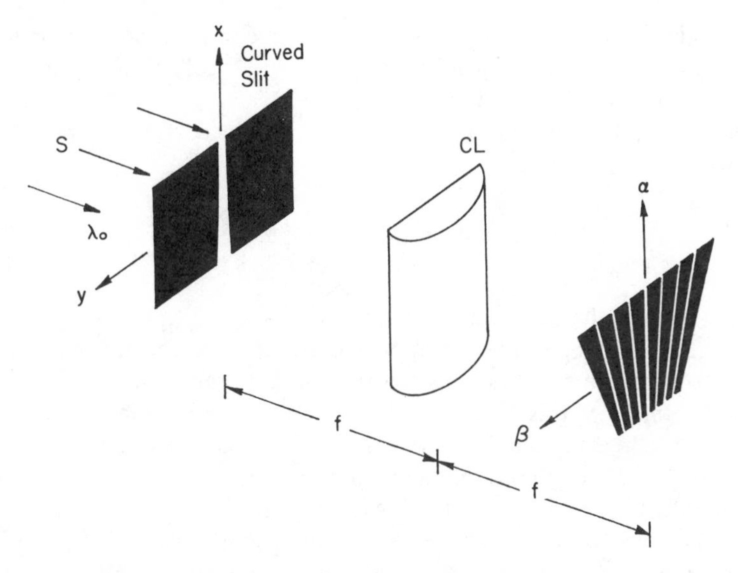

Figure 6.8. Generation of amplitude filter. *S*, monochromatic plane wave; *CL*, cylindrical transform lens.

aperture) in the Fourier plane. An amplitude filter can then be synthesized by simply recording this smeared sinc factor on a photographic plate. If the film gamma of the recorded plate is controlled to about unity (i.e., $\gamma = 1$), the amplitude transmittance of the recorded plate is equivalent to that of the desired fan-shaped amplitude filter. However, in practice, a fan-shaped amplitude filter is not that easy to synthesize for three primary reasons. First, if a very small source size is required for the filter synthesis, it usually takes a longer exposure time. Second, the spectral response of the recording plate is generally not uniform for all visible wavelengths. The recorded filter would produce uneven transmittance in the direction of the smeared color spectra. The effect of the transmittance variation of the filter would affect the fidelity of color reproduction and the degree of restoration. Third, it is difficult to synthesize the sidelobes of an amplitude filter, since the dynamic range of the photographic film is very limited.

There is an alternative technique for generating a fan-shaped amplitude filter with coherent illumination, as shown in Fig. 6.8. The purpose of using a curved-slit aperture is to accommodate the scale variation of the amplitude filter. The expression of the curved-slit aperture can be written as

$$d(x, y; \lambda) = \text{rect}\left(\frac{y}{w\lambda_0/\lambda}\right)\delta\left(x - \frac{\lambda f}{2\pi} p_0\right), \tag{6.27}$$

where λ_0 is the wavelength of the coherent source, w is the smeared length of the blurred image, f is the focal length of the cylindrical transform lens, and λ is the wavelength of the white-light source.

The corresponding Fourier transformation of the curved-slit aperture can be shown as

$$D(\alpha, \beta; \lambda) = \operatorname{sinc}\left(\frac{\pi w}{\lambda f}\beta\right) * \delta\left(\alpha - \frac{\lambda f}{2\pi}p_0\right), \tag{6.28}$$

where $*$ denotes the convolution operation.

It is clear now that a photographic recording of the spectra shown in Eq. (6.28) would produce a desirable fan-shaped amplitude filter for deblurring. In the synthesis of this broad-band amplitude filter, a HeNe laser, with a rotating ground glass to reduce the artifact noise, is used as a coherent source. A Kodak 131 plate is used for the recording plate, and a 6-min developing time in a POTA developer at 24°C is used to control the film gamma to about unity. The spectral wavelength limits are chosen from 4000 to 7000 Å. Within the dynamic range of the recording film, five sidelobes of $\sim$300–1 dynamic range (or a density range of 2.5) are recorded with good accuracy. The fan-shaped amplitude filter obtained is tested with satisfactory results.

6.2.3. Experimental Demonstrations

We provide a few experimental results of image deblurring utilizing a broad-band white-light source. In our experiments, a 75-W xenon arc lamp with a 200-μm pinhole is used as a broad-band white-light source. A phase grating of 130 lines/mm with 25% diffraction efficiency at each first-order diffraction is used at the input plane. An $f/8$ transform lens with 300-mm focal length is used for image Fourier transformation.

We first compare the effect of the broad-band deblurring with those the narrow-band and the coherent source techniques. For simplicity of illustration, we use a set of linear blurred letters of the alphabets as input objects, as shown in Fig. 6.9*a*. The smeared length is $\sim$0.5 mm. Figure 6.9*b* shows the deblurred image obtained with this broad-band deblurring technique, and the spectral bandwidth is $\sim$3000 Å under white-light illumination. Figure 6.9*c* is the result obtained with a narrow spectral band deblurring filter of $\sim\Delta\lambda = 100$ Å. Figure 6.9*d* is the deblurred image obtained with a HeNe coherent source. In comparing these examples, we see that the results obtained with the broad spectral band white-light source offer a higher quality deblurred image; the coherent artifact noise is substantially suppressed, and the deblurred image appears to be sharper than the one obtained with the narrow-band case.

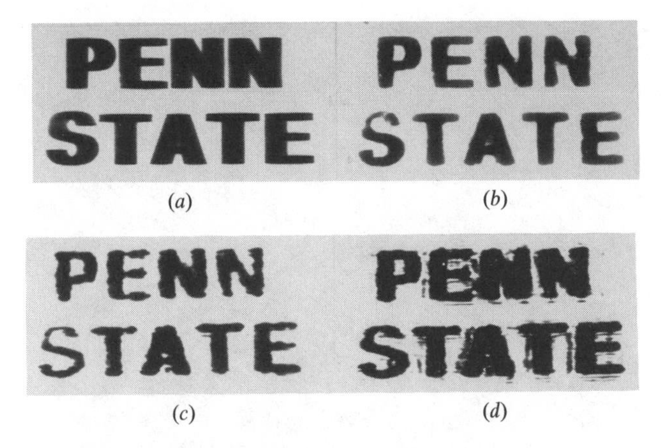

Figure 6.9. Smeared image restoration of the words PENN STATE. (*a*) Smeared image. (*b*) Deblurred image obtained with broad-band white-light source. (*c*) Deblurred image obtained with narrow spectral band white-light source. (*d*) Deblurred image obtained with coherent source.

We now experimentally demonstrate the capability of the white-light technique to deal with color images. Figure 6.10*a* shows a black-and-white version of a linear-motion blurred color picture of an F-16 fighter plane. The body of the plane is painted navy blue and white, the wings are mostly red, the tail is again navy blue and white, and the ground terrain is generally bluish-green. From this figure we see that the letters on the body, on one of the wings, and on this side of the tail are smeared beyond recognition. The details of the missiles at the tips of the wings are lost. The star symbols at the tail-end of the body and on the top of the wings are badly distorted. The features of ground terrain are obscured. Figure 6.10*b* shows, in black-and-white, the color image deblurring result that we have obtained with the proposed white-light deblurring technique. In this example, the letters USAF on the wing and YF-16 on the side of the tail can be clearly seen. The words U.S. AIR FORCE may be recognized. The star symbol on the wing can be clearly identified; however, the one on the body is rather obscured. Undoubtedly, the missiles at the tips of the wings can be seen, and the pilot in the cockpit is quite visible. The overall shape of the entire airplane is more distinctive than in the blurred example. Moreover the river, the highways, and the forests of the ground terrain are far more recognizable in this deblurred image. The color reproduction of the deblurred image is, for the most part, spectacularly faithful, and coherent artifact noise is virtually nonexistent. Some degree of color deviation, however, is inherent in the deblurred image. This is primarily

Figure 6.10. Color image deblurring. (*a*) A black-and-white picture of a smeared color image of an F-16 fighter plane. (*b*) A black-and-white picture of the deblurred color image.

due to chromatic aberration and to the antireflectance coating of the transform lenses. Nevertheless, these two drawbacks can be overcome by utilizing good-quality achromatic transform lenses.

6.3. COLOR IMAGE SUBTRACTION

Another interesting application of white-light optical signal processing is image subtraction. Image subtraction may be of value in many applications such as urban development, highway planning, earth resources studies, remote sensing, meteorology, automatic surveillance, inspection, and so on. Image subtraction may also be applied to communications, as a means of bandwidth compression. For example, it would be necessary to transmit only the differences between images in successive cycles, rather than the entire image in each cycle.

Optical image synthesis by complex amplitude subtraction was described by Gabor et al. [6.17]. The technique involves successive recordings of two or more complex diffraction patterns on a holographic plate and the subsequent reproduction of the composite holographic images. A few years later, Bromley et al. [6.18, 6.19] described a holographic Fourier subtraction technique, by which a real-time image and a previously recorded holographic image could be subtracted. Although good image subtraction was reported, it appears that the illumination for the holographic image reconstruction must be carefully aligned. In a more recent paper, Lee et al. [6.5] proposed a technique by which image subtraction and addition can be achieved by a diffraction grating technique. This technique involves the insertion of a diffraction grating in the spatial frequency domain of a coherent optical processor. The advantage of this technique is a real-time subtraction capability.

Since space does not permit us to review all the various techniques of image subtraction, we refer to the review paper by Ebersole [6.20]. However, most of the optical image synthesis involves a coherent source to carry out the image subtraction. But coherent sources also introduce artifact noise, which limits the subtracted image quality. In this section we apply the white-light processing technique to color image subtraction [6.21].

6.3.1. Broad-Band Image Subtraction

We now insert two color image transparencies, with a phase grating at the input plane P_1 of the white-light optical processor of Fig. 6.2. At the spatial frequency plane P_2, the complex light distribution for each wavelength λ of the light source may be described as

$$E(\alpha, \beta; \lambda) = S_1\left(\alpha - \frac{\lambda f}{2\pi} p_0, \beta\right) \exp\left(-i \frac{2\pi}{\lambda f} h_0 \beta\right) + S_2\left(\alpha - \frac{\lambda f}{2\pi} p_0, \beta\right) \exp\left(i \frac{2\pi}{\lambda f} h_0 \beta\right), \tag{6.29}$$

where $S_1(\alpha, \beta)$ and $S_2(\alpha, \beta)$ are the Fourier spectra of the input color images $s_1(s, y)$ and $s_2(x, y)$, $2h_0$ is main separation of the two color images s_1 and s_2, and p_0 is the angular spatial frequency of the input phase grating. Again we see that two input signal spectra disperse into rainbow colors along the α-axis of the spatial frequency plane.

For image subtraction, we insert a diffraction grating in the spatial frequency plane. Since the dispersed Fourier spectra vary with respect to the wavelength of the light source, we must insert a fan-shaped grating to compensate for the wavelength variation. If we let this fan-shaped grating be

$$H(\alpha, \beta; \lambda) = \left[1 + \sin\left(\frac{2\pi}{\lambda f} h_0 \beta\right)\right], \qquad \text{for all } \alpha, \tag{6.30}$$

then output image irradiance would be

$$\begin{aligned} I(x, y) \simeq \Delta\lambda[&|s_1(x, y - h_0)|^2 + |s_2(x, y + h_0)|^2 + \tfrac{1}{2}|s_1(x, y) - s_2(x, y)|^2 \\ &+ |s_1(x, y - 2h_0)|^2 + |s_2(x, y + 2h_0)|^2], \end{aligned} \tag{6.31}$$

where $\Delta\lambda$ is the spectral bandwidth of the white-light source. Thus the subtracted color image, that is, $|s_1(x, y) - s_2(x, y)|^2$, can be seen at the optical axis of the output plane. In practice, it is difficult to obtain a true white-light point source. However, this shortcoming can be overcome with the source encoding technique we have demonstrated as an example in Section 6.1.3.

To ensure a physically realizable source encoding function [see Eq. (6.9)], we let a reduced spatial coherence function with the point-pair coherence requirement be [i.e., Eq. (6.15)]

$$\Gamma(|y - y'|) = \frac{\sin[(N\pi/h_0)|y - y'|]}{N \sin[(\pi/h_0)|y - y'|]} \operatorname{sinc}\left(\frac{\pi w}{h_0 d}|y - y'|\right), \tag{6.32}$$

where $N \gg 1$ a positive integer, and $w \ll d$. Equation (6.32) represents a sequence of narrow pulses that occur at every $|y - y'| = nh_0$, where n is a positive integer, and their peak values are weighted by a broader sinc factor, as shown in Fig. 6.4. Thus a high degree of spatial coherence can be achieved at every point-pair between the two input color image transparencies. By applying the Van Cittert–Zernike theorem of Eq. (6.7), the corresponding source encoding function can be shown as

$$\gamma(|y|) = \sum_{n=1}^{N} \operatorname{rect} \frac{|y - nd|}{w}, \tag{6.33}$$

where w is the slit width, $d = (\lambda f/h_0)$ is the separation between the slits, f is the focal length of the achromatic collimated lens L_c, and N is the total number of the slits. Alternatively, Eq. (6.33) can be written in the following form:

$$\gamma(|y|) = \sum_{n=1}^{N} \operatorname{rect} \frac{|y - n\lambda f/h_0|}{w}, \tag{6.34}$$

for which we see that the source encoding function is a fan-shaped grating, as illustrated in Fig. 6.5. To obtain lines of rainbow color spectral light sources for the subtraction operation, we would utilize a linear extended white-light source with a dispersive phase grating, as illustrated in Fig. 6.3. In white-light image subtraction operation, a broad spectral band sinusoidal grating, such as

$$G = \frac{1}{2}\left[1 + \sin\left(\frac{2\pi\alpha h_0}{\lambda f}\right)\right], \tag{6.35}$$

should be used in the Fourier plane of Fig. 6.3. Thus with the prescribed broad-band source encoding mask of Eq. (6.34), and the fan-shaped sinusoidal grating of Eq. (6.35), as depicted in Fig. 6.3, the output image irradiance around the optical axis can be shown as

$$I(x, y) = K|s_1(x, y) - s_2(x, y)|^2. \tag{6.36}$$

Thus we see that a color subtracted image can readily be seen at the output plane.

6.3.2. Design Consideration for Source Encoding Mask

It is evident that complex image subtraction requires a destructive interference between the superposing images at the output plane. In order for such an interference process to take place, it is necessary that the two images be coherent and have a constant phase factor between them. Thus, as we have seen earlier, the design of the source encoding mask must establish a point-pair spatial coherence function and the grating filter must introduce a constant phase shift between the two images such that the complex amplitude subtraction operation can take place at the output plane. Let us refer to the broad-band image subtraction of Fig. 6.3; the source encoding mask [i.e., $\gamma(y)$] is illuminated by a linearly dispersed light source derived from a dispersive phase grating $T(x)$. Hence, for every given value of x, the source encoding mask is illuminated by a narrow spectral band (i.e., $\Delta\lambda$) of light. With the setup of Fig. 6.3, we see that the grating G is illuminated by the same spectral spread of light as the source encoding mask. Thus design of the grating G can be generated by the source encoding mask with the optical system of Fig. 6.3, but without the input image transparencies inserted.

In order to have a high contrast ratio for the subtracted image, the slit width of the source encoding mask should be made sufficiently narrow as compared with the spatial period of the encoding mask, as shown in Fig. 6.5. However, the narrower slit width would also reduce the encoded light illumination. As shown in Table 3.5 of Section 3.3, if the slit width is about 5% of

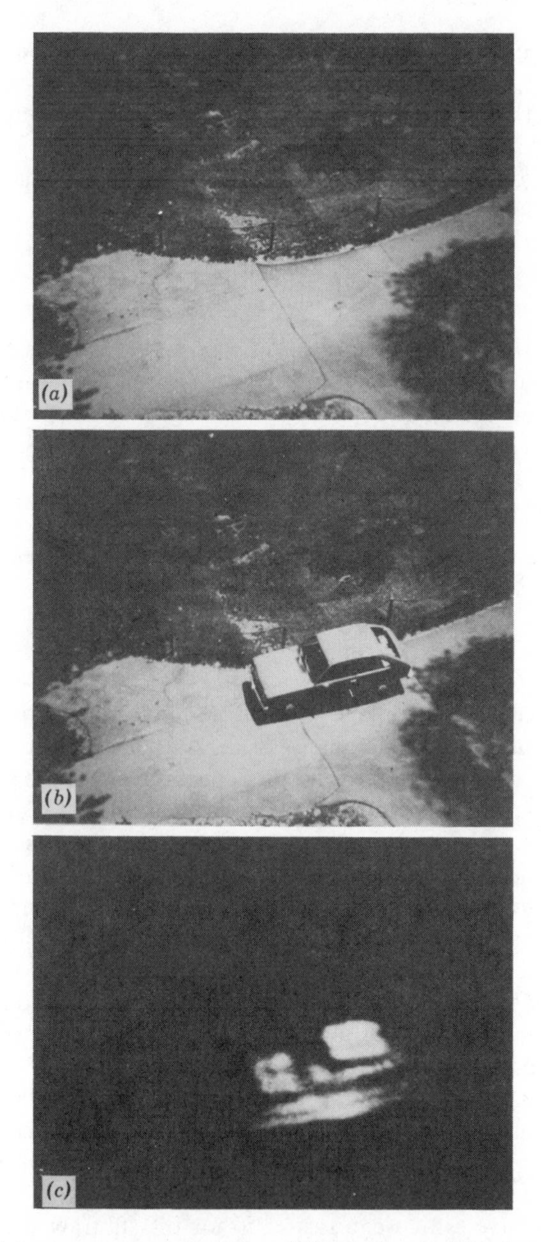

Figure 6.11. Color image subtraction: (*a*) and (*b*) Black-and-white pictures of input color images. (*c*) Black-and-white picture of output subtracted color image.

the spatial period, the system would provide a very high contrast subtracted image. Although, in principle, the source encoding mask may be generated by a computer technique, for simplicity we would synthesize the mask in the following way: First, by constructing a fan-shaped grating structure with thin copper wires guided by two separated threaded rods of different pitches. Second, the grating structure would then be photographically reduced by a high resolution camera (i.e., IC mask camera) to appropriate size. Thus a fan-shaped source encoding mask with specific slit width and spatial period for the rainbow color dispersion light source would be synthesized. In our experiment, the source encoding mask is about 19 mm long with the width increasing linearly from 0.6 to about 1.2 mm. The slit width is about 5 μm and the spatial period varies linearly from 32 μm at 3600 Å wavelength to 64 μm at 7200 Å wavelength.

6.3.3. Experimental Demonstrations

In our experiment, a xenon arc source with a 0.3 mm slit is used as a linear extended white-light source. The dispersive phase grating $T(x)$ used is about 136 lines/mm. The diameter of the achromatic transform lenses are about 70 mm and their focal length is about 381 mm.

Let us now provide an experimental result of color image subtraction obtained with the proposed white-light processing technique. Figure 6.11*a* and *b* show a set of black-and-white pictures of the input color image transparencies. The first figure shows a small paved road photographed from the rooftop of a campus building. The paved road appears light gray under the hot summer sun, and the lawn is, in general, green and yellowish. The second figure shows the same scene but with a compact car passing by. The car is orange-brown with brightness variation. Figure 6.11*c* shows a black-and-white version of the output color subtracted image obtained with this technique. From this figure, we see that the shape of the car can be easily recognized. In reality, the shade of the car (i.e., subtracted image) varies from yellow-brown to orange-brown and the shadow of the car is generally yellow-green. Although the resolution of the subtracted image is still below the generally acceptable standard, this drawback could be easily overcome by utilizing higher quality optics.

6.4. COLOR IMAGE RETRIEVAL

Archival storage of color films has long been an unresolved problem for film industries around the world. The major reason is that the organic dyes used in color films are usually unstable under prolonged storage. Thus a gradual color fading occurs in the film. Although there are several techniques available for preserving the color images, all of them possess certain definite drawbacks. One of the most commonly used techniques involves repetitive application of primary color filters to preserve the color images in three separate rolls of black-and-white film. To reproduce the original color image, a system with three primary color projectors should be used for these three rolls of films. These films should be projected in perfect unison so that the primary color images will be precisely recorded on a fesh roll of color film. However, this technique has two major disadvantages: First, the storage volume for each film is tripled. Second, the reproduction system is rather elaborate and expensive.

Historically, the use of monochrome transparencies to retrieve color images was first reported by Ives [6.22] in 1906. He introduced a slide viewer that produced color images by a diffraction phenomenon. Grating either of different spatial frequencies or of azimuthal orientation was used. More recently, Mueller [6.23] described a similar technique, employing a tricolor grid screen for image encoding. In decoding, he used three quasi-monochromatic sources for color image retrieval. Since then, similar work on color image retrieval has been reported by Macovski [6.24], Grousson and Kinany [6.25], and Yu [6.26]. However all of these techniques suffer a major common drawback, namely the unavoidable moiré fringe pattern in the retreived image. In this section, we describe a white light processing technique, for color image retrieval, for which the moiré fringe pattern can be avoided [6.27]. We show that this technique would be the most efficient and effective technique existing to date. This technique also allows direct viewing capability by simple white-light processing. This capability is particularly attractive for portable color image projection and home movie applications.

6.4.1. Spatial Encoding

We now describe a spatial encoding technique utilizing a white-light source. A color transparency is used as an object to be encoded; it is sequentially exposed with the primary colors of lights onto a black-and-white film, as illustrated in Fig. 6.12. The encoding onto the monochrome film takes place by spatial sampling of the primary color images of the color transparency, with a specific sampling frequency and a predescribed direction. In order to avoid the moiré fringe pattern in the retrieved color image, we propose to sample one of the primary color images in one independent spatial coordinate, and the remaining two primary color images in the other independent spatial coordinate, as shown in Fig. 6.13. Since any mixture of red with green or with blue produces a wide range of intermediate colors, we propose to encode the red image in one independent spatial coordinate direction, and

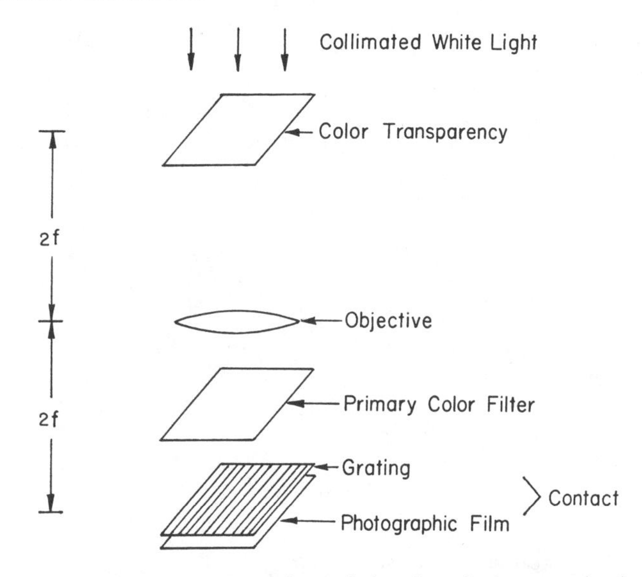

Figure 6.12. A spatial encoding technique for color image retrieval.

the blue and green images in the other remaining independent coordinate direction. Thus a small amount of color spread (i.e., color crosstalk) from blue to green (but not from green to blue) may not be avoided. However, this small amount of color spread will not cause significant degradation of the retrieved color image, for two main reasons: First, a slight mixture of blue into green will not produce significant color changes. Second, color transparencies do not, strictly speaking, exhibit natural colors; thus a small amount of color deviation would not be noticeable by human perception.

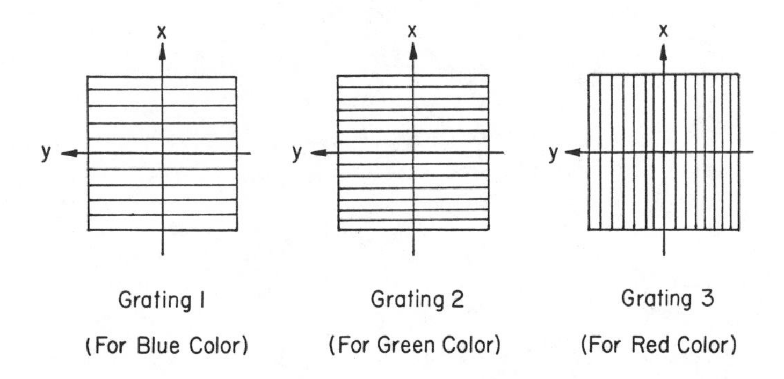

Figure 6.13. Three spatial sampling directions for three primary colors.

We now let the intensity transmittance of the encoded films be

$$\begin{aligned} T_n(x, y) = K\{&T_r(x, y)[1 + \operatorname{sgn}(\cos p_r y)] \\ &+ T_b(x, y)[1 + \operatorname{sgn}(\cos p_b x)] \\ &+ T_g(x, y)[1 + \operatorname{sgn}(\cos p_g x)]\}^{-\gamma}, \end{aligned} \tag{6.37}$$

where: $T_n(x, y)$ is the encoded black-and-white negative transparency; K is an appropriate proportionality constant; T_r, T_b, and T_g are the red, blue, and green color image exposures; p_r, p_b, and p_g are the respective carrier spatial frequencies; (x, y) is the spatial coordinate system of the encoded film; γ is the film gamma; and

$$\operatorname{sgn}(\cos x) \triangleq \begin{cases} 1, & \cos x \geqq 0, \\ -1, & \cos x < 0. \end{cases}$$

Instead of obtaining a positive image transparency, through a contact printing process, we bleach the encoded negative image film to obtain a phase object transparency [6.28, 6.29]. Let us assume that the bleached transparency is encoded in the linear region of the diffraction efficiency versus log exposure curve (i.e., the D–E curve). The amplitude transmittance of the bleached transparency can therefore be written as

$$t(x, y) = \exp[i\phi(x, y)], \tag{6.38}$$

where $\phi(x, y)$ represents the phase delay distribution, which is proportional to the exposure of the encoded film [6.30], such as

$$\begin{aligned} \phi(x, y) = M\{&T_r(x, y)[1 + \operatorname{sgn}(\cos p_r y)] \\ &+ T_b(x, y)[1 + \operatorname{sgn}(\cos p_b x)] \\ &+ T_g(x, y)[1 + \operatorname{sgn}(\cos p_g x)]\}, \end{aligned} \tag{6.39}$$

where M is an appropriate proportionality constant. Thus we see that a high diffraction efficiency spatially encoded transparency can be obtained with this spatial encoding process.

6.4.2. Color Image Reproduction

If we place the bleached encoded film of Eq. (6.38) in the input plane P_1 of a white-light optical processor, as illustrated in Fig. 6.14, then the first-order

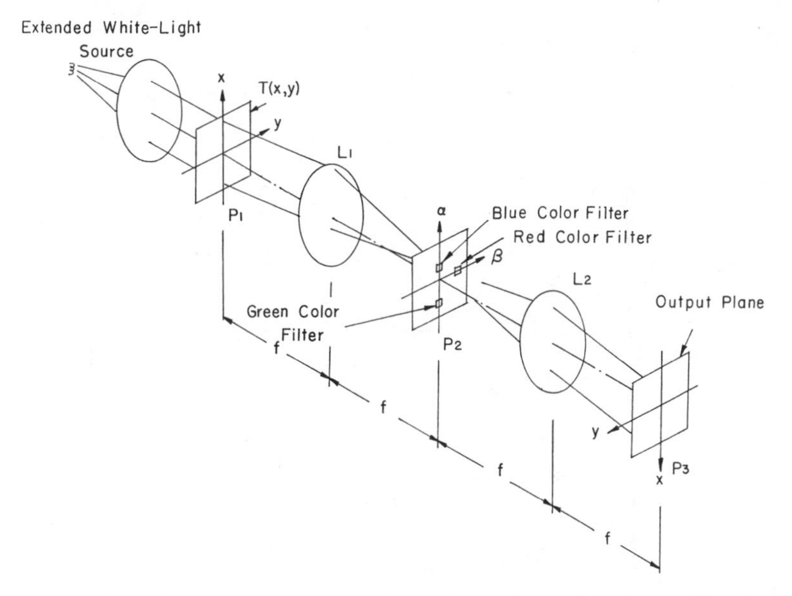

Figure 6.14. White-light color image retrieval. L; achromatic transform lenses; $T(x, y)$, phase encoded transparency.

complex light distribution, for every λ, at the spatial frequency plane P_2 can be shown as

$$\begin{aligned} S(\alpha, \beta; \lambda) \simeq\; & \hat{T}_r\left(\alpha, \beta \pm \frac{\lambda f}{2\pi} p_r\right) + \hat{T}_b\left(\alpha \pm \frac{\lambda f}{2\pi} p_b, \beta\right) \\ & + \hat{T}_g\left(\alpha \pm \frac{\lambda f}{2\pi} p_g, \beta\right) + \hat{T}_r\left(\alpha, \beta \pm \frac{\lambda f}{2\pi} p_r\right) * \hat{T}_b\left(\alpha \pm \frac{\lambda f}{2\pi} p_b, \beta\right) \\ & + \hat{T}_r\left(\alpha, \beta \pm \frac{\lambda f}{2\pi} p_r\right) * \hat{T}_g\left(\alpha \pm \frac{\lambda f}{2\pi} p_g, \beta\right) \\ & + \hat{T}_b\left(\alpha \pm \frac{\lambda f}{2\pi} p_b, \beta\right) * \hat{T}_g\left(\alpha \pm \frac{\lambda f}{2\pi} p_g, \beta\right), \end{aligned} \tag{6.40}$$

where: $\hat{T}_r$, $\hat{T}_b$, and $\hat{T}_g$ are the Fourier transforms of T_r, T_b, and T_g, respectively; $*$ denotes the convolution operation; and the proportional constants have been neglected for simplicity. We note that the last cross product term of Eq. (6.40) would introduce a moiré fringe pattern, which can be easily masked out at the Fourier plane. Thus by proper color filtering of the smeared Fourier spectra, as shown in Fig. 6.15, a true color image can be retrieved at the

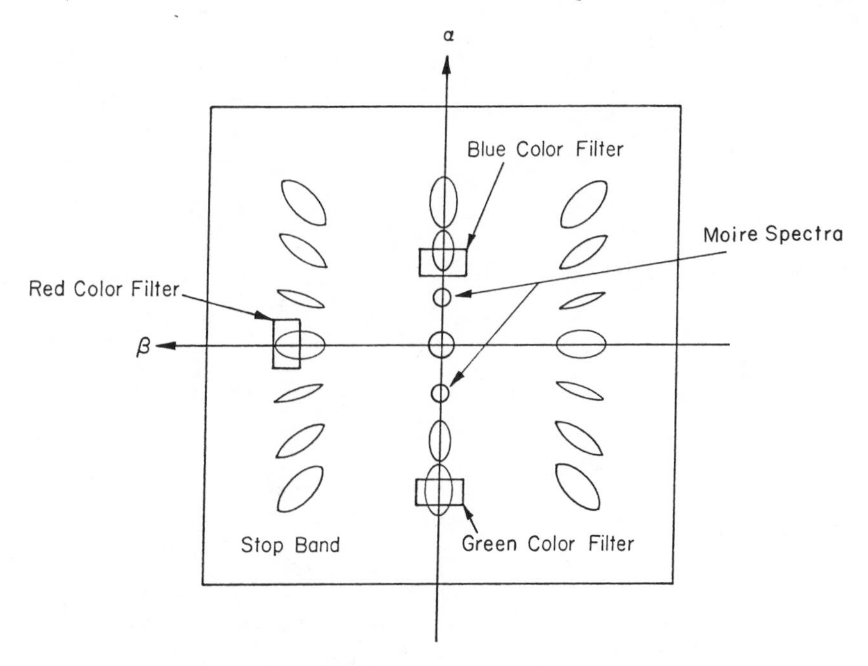

Figure 6.15. Spectral filtering at the Fourier plane.

output image plane P_3. The corresponding complex light field immediately behind the Fourier plane would be

$$S(\alpha, \beta) = \hat{T}_r\left(\alpha, \beta - \frac{\lambda_r f}{2\pi} p_r\right) + \hat{T}_b\left(\alpha - \frac{\lambda_b f}{2\pi} p_b, \beta\right) + \hat{T}_g\left(\alpha + \frac{\lambda_g f}{2\pi} p_g, \beta\right), \tag{6.41}$$

where λ_r, λ_b, and λ_g are the respective red, blue, and green color wavelengths. At the output image plane, the complex light distribution is

$$s(x, y) = T_r(x, y)\exp(iyp_r) + T_b(x, y)\exp(ixp_b) + T_g(x, y)\exp(-ixp_g). \tag{6.42}$$

The output image irradiance is therefore

$$I(x, y) = T_r^2(x, y) + T_b^2(x, y) + T_g^2(x, y), \tag{6.43}$$

which is a superposition of three primary encoded color images. Thus we see that a moiré-free color image can indeed be obtained.

6.4.3. Experimental Demonstration

In our experiment, we utilized three Kodak primary color filters, nos. 25, 47B, and 58, for the red, blue, and green color encoding and for the decoding process. The characteristics of these filters, illuminated by standard A illuminant, are specified by the "Commission Internationale de l'Eclairage" C.I.E. diagram, as shown in Fig. 6.16. The spatial encoding transparencies are made by Kodak technical pan film 2415 and Kodak microfilm 5460. The advantage of using Kodak film 2415 is that it is a high resolution film with a relatively flat spectral response, as shown in Fig. 6.17. The disadvantage is that this film is coated with a thin layer of dyed-gel backing, which introduces additional noise through the bleaching process. Although the microfilm used is a clear base film, however, it is a high gamma film and the spectral response decreases somewhat in the red color region, as shown in the same figure. In order to compensate this low spectral response, one would encode the red wavelength with a higher exposure. Needless to say, the resolution and contrast of the retrieval color image are also affected by the developing process of the film.

It should be emphasized that, to avoid the toe region of the D–E curve, the film should be preexposed. Otherwise, a low exposure nonlinear effect is introduced, which causes color unbalance in the retrieved image. The plots of

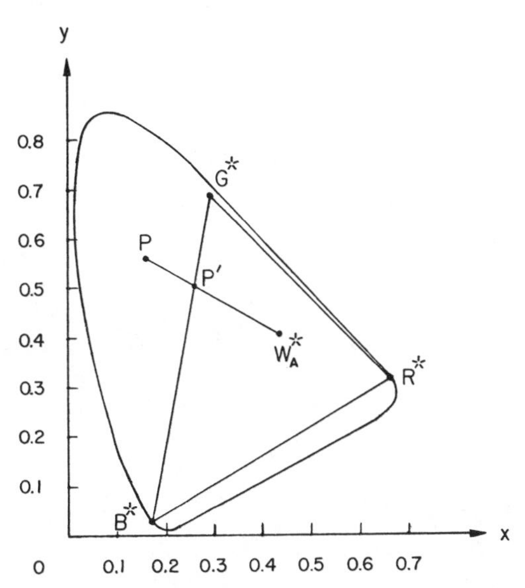

Figure 6.16. C.I.E. diagram, R^*, G^*, and B^* represent the locations of the red, green, and blue Kodak color filters nos. 25, 58, and 47B, respectively. W_A^* represents the standard A whitelight illuminant.

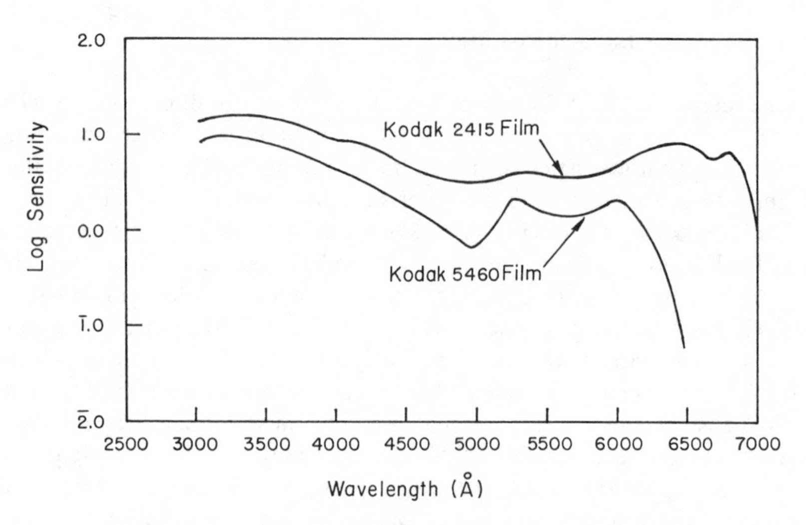

Figure 6.17. Spectral responses for Kodak films 2415 and 5460.

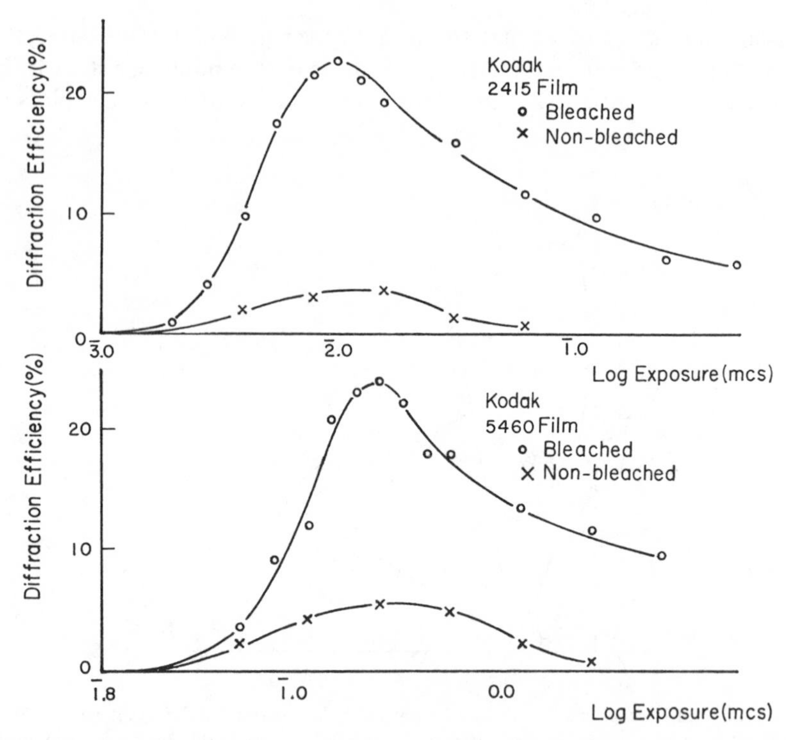

Figure 6.18. The plots of diffraction efficiency versus log exposure for Kodak 2415 and 5460 films. These plots were obtained with a sampling grating of 40 lines/mm.

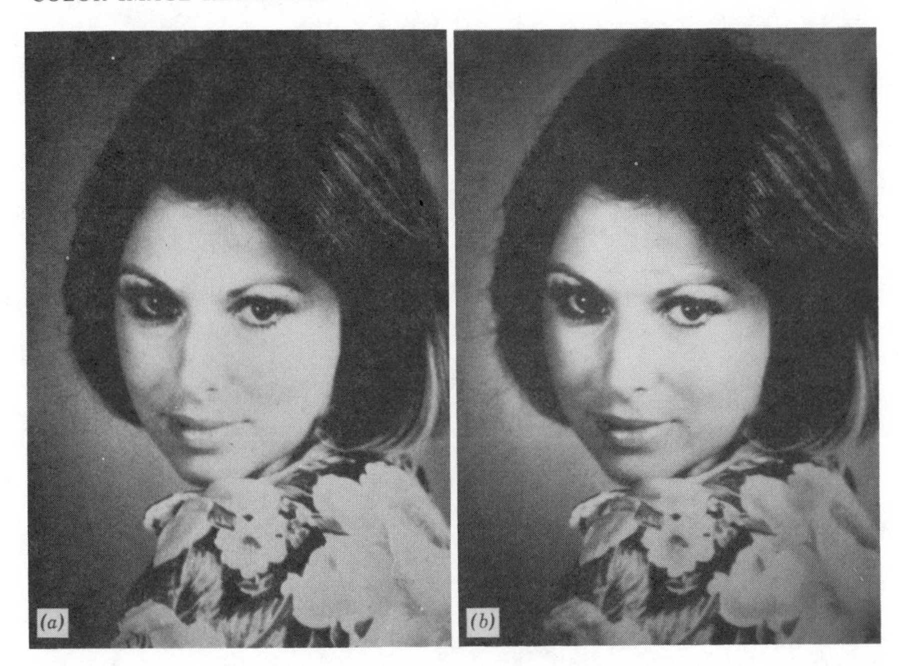

Figure 6.19. Black-and-white pictures of the color retrieved images. (*a*) Obtained with Kodak 2415 encoding film. (*b*) Obtained with Kodak 5460 encoding film.

diffraction efficiency versus exposure for Kodak 2415 and 5460 films with 40 lines/mm sampling frequency are plotted in Fig. 6.18. From this figure, we see that the bleached encoded films offer a higher diffraction efficiency; the optimum value occurs at exposures 8.50×10^{-3} mcs and 1.95×10^{-1} mcs, respectively, for Kodak films 2415 and 5460. With these optimum exposures, it is possible to optimize the encoding process as follows: First, by preexposing the film beyond the shoulder region. Second, by subdividing the remaining exposure, taking account of the film spectral response, into three parts for the red, green, and blue images.

For experimental demonstrations, we would like to provide two results obtained with Kodak 2415 and Kodak 5460 films, as shown in Fig. 6.19*a* and *b*, respectively. From these figures we see that moiré-free color images can indeed be obtained. We also see that the retrieved color image obtained with Kodak microfilm 5460 provides a higher image quality (i.e., higher resolution and lower noise level). The primary reason, as we have mentioned, is that commercially available Kodak 2415 is coated with a thin layer of dyed-gel backing, which causes an additional noise level through the bleaching process. For comparison, we also provide the original color transparency, shown

Figure 6.20. Black-and-white picture of the original color transparency.

in Fig. 6.20. By comparing Fig. 6.19 with Fig. 6.20, we see that the retrieved color images are spectacularly faithful, with virtually no color crosstalk. Although the resolution and contrast are still far below the acceptable stage for applications, these drawbacks may be overcome by utilizing a more suitable film, for which a research program is currently underway.

6.5. PSEUDOCOLOR ENCODING

Most of the optical images obtained in various scientific applications are gray-level density images, for example, the scanning electron microscopic images, multispectral band aerial photographic images, x-ray transparencies, and infrared scanning images. However, humans can preceive variations in color better than those in gray levels. In other words, a color-coded image can provide better visual discrimination.

In current practice, most of the pseudocoloring techniques are performed by a digital computer [6.31]. If the images are initially digitized, the computer technique may be a logical choice. However, for continuous tone images, an optical color encoding technique would be more advantageous for at least three major reasons: First, the technique can, in principle, preserve the spatial resolution of the image to be color coded. Second, the optical system is generally easy and economical to operate. Third, the cost of an optical pseudocolor encoder is generally less than that of its digital counterpart.

Mention must be made that density pseudocolor encoding by halftone screen implementation with a coherent optical processor was first reported by Liu and Goodman [6.32], and later with a white-light processor by Tai et al. [6.33]. Although good results have been reported, there is a spatial resolution loss with the halftone technique, and a number of discrete lines due to sampling are generally present in the color-coded image. A technique of density pseudocoloring through contrast reversal was reported by Santamaria et al. [6.34]. Although this technique has an advantage over the halftone technique, the optical system is more elaborate, and it requires both incoherent and coherent sources. Since a coherent source is utilized, the coherent artifact noise is unavoidable.

Recently a density pseudocolor encoding technique using white-light processing was reported by Chao et al. [6.35]. This technique offers the advantages of coherent noise reduction, no apparent resolution loss, versatility and simplicity of system operation, and low cost of pseudocoloring. Excellent results have been reported with this technique; however, pseudocolor encoding is primarily obtained by means of two primary colors. In a more recent paper by Mendez and Bescos [6.36], a two-step method for gray-level pseudocoloring with three primary colors, using a diffraction grating modulation method, was reported. Very good pseudocolor encoded images were shown in their article. However, their method may still suffer a major drawback; the moiré fringe pattern may not be avoided. Aside from the moiré fringes, they used a coherent image subtraction process to obtain the product (i.e., intermediate level) image. The coherent technique would introduce coherent artifact noise, and the subtraction method is rather cumbersome.

In this section we describe a white-light pseudocolor encoding technique with three primary colors [6.37]. We show that this technique is rather easy to use and is very cost effective. There is, however, one disadvantage of this technique: It is still not a real-time pseudocolor encoding method.

6.5.1. Pseudocoloring with Three Primary Colors

We now describe a white-light density pseudocolor encoding technique for monochrome images. We assume that a gray-level x-ray transparency (called a positive image) is available for pseudocoloring. By the contact printing process, a negative and a product (i.e., intermediate level) x-ray image transparencies can be made. Figure 6.21 shows a sketch of these normalized transmittances as function of gray scale. We now describe a spatial encoding technique to obtain a three-gray-level-image encoding transparency for the

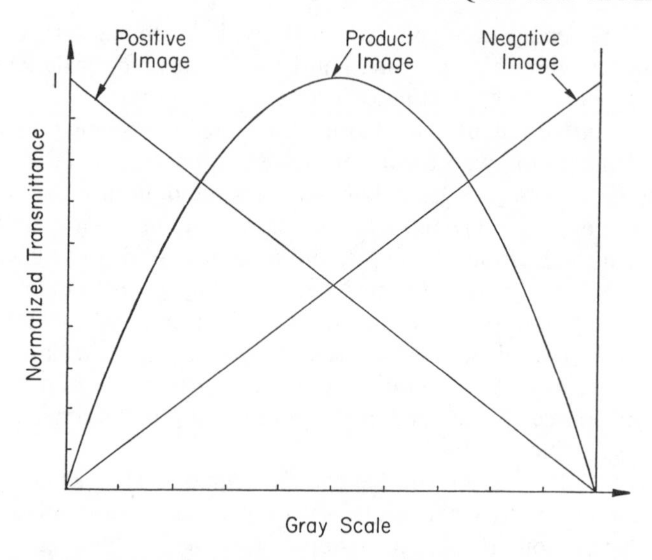

Figure 6.21. A normalized transmittance as a function of gray scale variation.

pseudocoloring. The spatial encoding is performed by respectively sampling the positive, the negative, and the product image transparencies onto a black-and-white photographic film, with specific sampling frequencies oriented at specific azimuthal directions. To avoid the moiré fringe pattern (see Section 6.4), we sample these three images in the orthogonal direction with different specific sampling frequencies, as shown in Fig. 6.22. The intensity transmittance of the encoded film can therefore be written as

$$\begin{aligned} T(x, y) = K\{ &T_1(x, y)[1 + \operatorname{sgn}(\cos p_1 y)] \\ &+ T_2(x, y)[1 + \operatorname{sgn}(\cos p_2 x)] \\ &+ T_3(x, y)[1 + \operatorname{sgn}(\cos p_3 x)]\}^{-\gamma}, \end{aligned} \tag{6.44}$$

where: K is an appropriate proportionality constant; T_1, T_2, and T_3 are the positive, the negative, and the product image exposures; p_1, p_2, and p_3 are the respective carrier spatial frequencies; (x, y) is the spatial frequency coordinate system of the encoded film; γ is the film gamma; and

$$\operatorname{sgn}(\cos x) \triangleq \begin{cases} 1, & \cos x \geqq 0, \\ -1, & \cos x < 0. \end{cases}$$

To obtain a surface relief phase object transparency, the encoded transparency is bleached with an R-10 formula [6.28, 6.29], and we assume that the

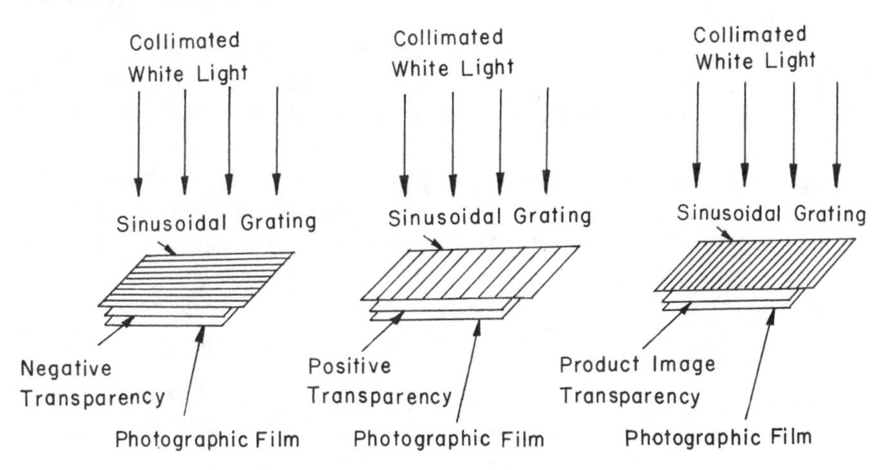

Figure 6.22. A spatial encoding technique.

bleached transparency is encoded in the linear region of the $D - E$ curve, as described in Section 6.4. The encoded phase transmittance is therefore

$$t(x, y) = \exp[i\phi(x, y)], \tag{6.45}$$

where

$$\begin{aligned}\phi(x, y) = M\{&T_1(x, y)[1 + \operatorname{sgn}(\cos p_1 y] \\ &+ T_2(x, y)[1 + \operatorname{sgn}(\cos p_2 x)] \\ &+ T_3(x, y)[1 + \operatorname{sgn}(\cos p_3 x)]\},\end{aligned} \tag{6.46}$$

where M is an appropriate proportionality constant. If we insert this encoded phase transparency at the input plane P_1 of a white-light optical processor, as illustrated in Fig. 6.16, the complex light distribution due to $t(x, y)$, for every λ, at the spatial frequency plane P_2 can be shown as

$$\begin{aligned}S(\alpha, \beta, \lambda) = {}&\hat{T}_1\left(\alpha, \beta \pm \frac{\lambda f}{2\pi} p_1\right) + \hat{T}_2\left(\alpha \pm \frac{\lambda f}{2\pi} p_2, \beta\right) \\ &+ \hat{T}_3\left(\alpha \pm \frac{\lambda f}{2\pi} p_3, \beta\right) + \hat{T}_1\left(\alpha, \beta \pm \frac{\lambda f}{2\pi} p_1\right) * \hat{T}_2\left(\alpha \pm \frac{\lambda f}{2\pi} p_2, \beta\right) \\ &+ \hat{T}_1\left(\alpha, \beta \pm \frac{\lambda f}{2\pi} p_1\right) * \hat{T}_3\left(\alpha \pm \frac{\lambda f}{2\pi} p_3, \beta\right) \\ &+ \hat{T}_2\left(\alpha \pm \frac{\lambda f}{2\pi} p_2, \beta\right) * \hat{T}_3\left(\alpha \pm \frac{\lambda f}{2\pi} p_3, \beta\right),\end{aligned} \tag{6.47}$$

where: $\hat{T}_1$, $\hat{T}_2$, and $\hat{T}_3$ are the smeared Fourier spectra of the positive, negative, and product images, respectively; $*$ denotes the convolution operation; and the proportional constants have been neglected for simplicity. Again, we see that the last cross product (i.e., the moiré fringe pattern) can be avoided by spatial filtering. Needless to say, by proper color filtering of the first-order smeared Fourier spectra a moiré-free pseudocolor encoded image can be obtained at the output plane P_3. The corresponding pseudocolor image irradiance is therefore

$$I(x, y) = T_{1r}^2(x, y) + T_{2b}^2(x, y) + T_{3g}^2(x, y), \tag{6.48}$$

where T_{1r}^2, T_{2b}^2, and T_{3g}^2 are the red, blue, and green intensity distributions of the three spatially encoded images.

6.5.2. Experimental Demonstrations

For experimental demonstration, Fig. 6.23*a* and *b* show a set of black-and-white pictures of color coded images of a woman's pelvis. The x-ray was taken following a surgical procedure. A section of the bone between the sacroiliac joint and spinal column has been removed. In Fig. 6.23*a*, the positive image is encoded in red, the negative image is encoded in blue, and the product image is encoded in green. By comparing the pseudocolor coded image with the original black-and-white x-ray picture, it appears that the soft tissues can be better differentiated in the color coded image, as demonstrated by the fact that the image contrast in the region containing the gastrointestinal tract is evidently superior in the color coded image. On the other hand, there seems to be a degradation in the resolution in the color coded image along edges of the hard tissues. This is perhaps caused by two factors: First, high frequency information may be eliminated due to the low spatial frequency encoding gratings (e.g., 40 and 26.7 lines/mm sampling gratings) employed. Second, the image may be smeared due to the film development process. These two problems can be easily corrected by selecting higher frequency encoding gratings and by gaining more experience in film processing.

Another point worth noting is that a reversal of the color encoding can be easily implemented, as shown in Fig. 6.23*b*, where the positive and negative images are encoded in blue and red while the product image remains in green. This possibility of switching colors could be beneficial because an image in different color combination may reveal subtle features that are otherwise undetected. For instance, the air pockets in the colon of the patient can be identified more easily with Fig. 6.23*b* than with Fig. 6.23*a*. Moreover, a wide variety of other pseudocolor encoded images can also be obtained by simply alternating the color filters in the Fourier plane of the white-light processor.

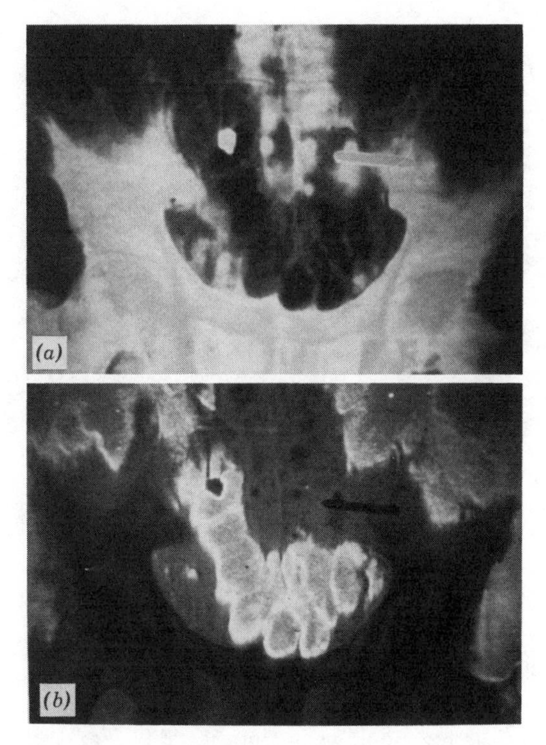

Figure 6.23. Black-and-white pictures of density pseudocolor images. (*a*) Positive image is coded in red, negative image is coded in blue, and product image is coded in green. (*b*) Positive image is coded in blue, negative image is coded in red, and product image is again coded in green.

Finally we stress that this white-light pseudocolor encoder offers several advantages over its digital counterpart. The encoder is far less expensive, and in principle the technique offers a higher color coded image resolution.

6.5.3. Application to Remote Sensing

We now illustrate an application of this proposed pseudocolor encoder to multispectral scanner Landsat data. Although five- or six-band composite pseudocoloring is achievable with this proposed white-light system, for simplicity three bands of multispectral scanner data were selected for composite pseudocoloring. These bands were from the blue-green (Band 4: 0.5–0.6 μm), red (Band 5: 0.6–0.7 μm), and reflected infrared (Band 7: 0.8–1.1 μm) spectral regions, as shown in Fig. 6.24. The scene is a 78 $\times$ 107 km subsample of Landsat scene 1440–15172 showing Southeastern Pennsylvania, as shown in the rectangular section of Fig. 6.25. If these bands of transparencies were

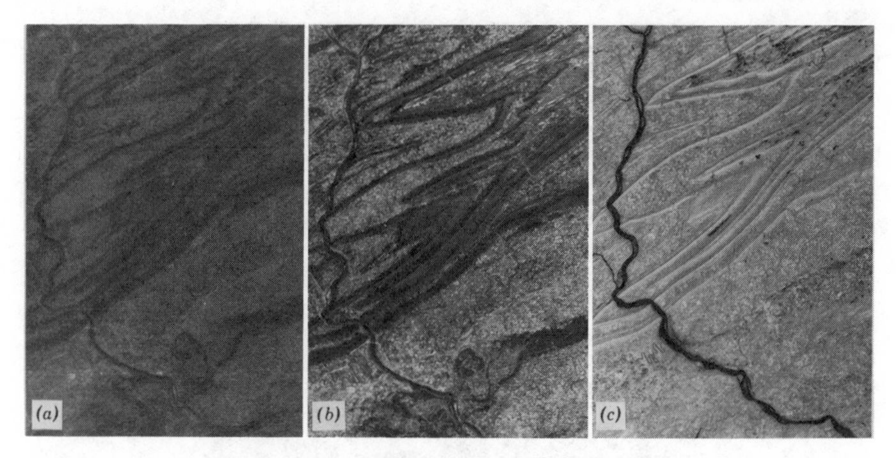

Figure 6.24. Three spectral band Landsat data. (*a*) Band 4: blue-green spectral region about 0.5–0.6 μm. (*b*) Band 5: red spectral region about 0.6–0.7 μm. (*c*) Band 6: reflected infrared about 0.8–1.1 μm.

spatially encoded onto a black-and-white transparency, as described in Section 6.5.1, then various composite false-color images of these spectral band data could be observed at the input image plane of the white-light processor of Fig. 6.16. In our experiment, we again utilized 26.7 and 40 lines/mm sampling gratings and Kodak 25, 47B, and 58 primary color filters for pseudocolor encoding. The results of the pseudocolor encoding of the Landsat multispectral scanner data are shown in Fig. 6.26. In Fig. 6.26*a*, where band 4 is encoded green and band 5 is encoded red, the Susquehanna River and small bodies of water are delineated in deep red. The islands in the Susquehanna River are easily distinguished. Strip mines are dark red, urban areas (e.g., Harrisburg) are medium red, and agricultural lands are light red, orange, and yellow. Forested areas are green.

When red-encoded band 5 is combined with blue-encoded band 7, as shown in Fig. 6.26*b*, the Susquehanna River is shown in violet. Small lakes and reservoirs appear as bluish hues. Some of the surface-mined areas appear as light violet along with some of the bare fields in the agricultural valleys. The forested regions are dark blue and the agricultural areas are red. Urban areas are not delineated.

Water appears as several shades of blue when band 4 is encoded green, and band 7 is encoded blue, as depicted in Fig. 6.26*c*. Northeast-southwest trending streams are also evident near the center of the image. Surface-mined areas are a much darker blue, and can be easily distinguished from water. Forested areas are most clearly distinguished in this image product as light green. The

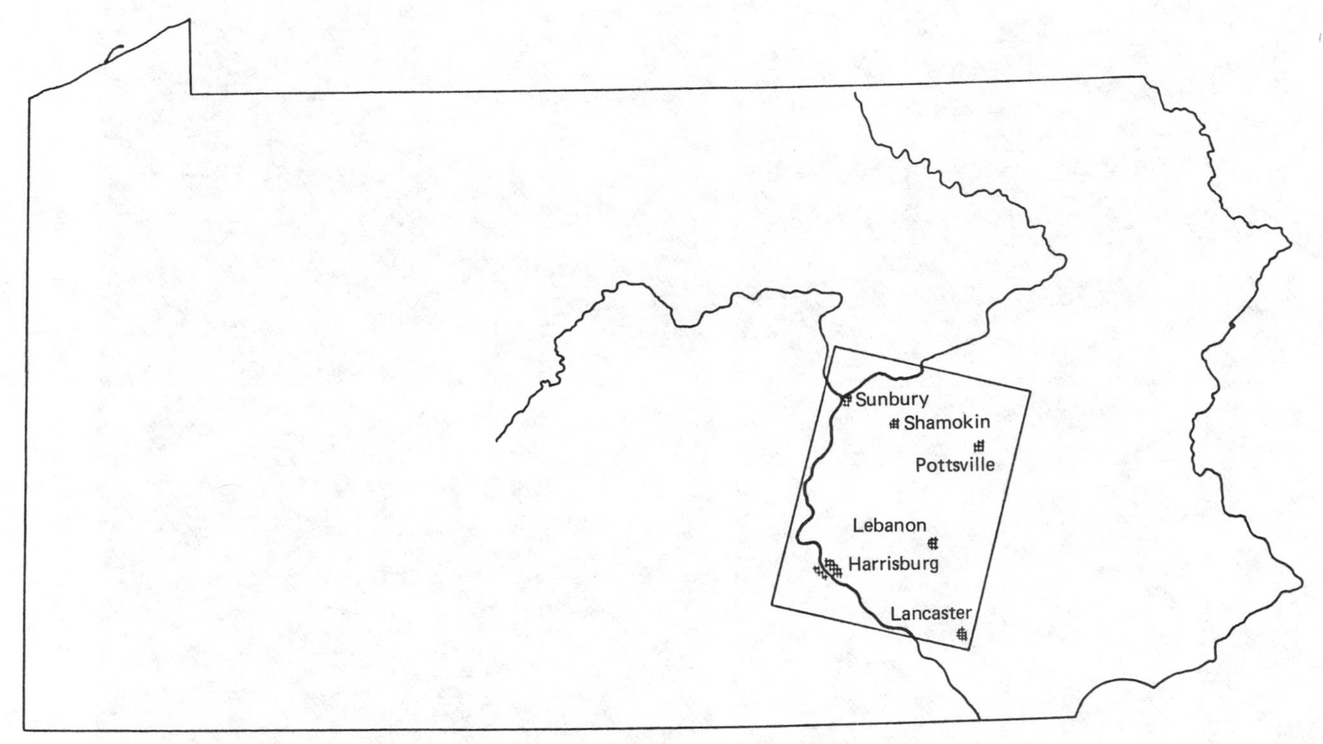

Figure 6.25. Location site of the Landsat data of Fig. 6.24.

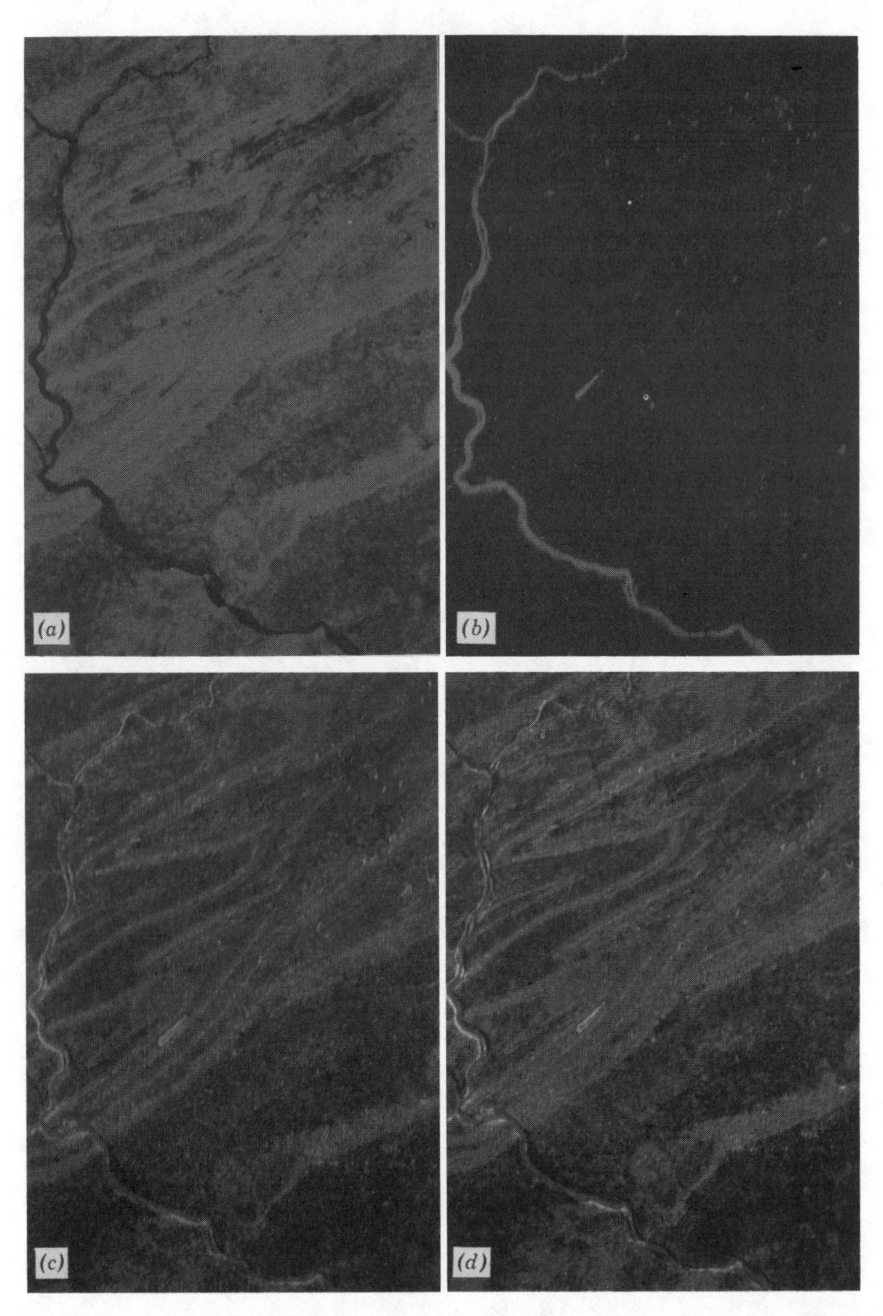

Figure 6.26. Black-and-white photographs of composite pseudo-color encoded images. (*a*) With band 4 encoded green and band 5 encoded red. (*b*) With band 5 encoded red and band 7 encoded blue. (*c*) With band 4 encoded green and band 7 encoded blue. (*d*) With band 4 encoded green, band 5 encoded red, and band 7 encoded blue.

dark green areas are agricultural regions. Urban areas and strip development along major highways appear as dark blue to black.

When all three bands of the Landsat data are encoded, as shown in Fig. 6.26*d*, the Susquehanna River appears as violet, and the other bodies of water as shades of blue. The surface mines and urban areas are dark red. The agricultural valleys are orange and the forested regions are green.

In ending this demonstration, we would note that the white-light pseudocolor encoder has great potential for application to multispectral band aerial photographic images. In principle, the resolution of the color-coded image can be as high as the original monochrome transparency. The white-light encoder is easy to construct and can be assembled for relatively low cost, less than $3000. Once the encoder is constructed, the system is also very easy to operate. Therefore, this system offers a new and inexpensive approach to the development of pseudocolor composites.

6.5.4. Application to Scanning Electron Micrograph

Here we discuss another interesting application of density pseudocolor encoding, the scanning electron micrograph. For the purposes of this discussion, photographically reproduced images of the microstructure of geological specimens obtained from a scanning electron microscope (SEM) are employed.

After the high energy electrons interact with a sample, the images generated by the electrons convey different information. Secondary electrons (SE) have high resolution and give useful information from rough surfaces. The composition of the target has very little influence on the SE image. On the other hand, the brightness of a backscattered electron (BSE) image is proportional to the average atomic number of the target imaged from a flat surface. McKinley et al. [6.38] have experimentally demonstrated that the backscatter coefficient η is related to the atomic number Z according to the following relationship:

$$\eta = 1.12 \times 10^{-3} Z^0 + 0.015 Z^1 - 1.75 \times 10^{-4} Z^2 + 7.92 \times 10^{-7} Z^3. \quad (6.49)$$

Figure 6.27 summarizes the experimentally observed data points and the empirically derived fit in Eq. (6.49). With most conventional backscatter detector systems commercially available today, the continuum in η expressed by the relationship η versus Z can be scaled into approximately 100 increments of gray levels, yielding a discrimination that is about at the limit of a single electron.

Two BSE images are presented. The first image, in Fig. 6.28, is a section of a microprobe standard in which a number of materials were cased into an

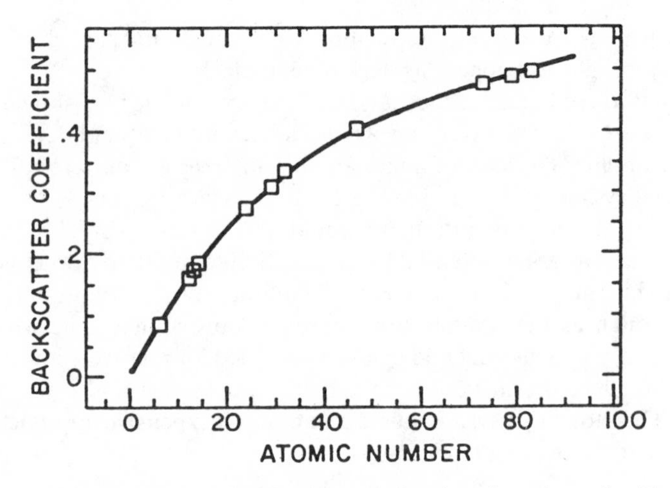

Figure 6.27. Backscatter efficiency *n* as a function of atomic number *Z*.

organic potting media, Spurrs, and a flat polished surface prepared. Figure 6.28*a* is the black-and-white backscatter image, Fig. 6.28*b* is the density pseudocolor encoded image obtained with the proposed white-light technique, and Fig. 6.28*c* is a map of the different phases. Table 6.2 summarizes the average electron densities of each of these materials. In Fig. 6.28*a*, the lowest electron density is represented by the potting medium at ≈ 6 and appears blue in the corresponding photograph in Fig. 6.28*b*. The silica, calcite, and alumina all possess an average electron density of 10 and, in principal, each should be represented by a single shade of yellow. However, in practice these three phases vary from a yellow to a yellow-blue to a blue-yellow. Comparing the varying gray levels in Fig. 6.28*a*, it becomes apparent that the pseudocolor encoding is in fact reproducing a variation in the gray levels of these three

Table 6.2. Electron Densities of Materials in Figure 6.28

Phase Composition	Average Electron Densities
Al_2O_3	10
SiO_2	10
$CaCO_3$	10
$NaKNbO_7$	16
FeS_2	19.3
Ti	22

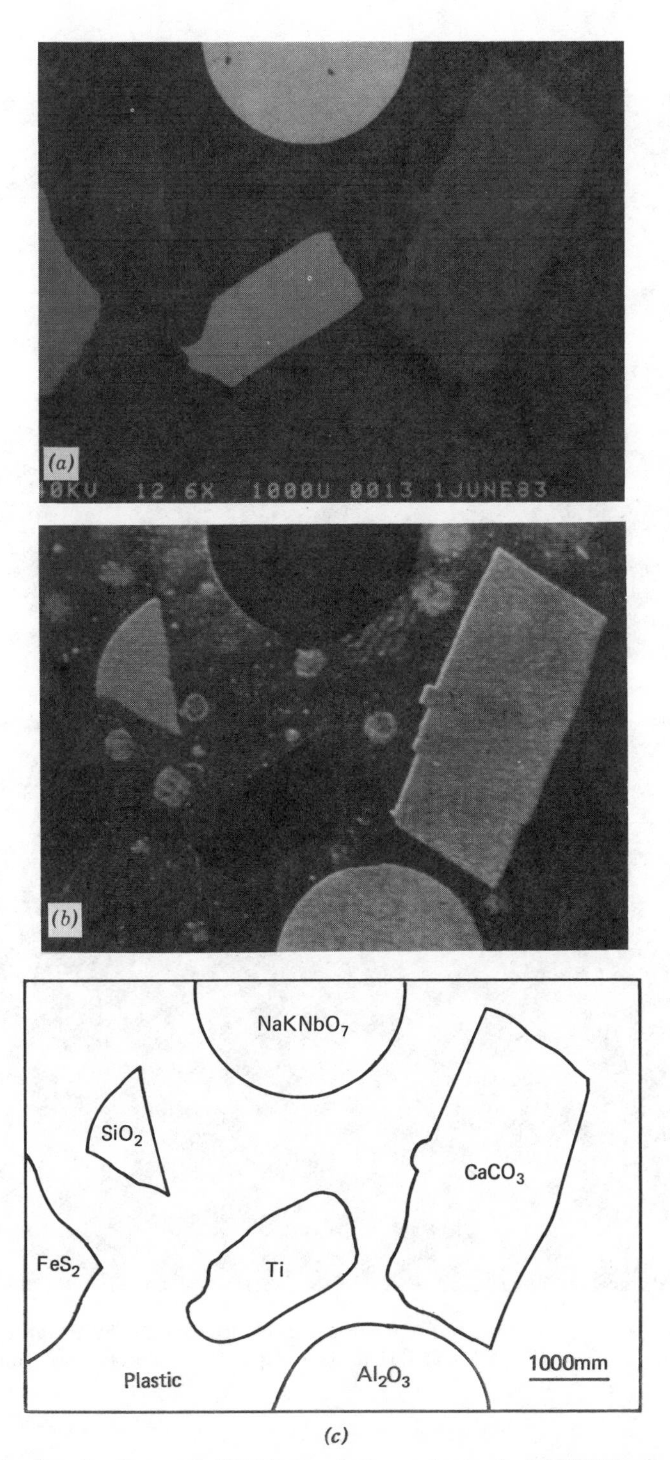

Figure 6.28. Pseudocolor encoded BSE image of microprobe standard. (*a*) BSE image of microprobe standard. (*b*) A black-and-white picture of a pseudocolor encoded BSE image. (*c*) Composition map of standard.

Figure 6.29. Pseudocolor encoded BSE image of a polished section of basalt. (*a*) BSE image of basaltic thick section. (*b*) A black-and-white photograph of a pseudocolor encoded BSE image.

phases, progressing from the lightest, calcite, to the darkest, alumina. Finally, the three phases, sodium potassium niobate, pyrite, and titanium metal, all appear as shades of red, representing the largest electron densities. Very little phase contrast appears in the pseudocolor encoded photograph because the original BSE image exposed these three phases equally. Clearly, the quality of the data that is recoverable is only as good as the quality of the initial black-and-white data set.

We now provide a more complex example, as displayed in Fig. 6.29*a*, which represents a BSE image of a polished section of basalt from the Devil's Pitchfork Body Crater in the Elegante Pincate Volcanic field of Sonora, Mexico. Figure 6.29*b* is the density pseudocolor coded image obtained with the proposed white-light technique. Figure 6.30 is a sketch of the backscatter image with energy dispersive spectra of the four principal mineral phases that constitute this sample. The dark blue pseudocolored lathlike crystals are a calcium-rich plagioslase feldspar, the large mass represented by the blue-yellow to yellow-red transition is in fact one pyroxene grain. Here too, as in the previous example, a comparison of the original black-and-white image indicates that there is a gradual transition in gray level from darker in the lower left to lighter in the upper right. Further, a careful examination of the upper right-hand portion of the image reveals a poorly polished surface that is manifested by the red to almost black mottling. The grain marked as an olivine in Fig. 6.30 exhibits a typical texture alteration. Olivines are chemically unstable and subsequent thermal cycling of the rock assists in the decomposition of the

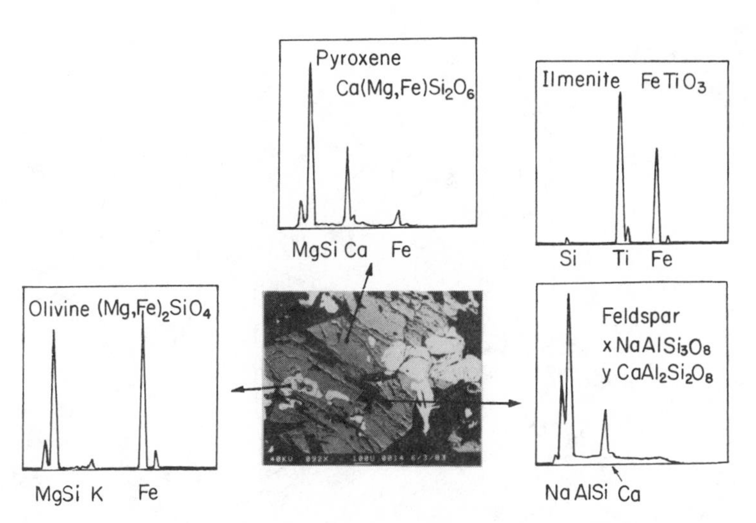

Figure 6.30. Sketch of polished basalt section in Figure 6.29 and energy dispersive x-ray spectra of the component mineral phases.

crystals. In the black-and-white image, only the significant embayment is readily observable, but in the density pseudocolor encoded image, the subtle alteration is marked. Finally, the titanium and iron-rich ilmenite grains appear as dark red and can only be recognized from the adjacent olivines by the smooth continuous surface. The adjacent olivine grains, in contrast, exhibit signs of alteration that break up the smooth surface appearance.

Finally, we would stress again that the human eye is much more sensitive to subtle changes in shades of color than to changes in levels of gray. We have demonstrated that, with this simple cost-effective system, pseudocolor encoding of black-and-white images is possible. Further, the technique is excellent for visual presentation where colors are easily correlated with the presence of mineral phases such as in the geologic specimen discussed above. The technique is general and can be applied to any photographically reproduced image in which information important to interpretation is contained in subtle gray-level changes. Unlike digital pseudocoloring techniques, this approach in principle is only limited in resolution by the resolution of the electron micrographs that are used.

6.5.5. A Programmable Pseudocolor Encoder

In this section, we propose a programmable white-light pseudocolor encoder, as schematized in Fig. 6.31. In contrast with the conventional optical signal processor, a programmable microprocessor can be utilized to control the mo-

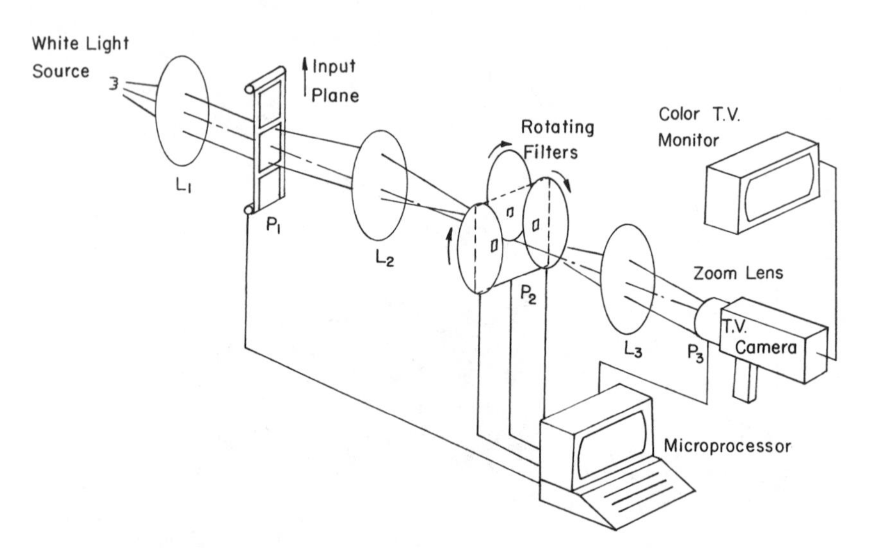

Figure 6.31. A programmable white-light pseudocolor encoder.

tion of the input encoded transparencies (e.g., as a movie projector), to select the spectral band filters for pseudocoloring, and to adjust the zoom lens of a TV camera for finer image display. Needless to say, a wide variety of pseudocolor encoded images in motion can be obtained for visualization, transmission, and storage. For example, let the input encoded transparencies be a sequence of x-ray images in motion. As this strip of encoded x-ray film is sequentially transported over the input plane of the proposed pseudocolor encoder, different shades of pseudocolor coded x-ray images in motion can be used for diagnostic analysis.

Since the optical pseudocolor encoder is a high resolution system, the detail of the color coded image can, in principle, be zoomed in for more detailed diagnosis. Thus it is apparent that the operator (e.g., a radiologist) can be seated in front of a color viewing screen. By controlling the equipped programmable computer, he or she can conveniently observe various color coded x-ray images in motion, as wished.

Furthermore, as we have previously mentioned, the overall system can be assembled for the relatively low cost of about $3000, in constrast with its digital counterpart, which would cost over $100,000. Once the system is constructed, it is as easy to operate as a home computer. The proposed system would offer a new and inexpensive alternative for high quality pseudocolor imaging.

6.5.6. Real-Time Pseudocolor Encoding

In the previous sections we have described a simple technique for white-light pseudocolor encoding through spatial image sampling. We have shown that the white-light pseudocoloring techniques are very simple and economical to operate. However, the techniques still suffer one major drawback; they are not real-time pseudocolor encoding techniques.

We now describe a real-time white-light pseudocoloring technique for spatial frequency and density encodings [6.39]. We stress that this real-time pseudocoloring technique may offer some advantages in some specific applications.

6.5.6.1. Spatial Frequency Pseudocoloring. In spatial frequency pseudocolor encoding, we place a gray-level image transparency $s(x, y)$ in contact with a two-dimensional high diffraction efficiency grating $T(x, y)$ at the input plane P_1 of a white-light optical processor, as shown in Fig. 6.32. For simplicity, we assume that the amplitude transmittance of the two-dimensional diffraction grating is

$$T(x, y) = 1 + \frac{1}{2}\cos p_0 x + \frac{1}{2}\cos q_0 y, \tag{6.50}$$

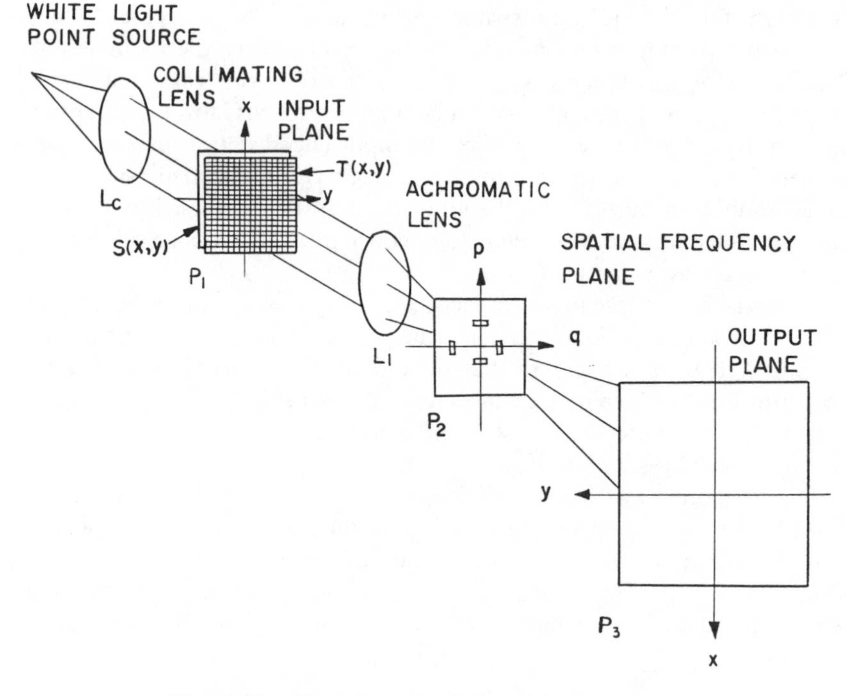

Figure 6.32. A real-time white-light pseudocolor encoder.

where p_0 and q_0 are the carrier spatial frequencies of the diffraction grating. The corresponding complex light distribution for a given wavelength λ at the spatial frequency plane P_2 is

$$E(\alpha, \beta; \lambda) = S(\alpha, \beta) + \frac{1}{4}\left[S\left(\alpha - \frac{\lambda f}{2\pi} p_0, \beta\right) + S\left(\alpha + \frac{\lambda f}{2\pi} p_0, \beta\right)\right.$$
$$\left. + S\left(\alpha, \beta - \frac{\lambda f}{2\pi} q_0\right) + S\left(\alpha, \beta + \frac{\lambda f}{2\pi} q_0\right)\right], \qquad (6.51)$$

where $S(\alpha, \beta)$ is the Fourier spectrum of input monochrome image $s(x, y)$. From this equation, we see that four first-order signal spectra are dispersed in rainbow color proportional to wavelength λ along the α- and β-axes. Since the spatial filtering is effective in the direction perpendicular to the color smeared spectral, we adopt one-dimensional spatial filters for pseudocolor encoding, as shown in Fig. 6.33. The complex light amplitude distribution immediately

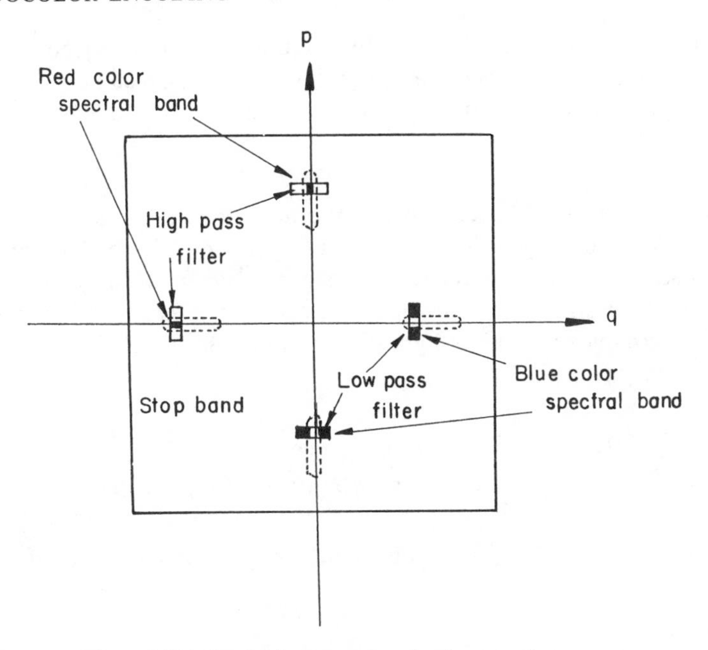

Figure 6.33. Spatial frequency pseudocolor encoding.

behind the spatial frequency plane is then

$$\begin{aligned} E(p, q; \lambda) = {} & S_r(p - p_0, q)H_1(q) + S_r(p, q - q_0)H_1(p) \\ & + S_b(p + p_0, q)H_2(q) + S_b(p, q + q_0)H_2(p), \end{aligned} \tag{6.52}$$

where S_r and S_b are the selected color-band image spectra (e.g., red and blue), H_1 and H_2 are the one-dimensional spatial filters, and (p, q) is the angular spatial frequency coordinate system of P_2. The output image irradiance can be shown as

$$\begin{aligned} I(x, y) \simeq {} & \Delta\lambda_r |\exp(ip_0 x)s_r(x, y) * h_1(y) + \exp(iq_0 y)s_r(x, y) * h_1(x)|^2 \\ & + \Delta\lambda_b |\exp(-ip_0 x)s_b(x, y) * h_2(y) + \exp(-iq_0 y)s_b(x, y) * h_2(x)|^2, \end{aligned} \tag{6.53}$$

where $\Delta\lambda_r$ and $\Delta\lambda_b$ are the color (e.g., red and blue) spectral bands of the signal spectra, and h_1 and h_2 are the corresponding impulse responses of H_1 and H_2. Thus we see that two spatial filtered images are incoherently superimposed to form a color encoded image at the output plane P_3 of the white-light image processor.

Figure 6.34 shows a black-and-white picture of a spatial frequency pseudocolor encoded radar image obtained by this processing technique, where the high spatial frequency contents are encoded in red and the low spatial frequency contents are encoded in blue.

6.5.6.2. Density Pseudocoloring. We now describe a real-time density pseudocolor encoding technique. In density pseudocoloring we insert two narrow strips of half-wave phase objects in the centers of the selected color-band image spectra to provide the image contrast reversal, as shown in Fig. 6.35. The complex amplitude light distribution immediately behind the spatial frequency plane is

$$E(p, q, \lambda) = S_r(p - p_0, q) + S_r(p, q - q_0) + S_g(p - p_0, q)H(q) + S_g(p, q - q_0)H(p); \tag{6.54}$$

where S_r and S_g are the selected color-band image spectra (e.g., red and green) and

$$H(q) = \begin{cases} -1, & q \simeq 0, \\ 1, & \text{otherwise}, \end{cases} \qquad H(p) = \begin{cases} -1, & p \simeq 0, \\ 1, & \text{otherwise}, \end{cases} \tag{6.55}$$

are the narrow strips of half-wave phase objects. At the output imaging plane P_3, the complex light amplitude distribution can be approximated by

$$g(x, y; \lambda) \approx [\exp(ip_0 x) + \exp(iq_0 y)]s_r(x, y) + [\exp(ip_0 x) + \exp(iq_0 y)]s_{gn}(x, y), \tag{6.56}$$

where $s_{gn}(x, y)$ is the (approximate) color contrast reversed image, that is,

$$s_{gn}(x, y) = s_g(x, y) - 2\langle s_g(x, y)\rangle, \tag{6.57}$$

where $\langle s_g(x, y)\rangle$ denotes the spatial ensemble average (i.e., the dc level) of $s_g(x, y)$. Since the two images s_r and s_{gn} are diffracted from two different color spectral bands of the light source, they are mutually incoherent. The output image irradiance is, therefore,

$$I(x, y) = \int |g(x, y; \lambda)|^2 \, d\lambda = \Delta\lambda_r I_r(x, y) + \Delta\lambda_g I_{gn}(x, y), \tag{6.58}$$

Figure 6.34. A black-and-white picture of a real-time spatial frequency pseudocoloring radar image. The high spatial frequency terrains are encoded in red and the low spatial frequency terrains are encoded in blue.

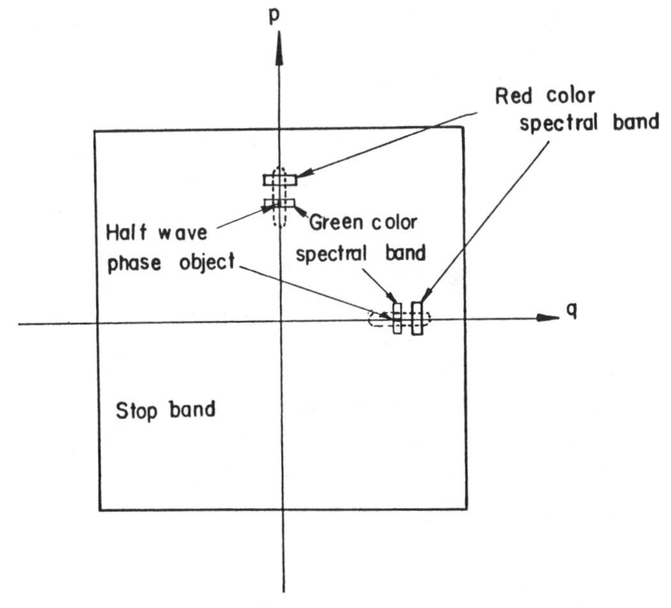

Figure 6.35. Contrast reversal density pseudocolor encoding.

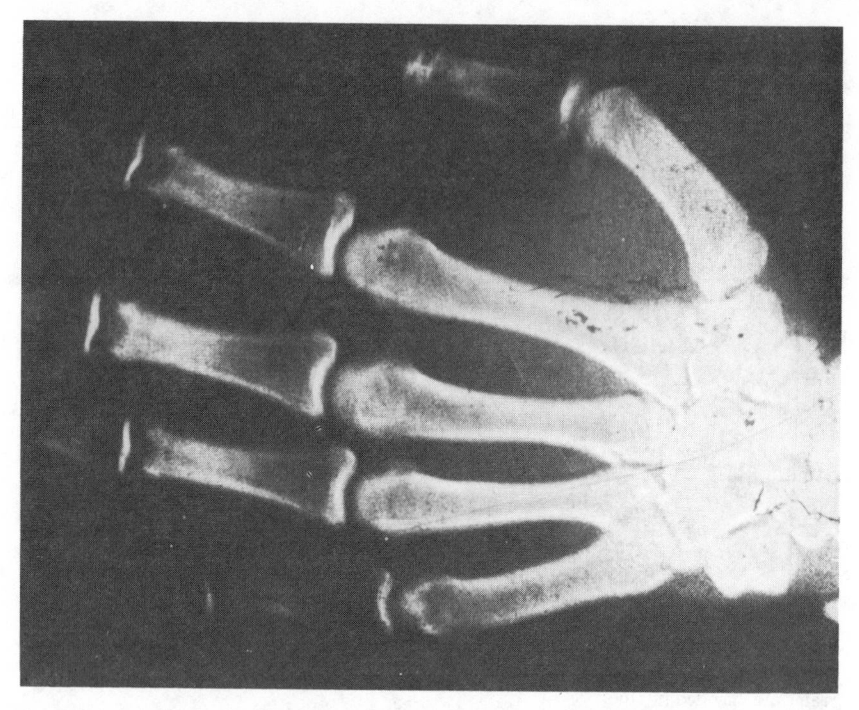

Figure 6.36. A black-and-white picture of a density pseudocolor encoding image of an x-ray transparency. In color the thicker bones are displayed in red and the fingers are green.

where $I_r(x, y)$ is a positive color (e.g., red) image irradiance, $I_{gn}(x, y)$ is an (approximate) contrast reversed or negative color (e.g., green) image irradiance, $\Delta\lambda_r$ and $\Delta\lambda_g$ are the narrow color spectral bands of the signal spectra (e.g., red and green). Thus we see that a density pseudocolor coded image is formed with the incoherent addition of a positive image in one color and a negative image in another.

Figure 6.36 shows a black-and-white picture of a real-time pseudocolor density encoded x-ray image of a hand. From this figure we see that a broad range of density pseudocolor encoded images can be obtained with this technique. Finally, we would note that, with this real-time white-light pseudocolor encoding technique, there is a freedom to select different color spectral bands for pseudocolor encoding. Thus in practice a wide range of different pseudocolor encoded images can easily be obtained.

6.5.7. Pseudocoloring Through Halftone Screen

It is also interesting to describe a technique utilizing a specially fabricated halftone screen to perform pseudocolor encoding by density [6.32, 6.33]. A halftone transparency of the original image is first obtained using a specially constructed halftone screen as described by Kato and Goodman [6.40]. The amplitude transmittance of the halftone image in one dimension can be described as

$$h(x) = \sum_{n=-\infty}^{\infty} \delta(x - nx_0) * \operatorname{rect}\left(\frac{x}{w}\right), \tag{6.59}$$

where $\delta(x)$ is the Dirac delta function, x_0 is the pulse period, w is the pulse width, and

$$\operatorname{rect}\left(\frac{x}{w}\right) = \begin{cases} 1, & \frac{x}{w} \leq \frac{1}{2}, \\ 0, & \frac{x}{w} > \frac{1}{2}. \end{cases}$$

Note that the pulse period x_0 is determined by the period of the halftone screen and that the pulse width w is a function of the density of the original image at x. The halftone image transparency is then inserted into the white-light processor shown in Fig. 6.37. The complex amplitude distribution at the back focal plane of the achromatic lens L_1 can be written as

$$E(\alpha, \lambda) = \exp\left[-iK_1\left(\frac{\alpha}{\lambda}\right)^2\right] \int h(x) \exp\left(-i\frac{2\pi}{\lambda f}\alpha x\right) dx, \tag{6.60}$$

where α is the spatial frequency coordinate, K_1 is an appropriate constant, and f is the focal length of the lens. The integration is performed over the spatial limits of the input transparency and the spectral width of the white-light source. For a given wavelength λ, the various orders of diffracted light at the spatial frequency plane are

$$E(\alpha, \lambda) = \exp\left[-iK_1\left(\frac{\alpha}{\lambda}\right)^2\right] \sum_{n=-\infty}^{\infty} \frac{\lambda^2 f^2}{\pi \alpha x_0} \sin\left(\frac{\pi \alpha w}{\lambda f}\right) \delta\left(\alpha - \frac{n\lambda f}{x_0}\right). \tag{6.61}$$

Thus we can see that, except for the zero-order term, the higher order diffractions are dispersed into rainbow colors along the α-direction at the spatial frequency plane P_2. Within the visible region of the diffracted light, there is no overlapping between diffraction orders up to the third order. For a given wavelength the irradiance of the nth diffraction order at $\alpha = n\lambda f/x_0$ is

$$I_n = \left(\frac{\lambda f}{n\pi}\right)^2 \sin^2\left(\frac{n\pi w}{x_0}\right). \tag{6.62}$$

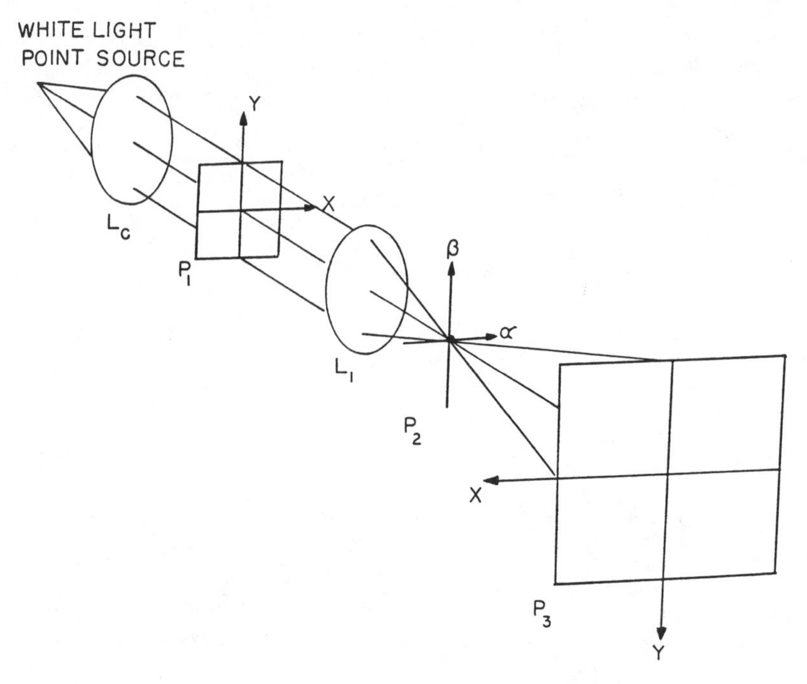

Figure 6.37. White-light optical processor. l_c, achromatic collimating lens; L_1, achromatic transform lens; P_1, halftone input transparency; P_2, spatial frequency plane; P_3, output image plane.

To perform pseudocolor encoding, we may select two different colors (e.g., red and green) from two different orders of diffraction by placing narrow slits at the appropriate positions to bandpass the desired colors. At the output plane P_3, we have a pseudocolor image formed by the addition of the intensities of the two filtered color images.

In our experiments a tungsten arc lamp is used as the white-light source and a specially fabricated one-dimensional multilevel halftone screen is used to produce the halftone input, as shown in Fig. 6.38. For simplicity, we use a single slit with adjustable width as the spatial filter. The slit is placed between the second and third orders, as shown in Fig. 6.39, such that only the red portion of the second-order diffraction and the blue-green portion of the third-order diffraction can pass through. In Fig. 6.40, we show a black-and-white picture of a pseudocolor encoded image obtained with this white-light processing technique.

Aside from the complication of generating a halftone image, this technique also suffers two major drawbacks, namely a spatial resolution loss and the

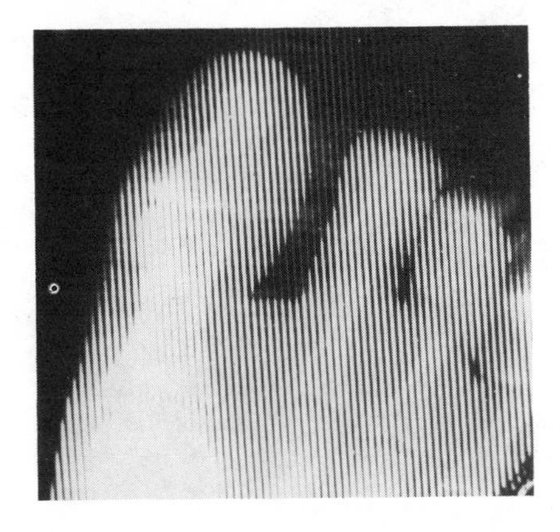

Figure 6.38. Halftone input transparency of an x-ray picture.

sampling lines generally present in the color coded image. However, this technique shows another method of producing a density pseudocolor coded image with simple white-light processing.

6.5.7.1. A Real-Time Technique. Here we propose a real-time pseudocoloring technique that utilizes a halftone screen method. The basic difference between this technique and the halftone encoder is the replacement of the halftone photographic image by a liquid crystal light valve (LCLV), as shown in Fig. 6.41. This substitution would allow the density pseudocolor encoding to be performed in real-time with a wide range of monochrome images.

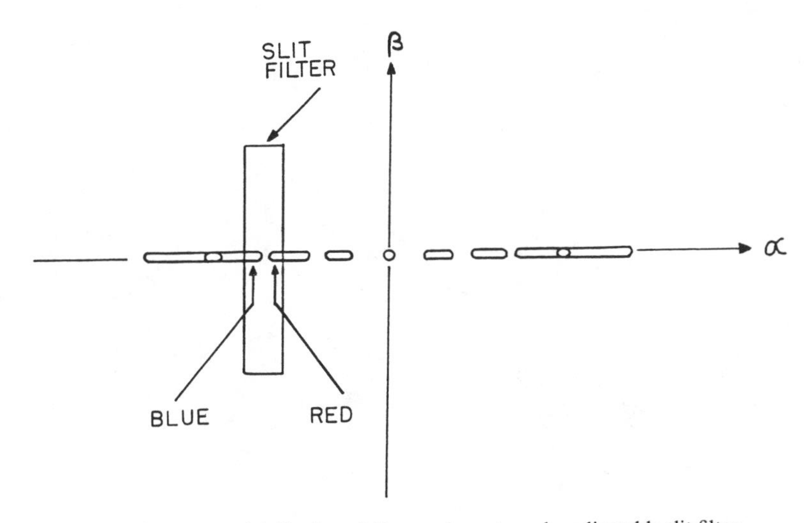

Figure 6.39. Spatial filtering of dispersed spectrum by adjustable slit filter.

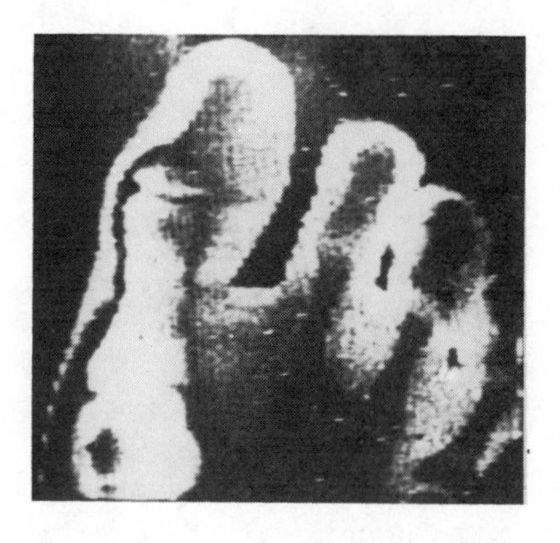

Figure 6.40. A black-and-white density color encoded image obtained with the halftone technique.

In order to obtain a halftone image [see Eq. (6.59)] that is suitable for the density pseudocoloring, it is necessary that the recording medium be a hard-clipping device. However, when an LCLV is utilized, the input-output intensity characteristic is nearly linear over a small range. This effect can be reflected in Eq. (6.59) by introducing a multiplicative term that represents the continuously varying intensity characteristic of the LCLV.

In pseudocoloring with a bandpass filter, as depicted in Fig. 6.41, two unfortunate characteristics would be introduced.

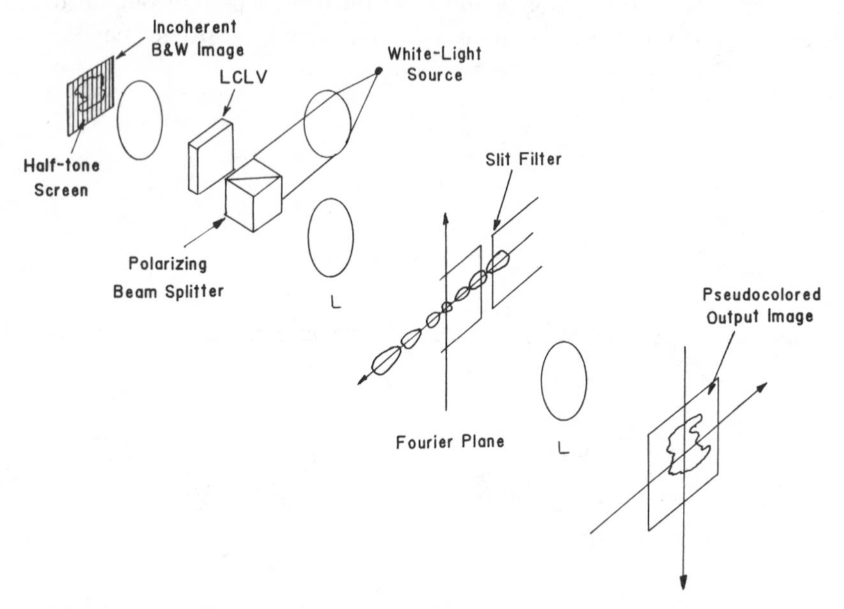

Figure 6.41. A real-time density pseudocolor encoding technique utilizing an LCLV.

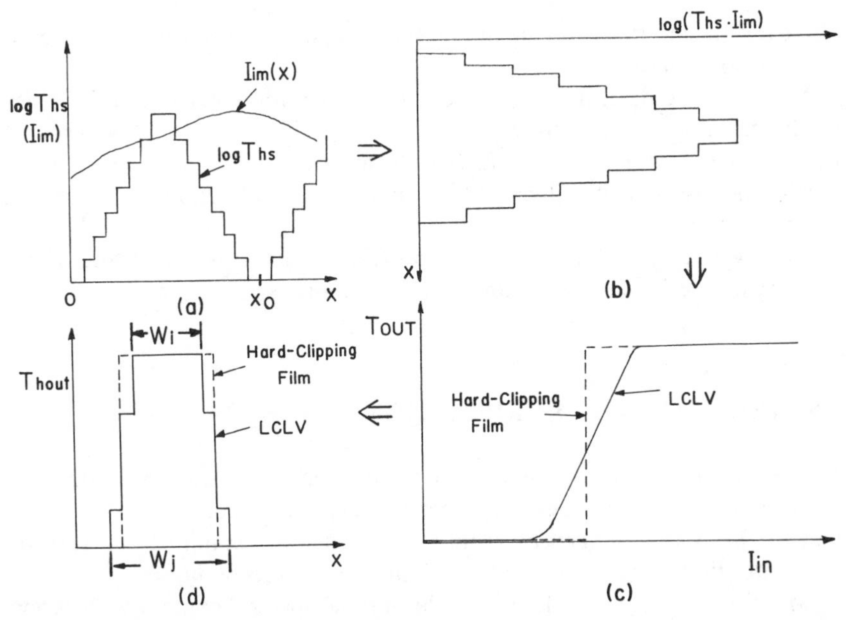

Figure 6.42. Transformation of halftone-LCLV encoding process: I_{im}, input image intensity; T_{hs}, halftone screen intensity transmittance; T_{out}, output intensity transmittance; T_{hout}, LCLV output encoded image. (*a*) Profiles of input irradiance and halftone intensity transmittance. (*b*) LCLV input intensity profile. (*c*) Transfer characteristic of the LCLV. (*d*) Resulting encoded output image.

First, if the density between adjacent levels in the halftone screen is small, the exposure range of the LCLV may extend over other density regions of the halftone screen. Thus within a single period of the screen, the effect can result in *m* overlapping concentric (along the *x*-axis) bars having widths w_i through w_j, where $0 < w_i < w_j < x_0$, as shown in Fig. 6.42. Since the contrast between the resulting adjacent regions is small, the additional contribution in the Fourier plane is distributed among *m* diffraction orders. The example illustrated in Fig. 6.42 shows the halftone encoding process within one screen cycle beginning with the input image intensity and halftone screen transmittance in Fig. 6.42*a*, the approximate corresponding LCLV input intensity profile in Fig. 6.42*b*, the LCLV input-output intensity characteristic in Fig. 6.42*c*, and the resulting LCLV output image in Fig. 6.42*d*. However, for most pseudocolor encoding applications, the number of halftone screen levels required to produce a satisfactory pseudocolored output image is quite small (e.g., three or four levels). The transmittance levels can then be chosen to give the appearance that the LCLV behaves more like a hard-clipping device. This also eases the limitation on maximum spatial frequency of the image imposed by the resolution of the LCLV.

Second, the output transmittance of the LCLV for a given input intensity is a function of the read-out wavelength. This problem can easily be resolved by inserting a color balancing filter in front of the white-light source or by placing neutral density filters at the appropriate diffraction orders in the Fourier plane.

Thus we see that the LCLV can possibly be used as a real-time hard-clipping device for color encoding with a white-light processor.

6.6. COLOR IMAGE CORRELATION

Unlike the human perception system, the conventional coherent optical image recognition system does not, in general, exploit the color or spectral content of the object under observation. Nonetheless, exploiting the spectral content of the image would further enhance pattern recognition.

Mention should be made that exploiting the spectral content with coherent processing has been reported: by Shi [6.41], using a volume hologram matched filter; by Case [6.42], Ishii and Murata [6.43], and Braunecker and Bryngdahl [6.44], utilizing a wavelength multiplexed matched filter; by Yu and Chao [6.45], with a grating based polychromatic coherent processor; and more recently by Warde et al. [6.46] and Ludman et al. [6.47], with a real-time joint spectral-spatial matched filtering method. In this section, we describe a white-light optical correlation technique that would exploit the spectral content of the object under observation [6.48]. In other words, the proposed white-light correlator is capable of recognizing multiple color objects of different shapes. The field of color-pattern recognition is very diverse and offers many applications. There are, however, two large categories where color is extremely important: natural color variations and objects deliberately colored for identification, for example in robotic vision applications.

6.6.1. Broad-Band Matched Filter Synthesis

We first describe a technique of generating a broad spectral band matched filter with white-light processing, as schematized in Fig. 6.43. A high diffraction efficient phase grating is used to disperse the spectral content of the light source into rainbow color at the back focal plane of the achromatic lens L_0. To alleviate the constraint of recording medium due to limited dynamic range, we spatially sample the smeared spectral light source with periodic pinholes such that a sequence of partially coherent sources for various spec-

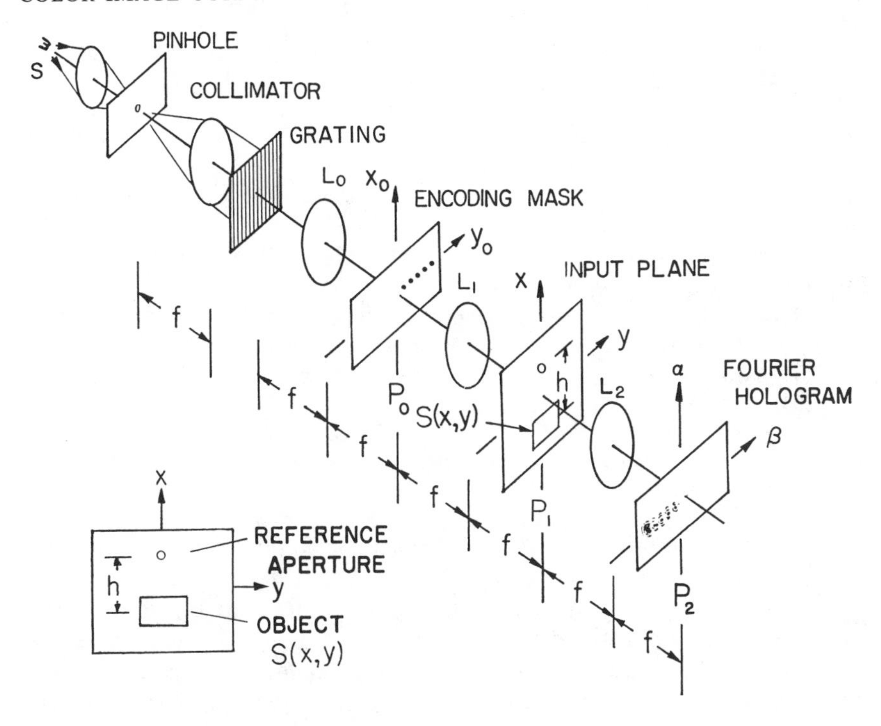

Figure 6.43. A technique of generating a broad spectral band Fourier hologram with a white-light source. *S*, white-light source.

tral colors can be obtained at the source plane P_0. The intensity distribution of these encoded sources, for various λ_n, would be

$$\gamma(x_0, y_0) = K \sum_{n=1}^{N} \operatorname{cir}_n\left[\frac{|x_0, y_0 - f v_0 \lambda_n|}{\Delta r_n}\right], \tag{6.63}$$

where K is a proportionality constant,

$$\operatorname{cir}_n\left[\frac{r}{\Delta r_n}\right] \triangleq \begin{cases} 1, & 0 \leqq r \leqq \Delta r_n \\ 0, & \text{otherwise,} \end{cases}$$

Δr_n is the radius of the nth sampling pinhole, f is the focal length of the achromatic lens, v_0 is the spatial frequency of the phase grating, λ_n is the main wavelength within the nth sampled source, and (x_0, y_0) is the spatial coordinate system of the source plane P_0. To achieve the required spatial coherence at the input plane P_1, it is apparent that the size of the sampling pinholes should be small compared with the overall input object, such as

$$\Delta r_n < \frac{f\lambda_n}{h}, \tag{6.64}$$

where h denotes the separation between the reference point and the input object $s(x, y)$, as shown in Fig. 6.43.

To maintain a high temporal coherence for matched filter synthesis, it is required that the wavelength spread $\Delta\lambda_n$ over the nth sampling aperture satisfy the following inequality:

$$\frac{\Delta\lambda_n}{\lambda_n} \ll \frac{\Delta\nu}{\nu_0}, \tag{6.65}$$

where $\Delta\nu$ is the spatial frequency bandwidth of the input object $s(x, y)$ and ν_0 is the spatial frequency of the phase grating.

Thus we see that a sequence of encoded partially coherent sources are illuminating the input plane P_1. Since the source encoding mask covers a wide range of the spectrum of the smeared white-light source, it is apparent that N discrete matched filters, for a broad spectral band, can be synthesized. Let us now assume that the input plane contains a target transparency and a reference pinhole, as described in the following equation:

$$f(x, y) = s\left(x + \frac{h}{2}, y\right) + \delta\left(x - \frac{h}{2}, y\right), \tag{6.66}$$

where $s(x, y)$ denotes an input object (i.e., target) function, $\delta(x, y)$ denotes the Dirac delta function, and h is the main separation between the target and the reference pinholes. It is therefore apparent that a broad spectral band matched filter can be interferometrically generated in the Fourier plane P_2. If we assume that the filter recording is in the linear region of the recording emulsion, the amplitude transmittance distribution of the matched filter can be approximated by

$$H(\alpha, \beta) \simeq \sum_{\substack{n=-N \\ n\neq 0}}^{N} \left\{K_1|S(\alpha, \beta + f\nu_0\lambda_n)|^2 + K_2|S(\alpha, \beta + f\nu_0\lambda_n)| \cdot \cos\left[\frac{2\pi h}{\lambda_n f}\alpha + \phi(\alpha, \beta + f\nu_0\lambda_n)\right]\right\}, \tag{6.67}$$

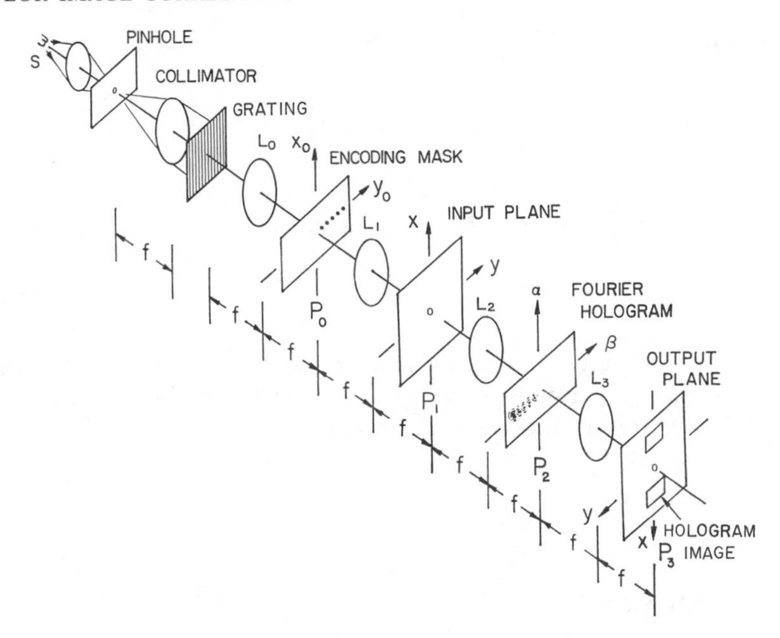

Figure 6.44. Fourier hologram image reconstruction with white-light processing.

where K_1 and K_2 are arbitrarily proportional constants, $S(\alpha, \beta; \lambda_n)$ is the Fourier spectrum of input target $s(x, y; \lambda_n)$, (α, β) is the spatial coordinate system of the Fourier plane, λ_n is the main wavelength of the nth sampled partial coherent source, f is the focal length of the achromatic transform lens L_2, and ν_0 is the spatial frequency of the grating.

It is interesting to note that if the broad spectral band matched filter $H(\alpha, \beta)$ (i.e., a broad-band Fourier hologram) is illuminated by a beam of encoded white-light source (shown in Fig. 6.44) such as

$$E(\alpha, \beta; \lambda) = C \sum_{\substack{n=-N \\ n \neq 0}}^{N} \operatorname{cir}_n[(\alpha, \beta + f\nu_0\lambda_n; \lambda)/\Delta\rho_n], \tag{6.68}$$

where C is a complex proportionality constant, $\Delta\rho_n = 1.22\lambda_n f/\Delta\gamma$, $\Delta\gamma$ is the radius of the input pinholes in Plane P_1, $\Delta\gamma = 1/\Delta\nu$, and $\Delta\nu$ is the spatial bandwidth of the object, then the complex light field at the output plane would be

$$g(x, y; \lambda) = \iint E(\alpha, \beta; \lambda)H(\alpha, \beta)\exp\left[-i\frac{2\pi}{\lambda f}(\alpha x + \beta x)\right]d\alpha\, d\beta. \tag{6.69}$$

By substituting Eqs. (6.67) and (6.68) into Eq. (6.69), we have

$$g(x, y; \lambda) = \sum_{n=1}^{N} C_1 \delta(x, y; \lambda) * s(x, y; \lambda_n) * s^*(-x, -y; \lambda_n)$$
$$+ \sum_{n=1}^{N} C_2 \delta(x, y; \lambda) * [s(x + h, y; \lambda_n) + s^*(-x + h, -y; \lambda_n)], \tag{6.70}$$

where C_1 and C_2 are complex constants, $*$ denotes the convolution operation, and the superscript $*$ represents the complex conjugation. The corresponding output irradiance can be shown as

$$I(x, y) = \int_{\Delta\lambda_n} |g(x, y; \lambda)|^2 \, d\lambda, \qquad \text{for } n = 1, 2, \ldots, N, \tag{6.71}$$

which is approximated by

$$I(x, y) \approx \Delta\lambda_n \sum_{n=1}^{N} \{K_1 s(x, y; \lambda_n) \circledast s^*(x, y; \lambda_n)$$
$$+ K_2[|s(x + h, y; \lambda_n)|^2 + |s(x - h, y; \lambda_n)|^2]\}, \tag{6.72}$$

where K_1 and K_2 are proportionality constants, and $\circledast$ denotes the correlation operation. From this equation we therefore see that two hologram images of $s(x, y)$ would be diffracted around $x = \pm h$ at the output image plane of Fig. 6.44.

6.6.2. Broad-Band Image Correlation

To illustrate the application of the broad-band matched filter of Eq. (6.67) to complex signal detection, we insert a signal target transparency $s(x, y)$ at the input plane of the white-light optical processor of Fig. 6.45. The complex light distribution impinging on the broad-band matched filter $H(\alpha, \beta)$ would be

$$E(\alpha, \beta; \lambda_n) = CS(\alpha, \beta + f v_0 \lambda_n), \tag{6.73}$$

which represents a smeared signal spectra over the broad-band matched filter, where λ_n is the main wavelength of this nth partial coherent source. Thus

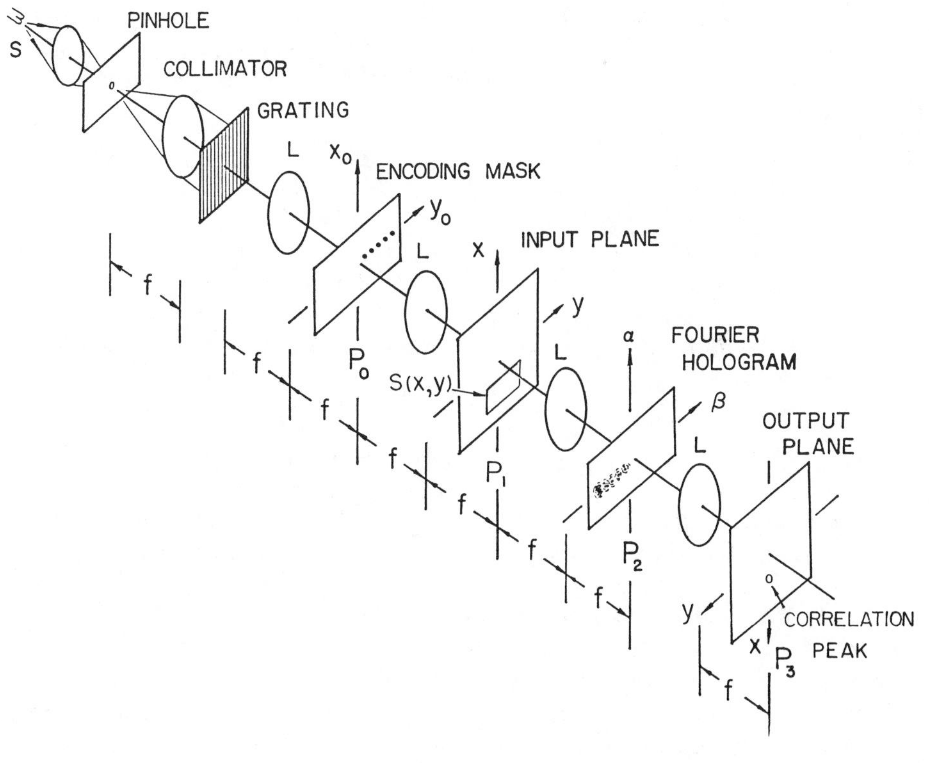

Figure 6.45. A white-light optical correlator.

the complex light distribution at the output plane P_3 would be

$$g(x, y; \lambda_n) = \iint E(\alpha, \beta; \lambda_n) H(\alpha, \beta) \exp\left[-\frac{2\pi}{\lambda f}(\alpha x + \beta x) \right] dx\, dy,$$

which can be shown to be

$$g(x, y; \lambda_n) = \sum_{n=1}^{N} C_1 s(x, y; \lambda_n) * s(x, y; \lambda_n) * s^*(-x, -y; \lambda_n)$$
$$+ \sum_{n=1}^{N} C_2 s(x, y; \lambda_n) * \left[s\left(x + \frac{h}{2}, y; \lambda_n \right) + s^*\left(-x + \frac{h}{2}, -y; \lambda_n \right) \right], \tag{6.74}$$

where C_1 and C_2 are complex proportionality constants, $*$ denotes the convolution operation, the superscript $*$ represents the complex conjugate, and λ_n is the main wavelength of the nth-order matched filter. The corresponding output irradiance is therefore

$$\begin{aligned} I(x, y) &= \int_{\Delta\lambda_n} |g(x, y; \lambda_n)|^2 \, d\lambda \\ &\simeq \sum_{n=1}^{N} \Delta\lambda_n \{K_1 s(x, y; \lambda_n) * s(x, y; \lambda_n) \circledast S^*(x, y; \lambda_n) \\ &\quad + K_2[s(x, y; \lambda_n) * s(x + h, y; \lambda_n) + s(x, y; \lambda_n) \circledast s^*(x - h, y; \lambda_n)]\}, \end{aligned} \tag{6.75}$$

where $\circledast$ denotes the correlation operation. From this equation, we see that an autocorrelation of the signal target is diffracted at $x = h$ at the output plane, as illustrated in Fig. 6.45. Since we utilized a broad spectral band white-light source, the spectral content of the target can be exploited.

6.6.3. Experimental Demonstrations

We now provide a couple of experimental results obtained with the proposed white-light optical correlator. In our experiments, a 75-W xenon arc lamp with a 1-mm square pinhole is used as a white-light source. A sinusoidal phase grating of about 110 lines/mm, with 25% diffraction efficiency for the first-order diffraction, is used to disperse the spectral content of the light source. The achromatic transform lenses used have a focal length of about 750 cm. The source encoding mask is composed of a set of periodic circular pinholes about 40 μm in diameter; diameter spaced about 2.5 mm apart. The separation between the reference aperture and the input target object is about 8 mm. The size of the reference aperture is roughly 300 μm.

We first show a broad-band Fourier hologram image reconstruction obtained with this technique. Figure 6.46*a* shows the continuous tone input object with a reference aperture. Figure 6.46*b* shows an enlarged photograph of a broad-band Fourier hologram generated with this technique. Figure 6.46*c* shows a twin hologram image reconstruction obtained at the output plane. The quality of this result is still far from acceptable; the image is, however, the first white-light Fourier hologram image reconstructed with this technique. Compared with the original object, the features of the input object image can readily be recognized.

Our next example shows the result obtained when this proposed technique was applied to character recognition. Figure 6.47*a* shows a Chinese character

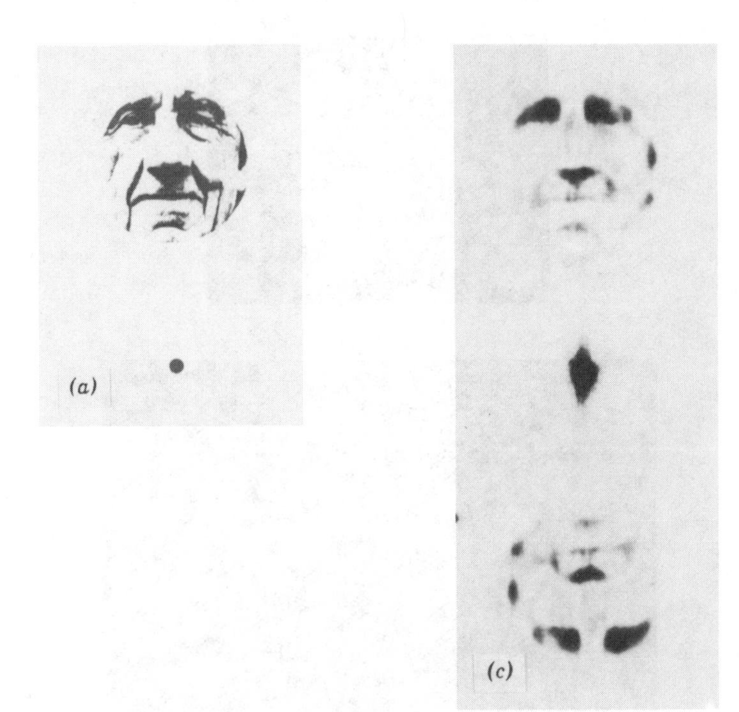

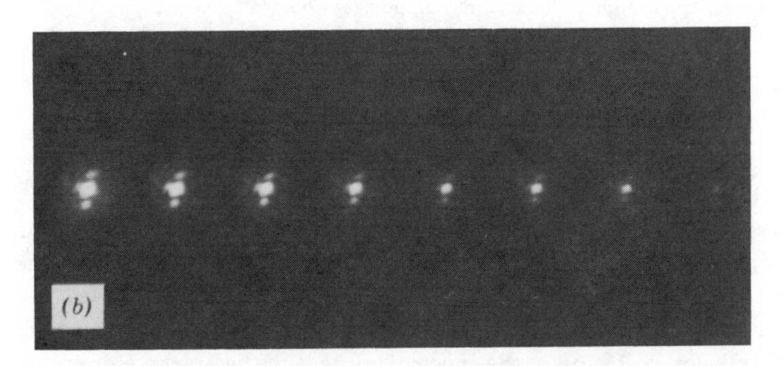

Figure 6.46. Fourier hologram image reconstruction. (*a*) Input object with a reference aperture. (*b*) A broad-band Fourier hologram. (*c*) Fourier hologram images.

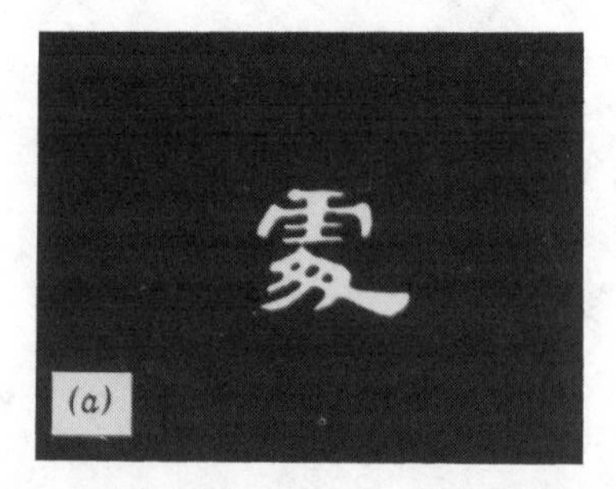

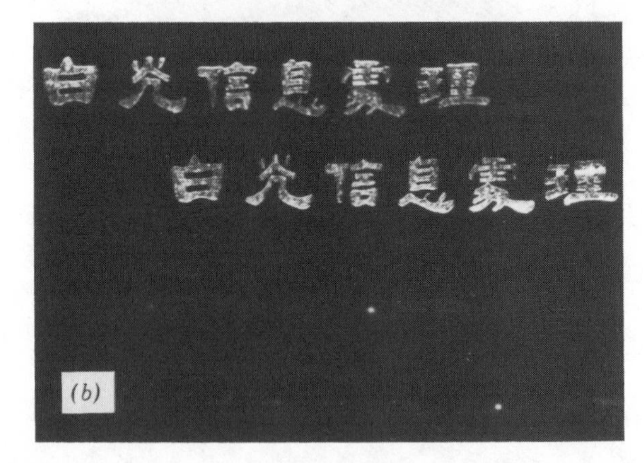

Figure 6.47. Application to character recognition. (*a*) Character to be recognized. (*b*) Character recognition through matched filtering. The correlation peaks identify where the character is located.

that we have used as a target object for the matched filter synthesis. Figure 6.47*b* shows the result obtained with this white-light correlator. Two autocorrelation peaks can readily be visualized in this figure, from which the locations of the character can be identified.

One of the important applications of the white-light optical correlator must be the exploitation of the spectral content of the object. Here we provide a result for color object recognition obtained with this technique. Figure 6.48*a* shows an input color transparency of an aerial photograph with a guided missile in flight. A broad-band matched filter of the missile is constructed, as described previously. Figure 6.48*b* shows a visible red correlation spot obtained at the output plane of the correlator. Thus the proposed white-light correlator has a unique feature for color pattern recognition based on both the spectral content and shape of an input object.

Figure 6.48. Color object identification. (*a*) An aerial photographic transparency. (*b*) The red correlation spot identifies the location and the spectral content of the missile.

REFERENCES

6.1. F. T. S. Yu, "Source encoding, signal sampling and spectral band filtering for partially coherent optical signal processing," *J. Opt.* **14**, 173 (1983).

6.2. S. T. Wu and F. T. S. Yu, "Source encoding for image subtraction," *Opt. Lett.* **6**, 652 (1981).

6.3. P. H. Van Cittert, "Die Wahrschcinlicke Schwingungs verteilung in einer von einer licht-quelle direkt Oden Mittels einer linse," *Physica* **1**, 201 (1934).

6.4. F. Zernike, "The concept of degree of coherence and its application to optical problems," *Physica* **5**, 785 (1938).

6.5. S. H. Lee, S. K. Yao, and A. G. Milnes, "Optical image synthesis (complex amplitude addition and subtraction) in real time by a diffraction-grating interferometric method," *J. Opt. Soc. Am.* **60**, 1037 (1970).

6.6. T. H. Chao, S. L. Zhuang, S. Z. Mao, and F. T. S. Yu, "Broad spectral band color image deblurring," *Appl. Opt.* **22**, 1439 (1983).

6.7. J. Tsujiuchi, "Correction of optical images by compensation of aberrations and spatial frequency filtering," in *Progress in Optics*, Vol. II, E. Wolf, Ed., North-Holland Publishing Company, Amsterdam, 1963.

6.8. G. W. Stroke and R. G. Zech, "A posteriori image-correcting deconvolution by holographic Fourier-transform division," *Phys. Lett.*, Ser. A **25**, 89 (1967).

6.9. A. W. Lohmann and D. P. Paris, "Computer generated spatial filters for coherent optical data processing," *Appl. Opt.* **7**, 651 (1968).

6.10. J. Tsujiuchi, T. Honda, and T. Fukaya, "Restoration of blurred photographic images by holography," *Opt. Commun.* **1**, 379 (1970).

6.11. J. L. Horner, "Optical spatial filtering with the least mean-square-error filter," *J. Opt. Soc. Am.* **59**, 553 (1969).

6.12. J. L. Horner, "Optical restoration of images blurred by turbulence using optimum filter theory," *Appl. Opt.* **8**, 167 (1970).

6.13. G. G. Yang and E. N. Leith, "An image deblurring method using diffraction gratings," *Opt. Commun.* **36**, 101 (1981).

6.14. S. L. Zhuang, T. H. Chao, and F. T. S. Yu, "Smeared photographic image deblurring utilizing white-light processing technique," *Opt. Lett.* **6**, 102 (1981).

6.15. F. T. S. Yu, S. L. Zhuang, and T. H. Chao, "Color-photographic-image deblurring by white-light processing technique," *J. Opt.* **13**, 57 (1982).

6.16. F. T. S. Yu, *Optical Information Processing*, Wiley-Interscience, New York, 1983, pp. 198–202.

6.17. D. Gabor, G. W. Stroke, R. Restrick, A. Funkhouser, and D. Brumm, "Optical image synthesis (complex amplitude addition and subtraction) by holographic Fourier transformation," *Phys. Lett.* **18**, 116 (1965).

6.18. K. Bromley, M. A. Monahan, J. F. Bryant, and B. J. Thompson, "Complex spatial filtering by holographic Fourier subtraction," *Appl. Phys. Lett.* **14**, 67 (1969)

6.19. K. Bromley, M. A. Monahan, J. F. Bryant, and B. J. Thompson, "Holographic subtraction," *Appl. Opt.* **10**, 174 (1971).

6.20. J. F. Ebersole, "Optical image subtraction," *Opt. Eng.* **15**, 436 (1975).

6.21. F. T. S. Yu and S. T. Wu, "Color image subtraction with extended incoherent sources," *J. Opt.* **13**, 183 (1982).

6.22. H. E. Ives, "Improvement in the diffraction process of color photography," *Br. J. Photog.* 609 (1906).

6.23. P. F. Mueller, "Color image retrieval from monochrome transparencies," *Appl. Opt.* **8**, 2051 (1969).

6.24. A. Macovski, "Encoding and decoding of color information," *Appl. Opt.* **11**, 416 (1972).

6.25. R. Grousson and R. S. Kinany, "Multi-color image storage on black and white film using a crossed grating," *J. Opt.* **9**, 333 (1978).

6.26. F. T. S. Yu, "White-light processing technique for archival storage of color films," *Appl. Opt.* **19**, 2457 (1980).

6.27. F. T. S. Yu, X. X. Chen, and S. L. Zhuang, "Progress report on archival storage of color films utilizing a white-light processing technique," *J. Opt.*, to be published.

6.28. J. Upatnieks and C. Leonard, "Diffraction efficiency of bleached, photographically recorded interference patterns," *Appl. Opt.* **8**, 85 (1969).

6.29. B. J. Chang and K. Winick, "Silver-halide gelatin holograms," *SPIE* **215**, 172 (1980).

6.30. H. M. Smith, "Basic holographic principles," in *Holographic Recording Materials*, H. M. Smith, Ed., Springer-Verlag, New York, 1977.

6.31. H. C. Andrews, A. B. Tescher, and R. P. Kruger, "Image processing by digital computer," **9**, 20 (1972).

6.32. H. K. Liu and J. W. Goodman, "A new coherent optical pseudocolor encoder," *Nouv. Rev. Opt.* **7**, 285 (1976).

6.33. A. Tai, F. T. S. Yu, and H. Chen, "White-light pseudocolor density encoder," *Opt. Lett.* **3**, 190 (1978).

6.34. J. Santamaria, M. Gea, and J. Bescos, "Optical pseudocoloring through contrast reversal filtering," *J. Opt.* **10**, 151 (1979).

6.35. T. H. Chao, S. L. Zhuang, and F. T. S. Yu, "White-light pseudocolor density encoding through contrast reversal," *Opt. Lett.* **5**, 230 (1980).

6.36. J. A. Mendez and J. Bescos, "Gray level pseudocoloring with three primary colours by a diffraction grating modulation method," *J. Opt.* **14**, 69 (1983).

6.37. F. T. S. Yu, X. X. Chen, and T. H. Chao, "White-light pseudocolor encoding with three primary colors," *J. Opt.* **15**, 55 (1984).

6.38. T. D. McKinley, K. F. J. Heinrich, and D. B. Wittry, *The Electron Microprobe*, Wiley and Sons, New York, 1966.

6.39. F. T. S. Yu, S. L. Zhuang, T. H. Chao, and M. S. Dymek, "Real-time white-light spatial frequency and density pseudocolor encoder," *Appl. Opt.* **19**, 2986 (1980).

6.40. H. Kato and J. W. Goodman, "Nonlinear filtering in coherent optical systems through halftone screen processes," *Appl. Opt.* **14**, 1813 (1975).

6.41. H. K. Shi, "Color-sensitive spatial filter," *Opt. Lett.* **3**, 85 (1978).

6.42. S. K. Case, "Pattern recognition with wavelength-multiplexed filters," *Appl. Opt.* **18**, 1890 (1979).

6.43. Y. Ishii and K. Murata, "Color-coded character-recognition experiment with wavelength-triplexed, reflection-type holographic filters," *Opt. Lett.* **7**, 230 (1982).

6.44. B. Braunecker and O. Bryngdahl, "Multiplex optical processing—Combination of polychromatic light and a dispersive element," *Opt. Commun.* **40**, 332 (1982).

6.45. F. T. S. Yu and T. H. Chao, "Color signal correlation detection by matched spatial filtering," *Appl. Phys.* **B32**, 1 (1983).

6.46. C. Warde, H. J. Caulfield, F. T. S. Yu, and J. E. Ludman, "Real-time joint spectral-spatial matched filtering," *Opt. Commun.* **49**, 241 (1984).

6.47. J. E. Ludman, B. Javidi, F. T. S. Yu, H. J. Caulfield, and C. Warde, "Real-time colored-pattern recognition," *SPIE* **465**, 143 (1984).

6.48. F. T. S. Yu and F. K. Hsu, "A white-light optical correlator," *Appl. Opt.*, to be published.

7

Applications

The use of optical technology for operations upon signals has grown rather rapidly in the past two and a half decades, since the work of Cutrona et al. [7.1]. This situation arises in part from the availability of new electro-optic devices and components, and in part from the increased realization that optical systems provide a powerful, versatile, and compact solution for many signal processing operations. There have been volumes of literature describing the basic techniques and applications of optical signal processing and computing [7.2–7.8]. In these publications, it has been shown that optics can be used to perform such operations as Fourier transformation, cross-correlation, spectrum analysis, convolution, image deblurring, image subtraction, pseudocolor encoding, wide-band processing, and many others. In fact, many of the linear signal processing operations can be carried out by optical techniques, with relatively mild constraints. The major underlying fact is that the optical processing technique offers the advantages of complex amplitude filtering, parallel processing operation, large capacity, high speed, multiwavelength diversity, and color. Otherwise the optical processing techniques would have been useless.

We have, in the previous chapter, shown some of the basic techniques and applications of the proposed white-light optical signal processor. In this chapter we continue to discuss some of their applications.

7.1. VISUALIZATION OF PHASE OBJECTS

The study of phase objects generally depends upon interferometric techniques [7.9, 7.10]. For slow object phase variation, the measurements can be very accurate with such techniques [7.11]. However, phase objects that contain fine structure, such as living cells, are exceedingly difficult, if not impossible, to study with the interferometric techniques. Frequently, the study of this type of phase object requires a special procedure to produce an image irradiance that is related to the phase of the object.

The Schlieren technique [7.12] and differential shearing interference microscopy [7.13, 7.14] both produce image irradiances that are proportional to the derivative of the object phase. Although the differential techniques provide visualization of shapes and sizes of the phase objects, they are not able to detect the details of phase variation between fringes. The total shearing interference microscope is, however, able to produce fringe patterns superimposed on a phase object, but the application is limited to evaluation of simple isolated phase structures, and it is difficult to apply to the more general case of phase variation.

There is a technique available for detecting a phase object whose image irradiance is proportional to the phase variation [7.15]. However, this technique also suffers one major drawback: The application is limited to very small phase variations in the order of a fraction of a wavelength.

The concept of detecting a large phase variation with differentiation and integration to obtain an image irradiance proportional to phase variation was proposed by DeVelis and Reynolds [7.12]. This concept has been subsequently tested by Spraque and Thompson [7.16] with a coherent optical processing technique. They achieved good results, obtaining an image irradiance that was proportional to the large object phase variation. However, the technique is rather elaborate and the phase detection is not a real-time operation. Moreover, the system utilizes a coherent source, and the annoying coherent artifact noise cannot be avoided. Although a multibeam interferometric technique [7.10] is able to produce sharper fringe patterns, it also suffers drawbacks. The technique still cannot provide the phase variation between the fringes with a single monotone fringe pattern. It is capable of detecting the phase variation between the fringes only by simultaneously changing the viewing angle and the reference beam, or the wavelength of the light source. In a recent paper, Roblin and El Sherif [7.17] describe an electronic scanning technique, based on phase modulation interferometry, for detecting phase variation between the fringes. But this scanning technique requires complicated electronic circuitry and the operation is rather cumbersome.

In Section 6.1 we have shown a technique of encoding an extended incoherent source for image subtraction. We extend the same basic optical processing concept to the detection of object phase variation with a pseudocolor encoding technique [7.18]. Since this technique utilizes incoherent sources, high quality color coded phase variation patterns can be visualized. We stress that this incoherent processing technique may alleviate certain drawbacks inherent to previous techniques, and the processing system is rather simple and economical to operate.

7.1.1. Object Phase Detection

Let us now describe an incoherent optical processing technique for object phase detection, as depicted in Fig. 7.1. From this figure we see that two

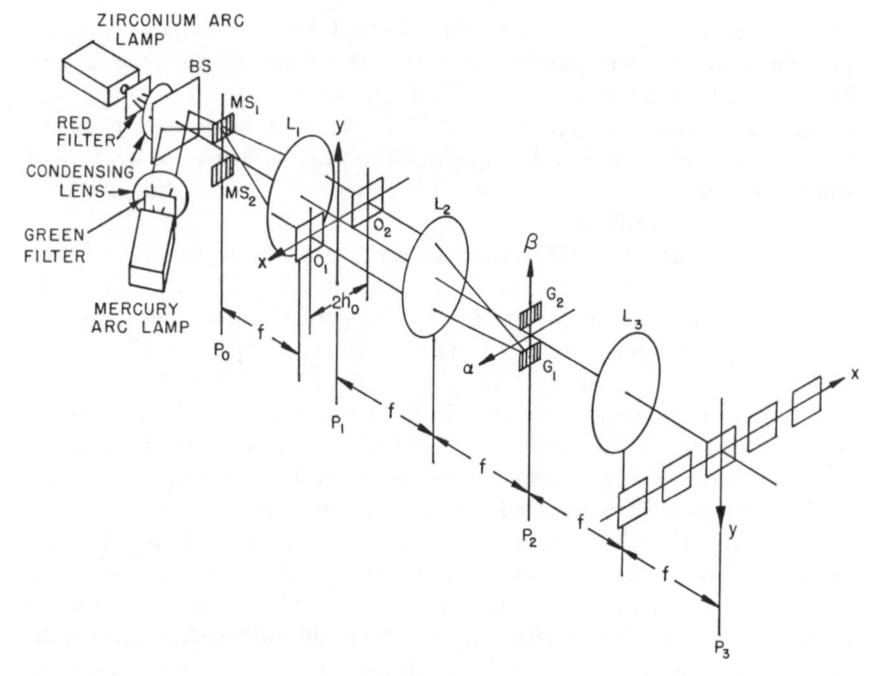

Figure 7.1. A multicolor incoherent processing system for detecting object phase variation. *BS*, beam splitter; MS_1 and MS_2, source encoding masks; L_1, L_2, and L_3, achromatic transform lenses; G_1 and G_2, diffraction gratings.

encoded extended incoherent sources for different colors of light (i.e., red and green) are used for the processing. The purpose of using a source encoding mask to achieve the point-pair spatial coherence requirement for an image subtraction operation has been discussed in the previous chapter.

In pseudocolor encoding we insert a phase object $\exp[i\phi(x, y)]$ in one of the open apertures in the input plane P_1. Two narrow strip sinusoidal gratings G_1 and G_2, with respect to the encoded red and green lights, are used in the spatial frequency plane P_2:

$$G_1 = 1 + \sin(h_0 p_r + \theta) \tag{7.1}$$

and

$$G_2 = 1 + \sin(h_0 p_g), \tag{7.2}$$

where $2h_0$ is the separation of the O_1 and O_2, $p_r = (2\pi\alpha/\lambda_r f)$ and $p_g = (2\pi\alpha/\lambda_g f)$ are the spatial frequencies of the gratings, λ_r and λ_g are the red and green wavelengths, α denotes the spatial coordinate in the same direction as p, f is the focal length of the achromatic transform lens, and θ is a phase factor that we have introduced. θ will play an important role in the color encoding process.

By a straightforward computation, the irradiance around the origin of the output image plane P_3 can be shown to be

$$\begin{aligned} I(x, y) &= I_r(x, y) + I_g(x, y) \\ &= K\{1 - \cos[\phi_r(x, y) + \theta]\} + K\{1 - \cos[\phi_g(x, y)]\}, \end{aligned} \tag{7.3}$$

where I_r and I_g denote the red and green image irradiances, $\phi(x, y)$ is the phase distribution of the object, θ is a constant phase factor that we have introduced (by shifting one of the gratings), and K is a proportional constant. The phase distributions $\phi_r(x, y)$ and $\phi_g(x, y)$ are slightly different because of different wavelength illuminations. From Eq. (7.3) we see that a broad range color coded phase fringe patterns can be visualized.

In color mixing analysis, we would use I_r and I_g to form a two-dimensional orthogonal coordinate system, as shown in Fig. 7.2. The variations of color mixing irradiance as a function of object phase variation θ for various values of θ are plotted. Since different wavelengths (i.e., red and green) are used in the color encoding, the color mixing curve is not generally enclosed for every 2π cycle as a function θ, as shown in the figure. In other words, the phase detection is more accurate within $-\pi \leqq \phi \leqq \pi$.

Consider a few cases of the color mixing procedure. For the example of $\theta = 0°$, the locus of the color mixing curve lies near the 45° angle of the (I_r, I_g) coordinates, as shown in Fig. 7.2*a*. There is no significant color change in this region to call for distinction of object phase variation. On the other hand, if $\theta = 180°$, the locus of the color mixing curve lies near the $-45°$ degree region (i.e., $I_r = 1 - I_g$) of the (I_r, I_g) coordinate system, and a broad range of color variation can be perceived. However, in this region it is difficult to distinguish a positive or a negative variation (i.e., $\pm\phi$). If we let $\theta = 90°$, the locus of the color mixing curve extends outward around the edges of the (I_r, I_g) coordinate system, as shown in Fig. 7.2*b*. Thus it not only provides a broad range color coded phase variation, but also establishes two different sequences of color coded bands so that the positive and negative phase variations can be detected. In other words, this color encoding technique is capable of detecting finer detail for phase variations, including positive and negative phases.

7.1.2. Experimental Demonstrations

In experiments a mercury arc lamp with a green filter (5461 Å) and a zirconium arc lamp with a red filter (6328 Å) are used for the color light sources. The

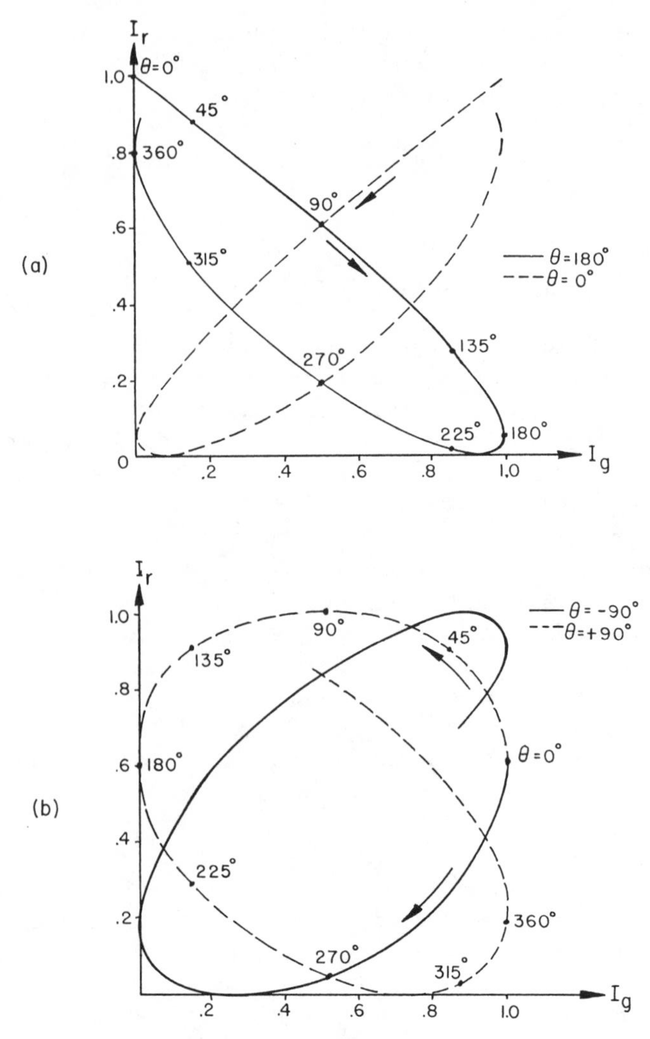

Figure 7.2. Color encoding mixing curves.

intensity ratio of these two color lights is adjusted to unity with a variable beam splitter.

Figure 7.3*a* shows a monotone fringe pattern of a phase object obtained with this incoherent processing technique. From this figure we see that detailed phase variation between the fringes is not perceptible. Figure 7.3*b* is obtained with a multicolor technique described earlier. The phase factor θ between the diffraction gratings G_1 and G_2 is set to about 90°. From this

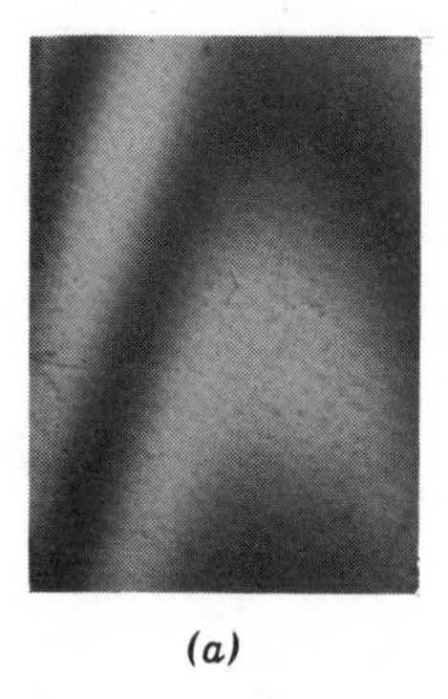

(a)

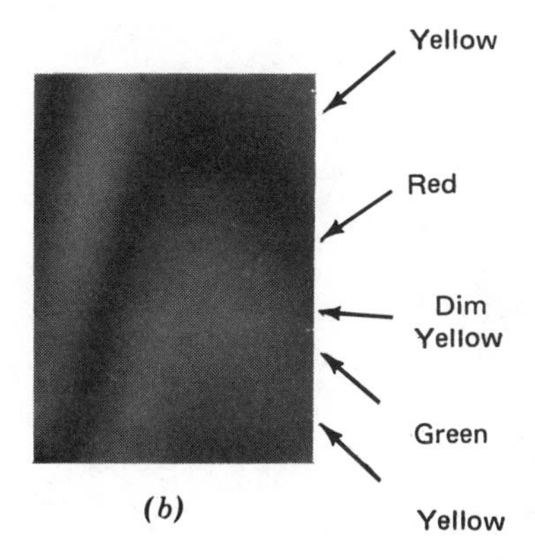

(b)

Figure 7.3. A color coded object phase variation. (*a*) Monotone phase fringe pattern. (*b*) A black-and-white picture of a color coded fringe pattern.

figure we see that a multicolor phase fringe pattern is formed so that a more detailed phase variation between fringes can be observed.

Between the two yellow bands there is a color variation from yellow to red to dim yellow to green and back to yellow again. The phase angle between these two color bands is 2π. These color variations correspond to the color mixing curve of Fig. 7.2*b*; the corresponding phase variations with respect to the color bands are $0° \rightarrow 180° \rightarrow 225° \rightarrow 315° \rightarrow 360°$, as indicated in Fig. 7.3*b*. Thus we see that there is a positive increasing phase variation from the top yellow band to the low yellow band. In another experiment, Figure 7.4*a* shows a monotone phase fringe pattern obtained with this technique from a phase object; again the detailed phase variation between the fringes cannot be determined. If we take the cross section along with line $A - A$ of the phase object between the two arrows, there are four possible phase variations that may be interpreted, as shown in Fig. 7.5. It may not be possible to retrieve the actual object phase variation with a monotone fringe pattern. However, with

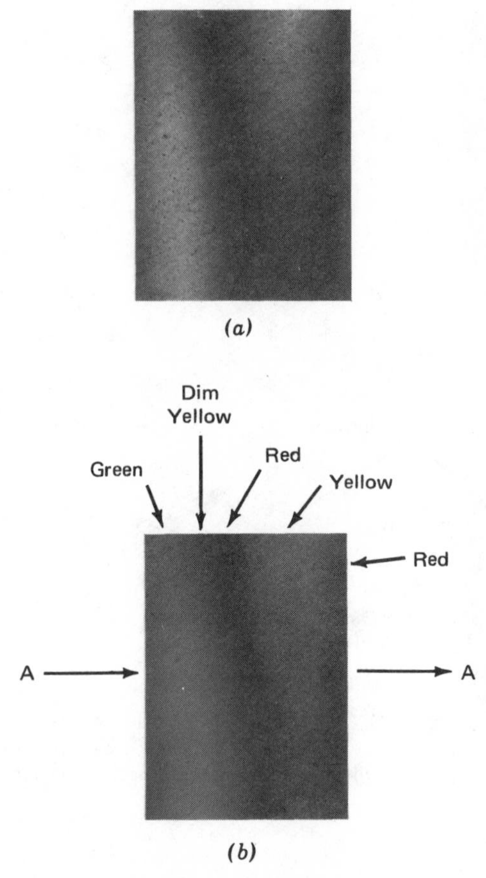

Figure 7.4. A color coded phase object (*a*) Monotone fringe pattern. (*b*) A black-and-white picture of a color coded fringe pattern.

the multicolor phase fringe pattern shown in Fig. 7.4*b*, we are able to identify the detailed phase variation between the fringes. If we take the same cross section $A - A$, a sequence of color bands from left to right, that is, green to dim yellow to red to yellow to red, can be observed. Corresponding to the color mixing curve of Fig. 7.2*b*, this sequence of color bands represents a sequence of phase angles of $315° \rightarrow 250° \rightarrow 180° \rightarrow 90° \rightarrow 180°$. Thus this region represents a phase depression object that corresponds to the case of Fig. 7.5*a*.

Object phase variation can be visually studied by a color coding technique with incoherent processing. This technique is accomplished with a complex image subtraction scheme using a partially coherent processing method. The

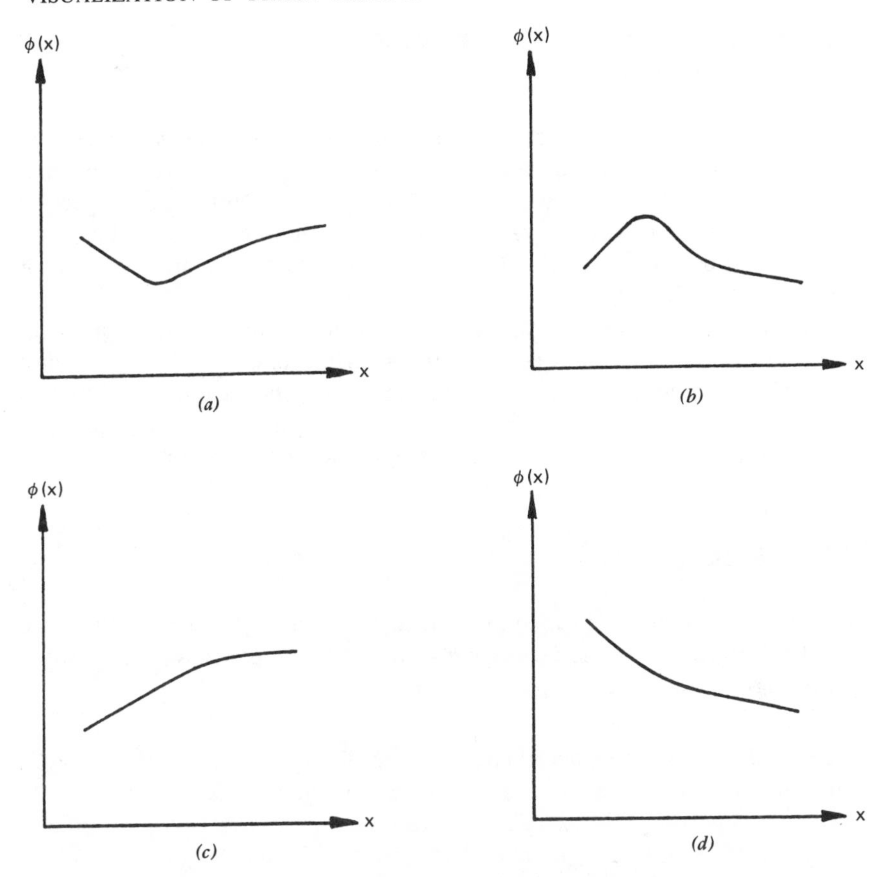

Figure 7.5. Four possible object phase variations for Fig. 7.4*a*.

object phase distribution within $-\pi \leqq \phi \leqq \pi$ can be detected with a color coded phase variation with lesser ambiguity. This technique offers more finely detailed object phase detection than other interferometric techniques. Since it uses inexpensive incoherent sources, the multicolor technique may offer a wider range of applications. In the case of microscopic phase object observation, the system can be designed to fit within a microscopic system. This technique may be used in other fields as well: study of birefringence and interference figures in polarizing microscopy and crystallography; distinction between compressed and stretched areas in photoelasticity; analysis of the aerodynamic pressure variations examined inside a wind tunnel by the stereoscopic technique; analysis of wavefronts to test the aberrations of an optical system; and others.

7.2. APPLICATIONS TO MICROCIRCUIT CHIP INSPECTION

In this section, we propose the use of a white-light optical image subtraction technique for automatic microcircuit chip inspection [7.19]. We believe that this proposed technique would have profound effect on automatic inspection schemes for fault detection and identification of microcircuitries. The technique could be used for rapid identification and inspection, and possibly for synthesis and fabrication.

We have, in Section 3.3, discussed the coherence requirement for image subtraction. With the application of Wolf's partial coherence theory, we have obtained the fundamental and second harmonic modulation transfer function (MTF) for the incoherent image subtraction system [i.e., Eqs. (3.44) and (3.45)]. We utilize these results to discuss the resolution limit of the proposed white-light image subtraction system.

7.2.1. Resolution Limit

It is known that the resolution limit is affected by the coherence and the space bandwidth product of the processing system. We discuss these effects in the following.

7.2.1.1. Effect Due to Coherence. We first discuss the effect on resolution limit due to temporal coherence. We note that for the case of strict spatial coherence, the slit width w of the source encoding mask should approach zero. The MTF of Eqs. (3.44) and (3.45) can be written as

$$\mathrm{MTF}(\nu) = \mathrm{sinc}\left(\pi\nu h \frac{\Delta\lambda}{\lambda}\right), \tag{7.4}$$

and

$$\mathrm{MTF}(2\nu) = \mathrm{sinc}\left(2\pi\nu \mathrm{h} \frac{\Delta\lambda}{\lambda}\right), \tag{7.5}$$

where $2h$ is the main separation of the two input object transparencies, λ is the center wavelength (equivalent to λ_n), and $\Delta\lambda$ is the corresponding wavelength spread (i.e., $\Delta\lambda_n$). The normalized MTF curves of the fundamental and second harmonic frequency are plotted in Fig. 3.7. It is obvious that the contrast of the subtracted image decreases monotonically as a function of the input object spatial frequency, for a given separation h. The MTF of the subtracted image also decreases as the spectral bandwidth of the light source. In other words, the quality of the subtracted image improves for narrower spectral bandwidth $\Delta\lambda$ and for lower spatial frequency of the object.

Let us define ν_c as the resolution limit of the subtracted image, where the MTF decreases to a minimum value C_m, as shown in Fig. 3.7. Figure 3.8 shows the functional relationship of the resolution limit ν_c and the spectral width $\Delta\lambda$ for various values of C_m. We now determine the spectral bandwidth requirement $\Delta\lambda$ from this figure. The relationship between the resolution limit ν_c, the spectral width $\Delta\lambda$, and the separation h of the input object transparencies is depicted in Fig. 3.9. We see that the resolution limit decreases monotonically as the spectral bandwidth and the separation of input objects increase. In order to have a feeling of magnitude, Table 7.1 illustrates the relation of ν_c, h, and $\Delta\lambda$, where the center wavelength λ is assumed to be 5461 Å and the MTF(ν) = 0.3. Thus we see that, to obtain a high resolution of the subtracted image, a narrow spectral bandwidth $\Delta\lambda$ is required.

Table 7.1. Effect of Resolution upon $\Delta\lambda$ and h, $\lambda = 5461$ Å

	ν_c(lines/mm)			
h(mm)	$\Delta\lambda = 50$ Å	$\Delta\lambda = 25$ Å	$\Delta\lambda = 10$ Å	$\Delta\lambda = 5$ Å
6.0	14	30	68	137
4.0	20	41	102	205
2.0	41	82	205	410

We now consider the effect of resolution due to spatial coherence. For $\Delta\lambda \to 0$, the MTF of Eqs. (3.44) and (3.45) can be written as

$$\mathrm{MTF}_1 = \frac{(2 - C_0^2)[1 - 2\,\mathrm{sinc}(2\pi w \nu_0)]}{10 - C_0^2 - 8\,\mathrm{sinc}(2\pi w \nu_0)}, \tag{7.6}$$

and

$$\mathrm{MTF}_2 = \frac{2 - C_0^2}{10 - C_0^2 - 8\,\mathrm{sinc}(2\pi w \nu_0)}, \tag{7.7}$$

where $\nu_0 = h/(\lambda f)$. From these equations we see that the MTFs are independent of the object's spatial bandwidth; however, they are functions of the slit width w. The corresponding MTF versus the separation h is plotted in Fig. 3.10. Thus to obtain a high contrast subtracted image, the separation h should be small. The relationship between the MTF, the object separation, and the slit width is tabulated in Table 7.2, where the focal length of the achromatic transform lens is assumed to be 300 mm. From this table, we see that, to maintain an adequately high MTF, a very narrow source is required. However, this problem can be alleviated with a source encoding technique, as illustrated in the following.

Table 7.2. Source Size Requirement, $\lambda = 5461$ Å

	w(mm)		
MTF	$h = 5.0$ mm	$h = 7.5$ mm	$h = 10.0$ mm
0.1	0.0076	0.005	0.0038
0.3	0.0052	0.0035	0.0026
0.6	0.0031	0.0021	0.0015

A multislit source encoding mask is used for this purpose. The spatial period of the encoding mask is precisely equal to the sinusoidal grating G (i.e., $d = 1/v_0$, where v_0 is the spatial frequency of the grating). The spatial coherence function, at the input plane, is governed by the ratio of the slit width to the spatial period of the encoding mask, that is, by w/d. If the ratio of w/d is adequately small, a high degree of point-pair spatial coherence at the input plane is achievable. The dependence of the MTF upon w/d is shown in Fig. 3.11. It is apparent that the subtraction operation ceases when the MTF approaches zero, that is, as $w/d \to 1$. In order to see this effect due to spatial coherence, a few numerical examples derived from Fig. 3.11 are given in Table 7.3.

7.2.1.2. Effect Due to Space Bandwidth Product. Since the microcircuit chip under inspection is a high-spatial-frequency-type object, the transform lenses L_1 and L_2 are required to be highly diffraction limited. Here are a few commercial microscopic objectives suitable for this purpose, as tabulated in Table 7.4. The size requirement of the microcircuit chip under inspection [i.e., $S = h = \phi/(4\sqrt{2})$] and the resolution limit of the processing system can be calculated if a specific microscopic objective lens is selected. Examples of the subtraction capability of the proposed optical system are tabulated in Table 7.5.

Table 7.3. Spatial Coherence Requirement for Various w/d, $\lambda = 5461$ Å

w/d	0.05	0.10	0.20	0.30
MTF	0.85	0.57	0.18	0.006

Table 7.4. Some Commercially Available Microscopic Objectives

Number	Magnification	Numerical Aperture	Field of View ϕ(mm)	Resolution Limit R(lines/mm)	Space Bandwidth Product ϕR
1	1	0.12	50	200	11000
2	1/5	0.2	15	370	5550
3	1/10	0.3	8	550	4400

Table 7.5. Image Subtraction Processing Capability $\Delta\lambda = 10$ Å, $w = 0.003$ mm.

Number	S(mm)	R(lines/mm)
1	9	220
2	2.7	370
3	1.4	550

For instance, if the spectral bandwidth is 10 Å, the slit size of the source encoding mask is 3 μm, and the number 1 microscopic objective lens is used, then the optical processor has the capability of observing a microcircuit chip width as large as 9 mm with a spatial resolution of 220 lines/mm. The capability is, of course, dependent upon the minimum detectable value of MTF, which is less than 7% in the above example.

7.2.2. Experimental Demonstrations

In experimental demonstrations, we insert a standard simulated microcircuit chip transparency in one of the input apertures of the processor, as shown in Fig. 7.6*a*, and a defective or faulty one in the other aperture, as shown in Fig. 7.6*b*. By the comparison of these two figures, we see that there are several links missing in Fig. 7.6*b*. Figure 7.6*c* shows the subtracted image obtained at the output plane of the proposed processor. Since we have utilized an incoherent source, the coherent artifact noise in the subtracted image is substantially suppressed.

Another interesting experimental demonstration for the inspection of an Integrated Circuit (IC) mask [7.20] utilizing a simple white-light processor is shown in Fig. 7.7. Since the white-light source radiates all the visible wavelengths, it is possible to identify the defects or faulty cracks through color

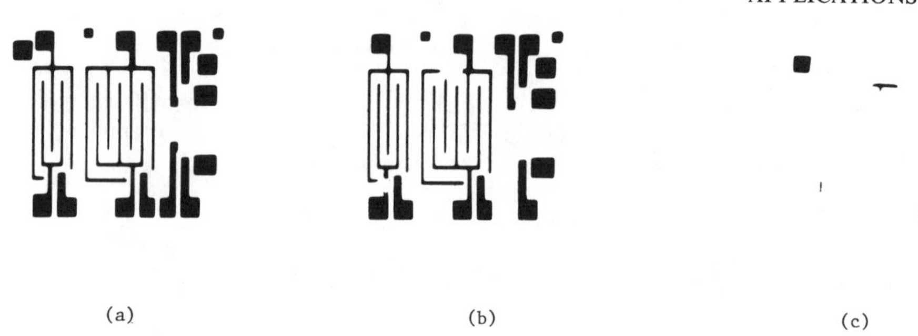

(a) (b) (c)

Figure 7.6. IC chip inspection. (*a*) and (*b*) are input objects. (*c*) Subtracted image.

encoded spatial frequency filtering. To demonstrate the defect detection, we assume that the Fourier spectra distribution of the IC mask is known *a priori.* We note that the scale of the signal spectrum is proportioned to the wavelength of the light source; a red and green spatial filter, as shown in Fig. 7.8, is utilized in our experiment. Since the faulty cracks are generally high spatial frequency signals, they would be expected to be diffracted outside the red region of Fig. 7.8*a*. Thus the defects or faulty cracks would form green image signals superimposed on the red IC mask, as shown in Fig. 7.8*b*. From this figure, we see that the defects and faulty cracks can be identified through the color coded image.

In concluding this section, we would like to point out that it is feasible to utilize an incoherent, or white-light, optical processing technique for micro-

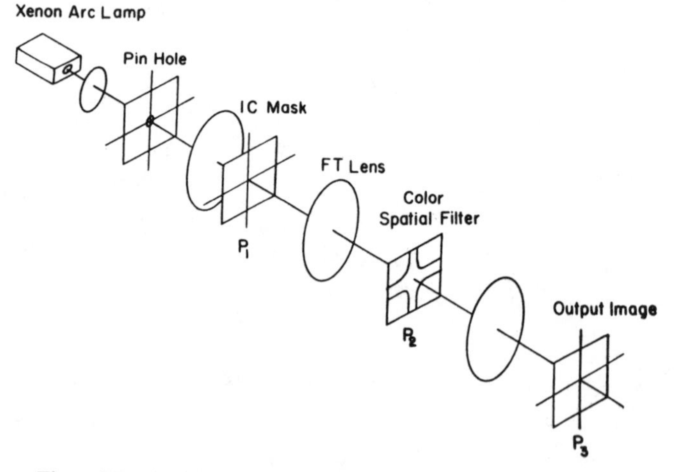

Figure 7.7. A white-light optical processor for IC mask inspection.

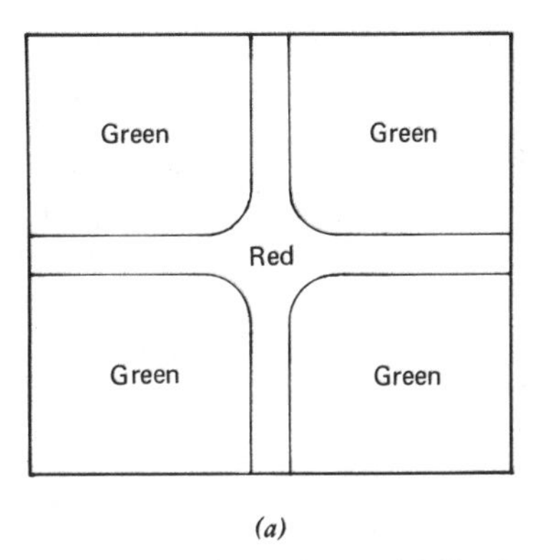

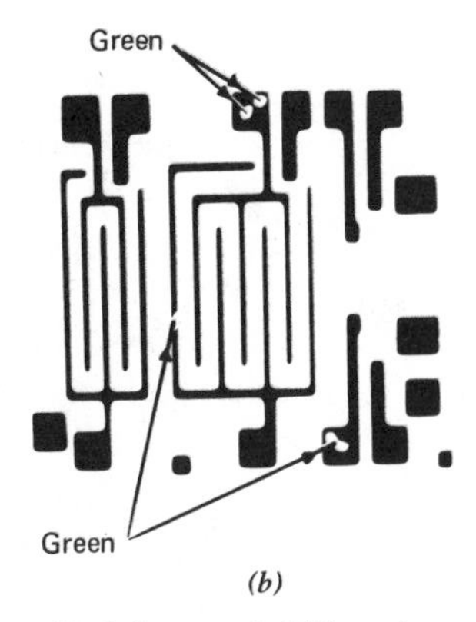

Figure 7.8. (*a*) Color spatial filter for IC mask inspection. (*b*) Color encoded IC mask.

circuit inspection. The techniques are capable of improving the reliability, efficiency, and cost for large scale automatic precision inspection.

7.3. GENERATION OF SPEECH SPECTROGRAM

Since the development of the first electronic sound spectrograph by Bell Telephone Laboratories nearly four decades ago [7.21], great strides have been made in the field of speech analysis. The electronically generated sound spectrograph has been widely used for various applications including such areas as linguistics, phonetics, speech synthesis, and speech recognition. However, an optical system utilizing the Fourier transform properties of the positive lens offers a viable alternative to the electronic system. The multichannel optical processor suggested by Cutrona et al. [7.1] was found to be able to efficiently utilize the two-dimensional nature of an optical system. Moreover, by sacrificing the multichannel capability of such a processor, Thomas [7.22] pointed out the feasibility of generating a near real-time spectrum analysis for a large space-bandwidth signal. Later, Yu [7.23] reinforced Thomas' concept in the synthesis of a coherent optical sound spectrograph. He pointed out that a near real-time optical sound spectrograph can be designed and constructed at a competitively low cost [7.24].

In this section, we describe a technique for generating multicolor speech spectrograms with a white-light optical processor [7.25]. This technique utilizes a cathode ray tube (CRT) scanner density modulator to convert a temporal signal to a spatial signal that is suitable for white-light signal processing. To obtain a color coded speech spectrogram, a dispersive element such as a grating (or prism) can be used at the input plane of the processor. In frequency color encoding, a narrow slit is placed over the smeared color signal spectra at the Fourier plane so that a frequency color encoded spectrogram can be recorded. In the following, we describe the basic performance of this white-light optical spectrum analyzer as applied to speech signals. The frequency resolution limit of the system, as well as the frequency scaling of the spectrogram, are given. Experimental demonstrations of the color-coded speech spectrograms, obtained with the white-light processing technique, are provided.

7.3.1. Temporal-Spatial Signal Conversion

It is well known that an optical processor is capable of performing two-dimensional spatial Fourier transformation. However, the processing of time signals by optical means necessitates the transformation of the signals to a two-dimensional spatial format. Generally speaking, there are two ways of performing this conversion, namely, the density and the area modulation techniques. Nevertheless, the density modulation method is the simplest and most commonly used technique in practice [7.26]. We use a density modulation technique for temporal-to-spatial signal conversion and also show that the temporal-spatial formats obtained from this technique are very suitable for color coded spectrogram generation with a white-light source.

The conversion can be obtained by displaying the time signal with a CRT scanner and then recording it on a moving photographic film. In other words, the time signal is first applied to the Z-axis of a high resolution CRT scanner to produce an intensity modulated electron beam such that the fluorescence light intensity varies linearly with the input signal voltage. Since the signal voltage is usually a bipolar function and the light intensity is a positive real quantity, an appropriate bias should be added to the time signal to produce a density modulated signal. To ensure a linear density modulated signal, the input signal should be scaled down within the linear region of the CRT scanner. Furthermore, a ramp function is fed into the Y-axis of the CRT scanner to produce the vertical sweep. In order to maintain the uniformity of sweep intensity, the scanning rate should be set considerably higher than the maximum frequency content of the input signal. Thus the signal would be displayed on the CRT screen as a uniform intensity scan at each vertical sweep. Consequently, a spatially recorded format, whose amplitude transmittance is linearly proportional to the intensity of the CRT display, can be produced by imaging the CRT display onto a moving photographic film. Figure 7.9*a* shows the schematic diagram of the conversion process. The recorded format, as typified in Fig. 7.9*b*, can then be inserted into the input plane of the optical processor for spectrogram generation.

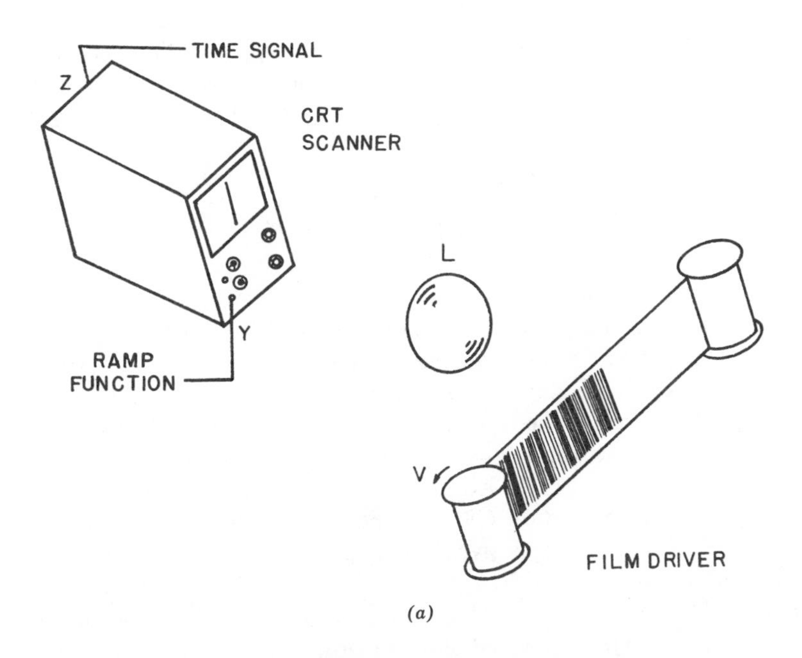

(a)

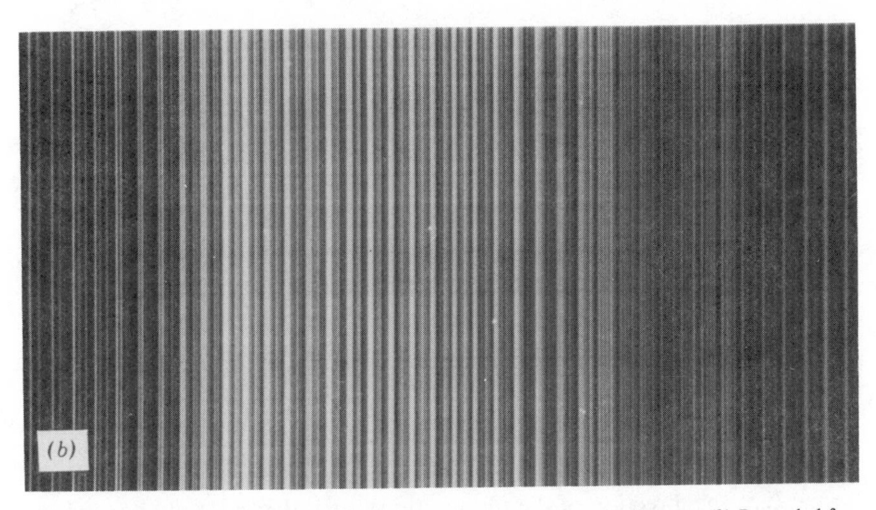

Figure 7.9. A temporal-to-spatial signal conversion. (*a*) Schematic diagram. (*b*) Recorded format. *L*, imaging lens.

In the time-spatial conversion recording, the speed of the recording film limits the highest frequency content of the signal to be recorded. For instance, if the highest frequency content of the signal is ν cps, then the speed of the film motion should be

$$v \geqq \frac{\nu}{R}, \tag{7.8}$$

where R is the spatial frequency limit of the film. Thus it is apparent that to resolve the highest frequency content of the recorded format, the speed of the recording film should be adjusted higher than the ratio of ν/R. Unfortunately, the finite optical processing window would set the speed limit of the recording film. A higher speed of the recording film would result in a lower frequency resolution of the spectrogram generation, thus corresponding to a wide-band spectrogram analysis [7.27]. Although this drawback may be alleviated by utilizing a higher resolution CRT scanner, in practice, a higher resolution CRT tends to be more expensive. Moreover, the frequency resolution of the optical processor is limited by the source size as well as by the point spread function of the imaging lens, and of course by the width of the optical window in the processor. Thus the speed of the recording film should be restricted by the following inequality:

$$v \leqq W\nu_l, \tag{7.9}$$

where W is the width of the optical window of the processor and ν_l is the lowest frequency limit.

7.3.2. System Analysis

Let us now discuss a white-light optical spectrum analyzer as depicted in Fig. 7.10, where L_1, L_2, and L_3 are achromatic Fourier transform lenses. If the recorded format of a time-spatial signal, as described in the previous section, is loaded in a linear motion film transport at the input plane of the processor, a slanted set of rainbow smeared Fourier spectra in the Fourier plane can be observed. We note that the effect of the phase grating at the input plane is to disperse the signal spectra in a direction perpendicular to the recorded input format (i.e., the direction of the film motion), such that a set of nonoverlapping slanted (or fan-shaped) rainbow color smear spectra can be displayed at the Fourier plane. It is now apparent that, if a slanted narrow slit is properly utilized at the Fourier plane, as illustrated in the figure, a frequency color coded spectrogram can be recorded at the output plane, with a moving color film. The color coded spectrographic signal can also be picked up by a color

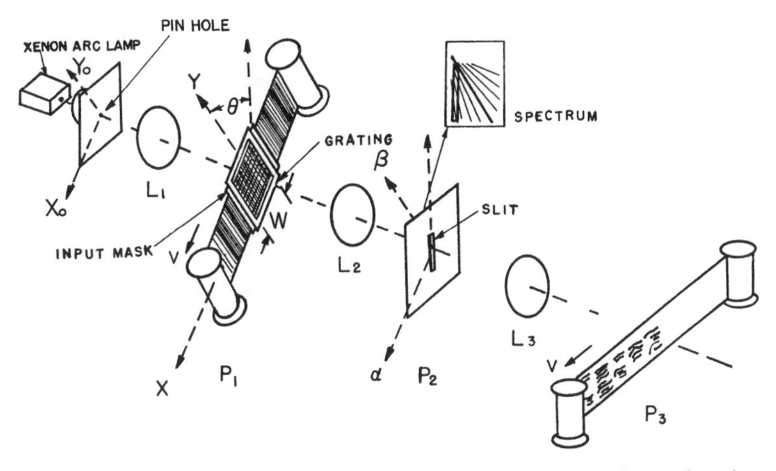

Figure 7.10. A white-light optical sound spectrograph. L, achromatic transform lens.

TV camera for display, storage, transmission, and further processing by electronic or digital system. We further note that, by simply varying the width of the input optical window, one would expect to obtain the so called *wide-band* and *narrow-band* spectrograms. In other words, if a broader optical window is utilized for the spectrogram generation, then a high spectral resolution, corresponding to a narrow-band spectrogram, can be obtained. However, by increasing the spectral resolution of the spectrogram, we would expect to lose the spatial (i.e., time) resolution. Conversely, if the optical window is narrower, which corresponds to a wide-band analysis, then a loss of spectral resolution is expected. However, the loss of the spectral resolution improves the time resolution of the spectrogram. This time-bandwidth relationship is in fact the consequence of the Heisenberg's uncertainty relation in quantum mechanics [7.28].

For simplicity in analysis, we let the light source be a square white-light source. The source irradiance may be written as

$$\gamma(x_0, y_0; \lambda) = S(\lambda)\mathrm{rect}\left[\frac{x_0}{b}\right]\mathrm{rect}\left[\frac{y_0}{b}\right], \tag{7.10}$$

where $S(\lambda)$ is the spectral distribution of the light source, b denotes the size of the square source, and

$$\mathrm{rect}\left[\frac{\chi}{b}\right] = \begin{cases} 1, & |\chi| \leqq b/2 \\ 0, & \text{otherwise.} \end{cases}$$

With reference to Wolf's partial coherence theory [7.9], the mutual coherence function at the input plane P_1 of the spectrum analyzer of Fig. 7.10 would be

$$\Gamma(x_1, y_1; x_2, y_2; \lambda) = \iint \gamma(x_0, y_0; \lambda) \cdot \exp\left\{-i\frac{2\pi}{\lambda f}[x_0(x_1 - x_2) + y_0(y_1 - y_2)]\right\} dx_0\, dy_0, \tag{7.11}$$

where the integration is over the source irradiance. By carrying out the integration, the mutual coherence function at plane P_1 becomes

$$\Gamma(x_1, y_1; x_2, y_2; \lambda) = KS(\lambda)\text{sinc}\left[\frac{\pi b(x_1 - x_2)}{\lambda f}\right]\text{sinc}\left[\frac{\pi b(y_1 - y_2)}{\lambda f}\right], \tag{7.12}$$

where

$$\text{sinc}[\pi\chi] \triangleq \frac{\text{sinc}(\pi\chi)}{\pi\chi},$$

and K is an appropriate constant.

Now, if the time-spatial signal format is inserted at the input plane P_1 of the processor, the mutual coherence function behind the input plane would be

$$\Gamma'(x_1, y_1; x_2, y_2; \lambda) = \Gamma(x_1, y_1, x_2, y_2; \lambda)\exp[iq_0(y_1 - y_2)] \cdot t(x_1)t^*(x_2)\text{rect}\left[\frac{x_1}{W}\right]\text{rect}\left[\frac{x_2}{W}\right], \tag{7.13}$$

where the exponential factor represents the phase transform of the one-dimensional phase grating, q_0 is the angular spatial frequency of the grating, $t(x)$ is the amplitude transmittance function of the recorded format, and W is the width of the input optical window as shown in the figure. Thus, it is apparent that the mutual coherence function arriving at the Fourier plane P_2 would be

$$\Gamma(p_1, q_1; p_2, q_2; \lambda) = \iiiint \Gamma'(x_1, y_1; x_2, y_2; \lambda) \cdot \exp\{-i(x_1 p_1 + y_1 q_1 - x_2 p_2 - y_2 q_2)\} dx_1\, dy_1\, dx_2\, dy_2, \tag{7.14}$$

where: the proportional constant is ignored for convenience; (p, q) is the angular spatial frequency coordinate system, which can be written as $p \triangleq 2\pi\alpha/f\lambda$, $q \triangleq 2\pi\beta/f\lambda$; and (α, β) is the spatial coordinator system of the Fourier plane P_2.

By properly substituting Eq. (7.13) into Eq. (7.14) and integrating over the y_1 and y_2 variables, we have

$$\Gamma(p_1, q_1; p_2, q_2; \lambda) = S(\lambda)\operatorname{rect}\left[\frac{\lambda f(q_1/2 + q_2/2 - q_0)}{2\pi b}\right] \cdot \delta(q_1 - q_2)$$
$$\cdot \iint \operatorname{sinc}\left[\frac{\pi b(x_1 - x_2)}{\lambda f}\right] t(x_1)t^*(x_2)$$
$$\cdot \operatorname{rect}\left[\frac{x_1}{W}\right]\operatorname{rect}\left[\frac{x_2}{W}\right]\exp\{-i(x_1p_1 - x_2p_2)\}dx_1\, dx_2. \tag{7.15}$$

For simplicity of illustration, we assume that the input format is a single sinusoidal signal, that is,

$$t(x) = 1 + \cos(p_0x), \qquad \text{for all } y, \tag{7.16}$$

where p_0 is the spatial frequency of the sinusoid. Since the cosine function of Eq. (7.16) can be written into two exponential functions, we focus our attention merely on one of the diffraction orders, that is

$$t(x_1)t^*(x_2) = \exp\{ip_0(x_1 - x_2)\}. \tag{7.17}$$

By substituting Eq. (7.17) into Eq. (7.15), we obtain

$$\Gamma(p_1, q_1; p_2, q_2; \lambda) = S(\lambda)\operatorname{rect}\left[\frac{\lambda f(q_1/2 + q_2/2 - q_0)}{2\pi b}\right]\delta(q_1 - q_2)$$
$$\cdot \iint \operatorname{sinc}\left[\frac{\pi b(x_1 - x_2)}{\lambda f}\right]\exp[ip_0(x_1 - x_2)]$$
$$\cdot \operatorname{rect}\left[\frac{x_1}{W}\right]\operatorname{rect}\left[\frac{x_2}{W}\right]\exp[-i(x_1p_1 - x_2p_2)]dx_1\, dx_2. \tag{7.18}$$

Since the interest is centered at the output irradiance, by letting $p_1 = p_2 = p$ and $q_1 = q_2 = q$, the irradiance at the Fourier plane would be

$$I(p, q; \lambda) = S(\lambda)\text{rect}\left[\frac{\lambda f(q - q_0)}{2\pi b}\right] I(p), \tag{7.19}$$

where

$$I(p) \triangleq \iint \text{sinc}\left[\frac{\pi b(x_1 - x_2)}{\lambda f}\right] \exp[ip_0(x_1 - x_2)] \cdot \text{rect}\left[\frac{x_1}{W}\right] \text{rect}\left[\frac{x_2}{W}\right] \exp[-i(x_1 - x_2)p] dx_1\, dx_2. \tag{7.20}$$

To investigate the variation of the output irradiance due to the width of the input optical window, Fig. 7.11 shows a plot of the normalized intensity variation of Eq. (7.20), for two values of W. From this figure, we see that a narrower spread of output irradiance corresponds to a broader optical window W at the input plane. Thus a narrow-band spectrogram can be generated in this manner. On the contrary, if a narrower optical window is utilized,

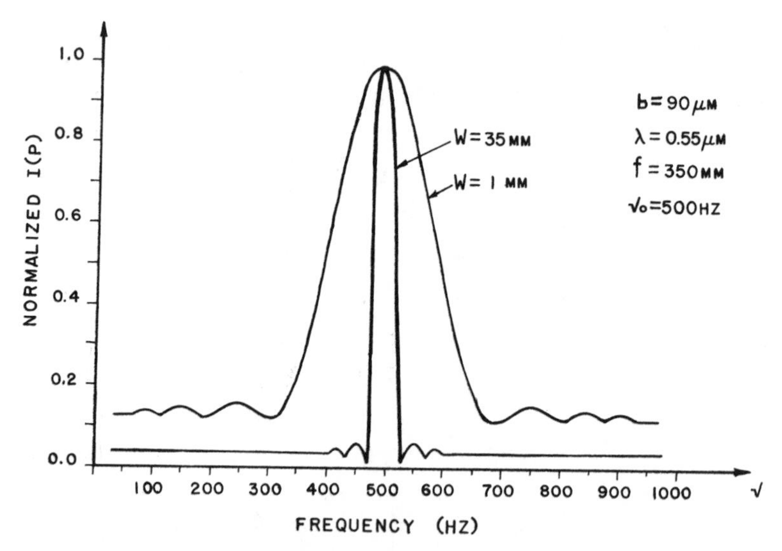

Figure 7.11. Normalized output spectral distribution as a function of input frequency for two sizes of optical windows.

then a broader spread of the output irradiance is expected. Thus a wide-band spectrogram may be generated with a narrow optical window.

In order to have a better understanding of Eq. (7.9), two extreme cases of the light source are discussed in the following.

1. If we assume that the broad-band light source is vanishingly small (i.e., $b \to 0$), then

$$\lim_{b \to 0} \operatorname{sinc}\left[\frac{\pi b(x_1 - x_2)}{\lambda f}\right] = 1$$

and Eq. (7.19) takes the form

$$I(p, q; \lambda) = S(\lambda)\operatorname{rect}\left[\frac{\lambda f(q - q_0)}{2\pi b}\right] W^2 \operatorname{sinc}^2[\pi W(p - p_0)]. \tag{7.21}$$

This result describes a *completely spatially coherent* illumination.

2. On the other hand, if the light source becomes infinitely large, (i.e., $b \to \infty$), then

$$\lim_{b \to \infty} \operatorname{sinc}\left[\frac{\pi b(x_1 - x_2)}{\lambda f}\right] = K\delta(x_1 - x_2),$$

where K is an appropriate constant, and Eq. (7.19) becomes

$$I(p, q; \lambda) = S(\lambda)\operatorname{rect}\left[\frac{\lambda f(q - q_0)}{2\pi b}\right] W^2. \tag{7.22}$$

This result corresponds to a *completely spatially incoherent* illumination. Nevertheless, the proposed white-light spectrum analyzer, as can be seen, is operating in a partially coherent mode.

7.3.3. Frequency Calibration and Resolution

We now discuss the frequency calibration of the proposed white-light optical sound spectrograph. The accuracy of the frequency measurement depends on the width of the input optical window, as well as on the width of the sampling slit at the Fourier plane. By adjusting these parameters, a properly calibrated speech spectrogram can be generated at the output plane.

Let us assume that the transform lenses are achromatic (i.e., $\partial f/\partial \lambda = 0$), and that the angular spatial frequency of the input phase grating is q_0. Thus we see that the Fourier spectrum of the phase grating would be diffracted at

$$\beta = \left(\frac{\lambda f}{2\pi}\right) q_0, \tag{7.23}$$

where f is the focal length of the achromatic transform lens. Since Eq. (7.23) is linearly proportional to the wavelength of the light source, a rainbow color of smeared spectra can be observed along the β-axis in the Fourier plane P_2. For simplicity, we assume that the input format is a single sinusoid of spatial frequency p_0; then the Fourier spectra points are located at

$$\alpha = \pm \frac{\lambda f}{2\pi} p_0. \tag{7.24}$$

For convenience, we use the positive value of Eq. (7.24) in our discussion. By taking the ratio of Eqs. (7.23) and (7.24), we have

$$\alpha = \frac{p_0}{q_0} \beta, \tag{7.25}$$

where we assume that $p_0 \ll q_0$; q_0 is the spatial frequency of the phase grating. Thus we see that the frequency locus of the input signal can be properly traced out in the Fourier plane. In view of $q_0 \ll p_0$, we further see that the spectrum of the input signal would disperse into rainbow color at a slanted angle in the Fourier plane. It is therefore apparent that, if a narrow masking slit is properly utilized in the Fourier plane, a frequency color coded spectrogram can be obtained at the output plane.

Now let us suppose that the masking slit can be described as

$$\beta = \frac{\lambda_0 f}{2\pi} q_0 + \alpha \tan \gamma, \tag{7.26}$$

where λ_0 denotes the upper or lower limit of the source wavelength, depending on the slit orientation and $\tan \gamma$ is the slope of the slit. By substituting Eqs. (7.23) and (7.24) into Eq. (7.26), we have

$$p_0 = \frac{q_0(\lambda - \lambda_0)}{\lambda \tan \gamma},$$

which can be written as

$$\nu = \frac{p_0 v}{2\pi} = \frac{q_0(\lambda - \lambda_0)v}{2\pi\lambda \tan \gamma}, \tag{7.27}$$

where ν is the time frequency of the input signal. Thus we see that the temporal frequency of the input signal and the spectral wavelength of the light source form a nonlinear function, as plotted in Fig. 7.12*a*. Figure 7.12*b* shows

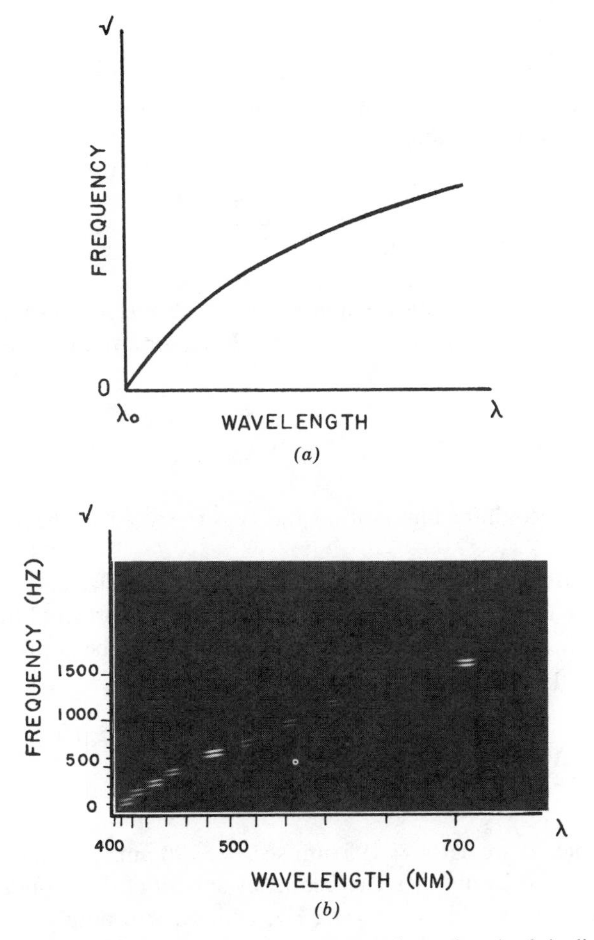

Figure 7.12. Variation of input frequency versus spectral wavelength of the light source. (*a*) Calculated result. (*b*) Experimental result.

the experimental result to confirm this relationship. Nevertheless, this nonlinearity of the frequency–spectral wavelength relationship can be linearized by using an appropriate curved slit instead of the linear one in the Fourier plane.

We now discuss the frequency resolution of the proposed system. We note that the frequency resolution is primarily limited by the width of the input optical window, the source size, and the width of the sampling slit in the Fourier plane. In order to investigate these effects, we assume that the size of the white-light source is b. Then the size of a point spread image would be

$$b' = \frac{bf}{f'}, \tag{7.28}$$

where f' and f are the focal lengths of collimating and achromatic Fourier transform lenses, respectively. Thus the width of the point spread image at the spatial frequency plane would be

$$\Delta\nu_1 = \frac{b'}{\lambda f} v = \frac{bv}{\lambda f'}. \tag{7.29}$$

Since the width of input optical window affects the spectral resolution of the processor, the width of the spectral lines in Hertz, for a point source, is

$$\Delta\nu_2 = \frac{v}{W}, \tag{7.30}$$

where v is the speed of the film motion and W is the width of the input optical window.

For simplicity, we treat these two factors (i.e., source size and the width of optical window) independently. It is apparent that the spread of the smeared spectral line (i.e., frequency resolution) of the system can be approximated by the following equation:

$$\Delta\nu = [(\Delta\lambda)^2 + (\Delta\nu_2)] = v\left[\left(\frac{b}{\lambda f'}\right)^2 + \left(\frac{1}{W}\right)^2\right]^{1/2}. \tag{7.31}$$

As an example, if we let $v = 195$ mm/sec, $b = 90\ \mu\text{m}$, $\lambda = 0.55\ \mu\text{m}$, $W = 35$ mm, and $f' = 762$ mm, then the frequency spread of the proposed optical spectrum analyzer would be $\Delta\nu = 42$ Hz, which corresponds to a narrow-band analysis (e.g., narrow-band speech spectrogram). However, we see that the overall output resolution of the spectral spread is determined by the

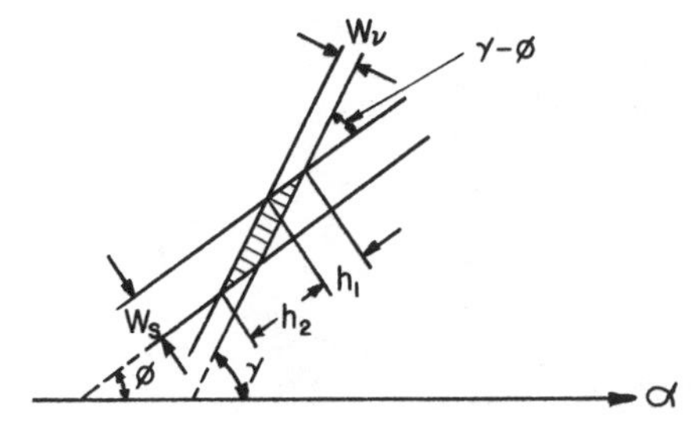

Figure 7.13. Determination of the output spectral resolution.

width and orientation of the sampling slit at the Fourier plane, as illustrated in Fig. 7.13. Thus the effective output frequency resolution can be shown as

$$\Delta\nu_e = \varepsilon(h_1 + h_2) = \left\{\frac{W_v}{\sin(\gamma - \phi)} + \frac{W_s}{\tan(\gamma - \phi)}\right\}, \qquad (7.32)$$

where W_v and W_s are the widths of the smeared frequency spectral line and the sampling slit, ε is an appropriate conversion factor, γ and ϕ are defined in the figure.

From this equation we see that the effective output frequency resolution is proportional to the diagonal region of the spectral line that intersects the sampling slit, as shown in the shaded area of the figure. Therefore, the overall output frequency resolution is somewhat lower than the width of the smeared spectral line, that is,

$$\Delta\nu_e \geqq \Delta\nu. \qquad (7.33)$$

This is the price that we paid for the color encoding. Nevertheless, the price is considered small when compared with the advantage we have gained from the white-light processing.

7.3.4. Experimental Demonstrations

In our experiments, a 75-W xenon arc lamp with a 90-μm pinhole was used as a white-light source. A phase grating of 80 lines/mm was used as a dispersive element at the input plane. A narrow slit of about 70 μm is inserted in the Fourier plane for color encoding. The focal length of the achromatic transform lens used was about 350 mm.

In an experimental demonstration, we show two sets of color coded speech spectrograms obtained with this technique, as depicted in Figs. 7.14 and 7.15. The frequency contents of these spectrograms are encoded in rainbow color from red to blue for upper to lower frequencies. These spectrograms were

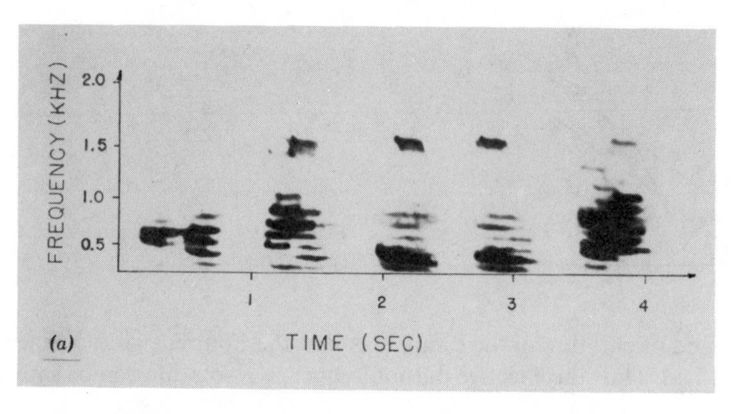

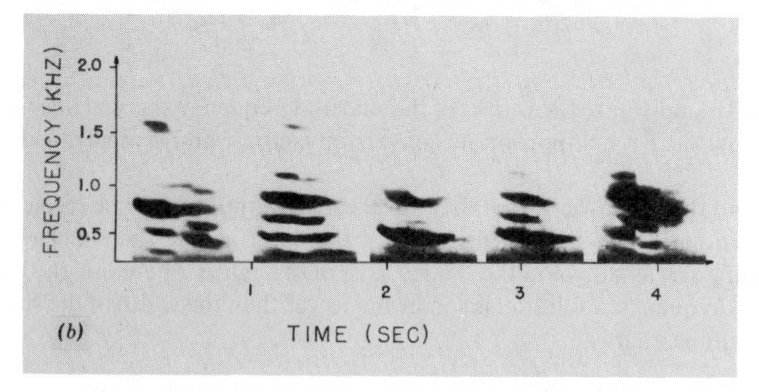

Figure 7.14. Black-and-white pictures of narrow-band color coded speech spectrograms (about 45-Hz bandwidth). (*a*) Obtained with a male voice. (*b*) Obtained with a female voice.

produced by a sequence of English words. "Testing, one, two, three, four," spoken by a male and a female voice, respectively. Figure 7.14 shows a set of narrow-band speech spectrograms, representing a 45-Hz bandwidth resolution. And Fig. 7.15 illustrates a set of wide-band speech spectrograms with a 300-Hz bandwidth analysis. From these sets of spectrograms we see that the basic structure of the spectrographic contents are preserved. The characteristics of the formant variation can readily be identified. Since Fig. 7.14*a* is produced by a low-pitched voice (i.e., male), the spectrogram shows more abundant harmonics than does that of the high-pitched voice of Fig. 7.14*b*.

It is worth mentioning that, although the narrow-band spectrogram is capable of resolving the spectral contents of a speech, it loses the time striation, as can be seen in Fig. 7.14. On the other hand, the wide-band spectrograms of Fig. 7.15 are capable of handling the time resolution, but they fail to

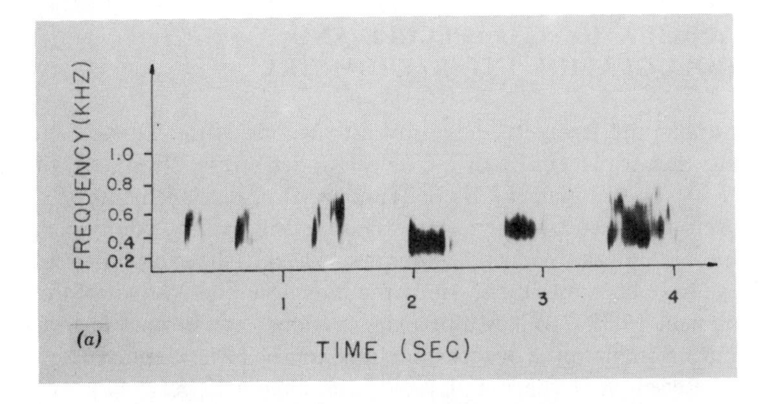

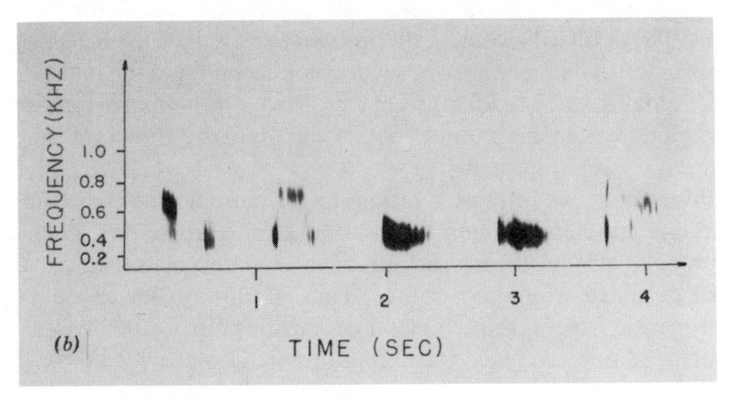

Figure 7.15. Black-and-white pictures of wide-band color coded speech spectrograms (about 300-Hz bandwidth). (*a*) Obtained with a male voice. (*b*) Obtained with a female voice.

resolve the spectral lines. We note that the trading of time and frequency resolutions between the wide and narrow spectrograms is essentially the well-known $\Delta\nu\Delta t$ concept in communication theory (i.e., $\Delta\nu\Delta t \geqq 1$), which is related to the famous Heisenberg's uncertainty principle in quantum mechanics.

In concluding this section we stress that, due to the very limited resolution of the CRT scanner used in our experiments, the results that we have obtained only extended to 1.6 kHz. This limitation is more pronounced for wide-band spectrographic analysis, due primarily to the low-light performance of the narrower optical window used at the object plane. However in our experience, if a higher resolution CRT scanner is utilized, the frequency range can easily be pushed up to 4 kHz. This frequency range is commonly used for most telephone systems.

7.4. APPLICATION TO SPECKLE AND HOLOGRAPHIC INTERFEROMETRY

Holographic interferometry has proven to be one of the valuable means of studying the displacement of an object under stress, strain, or vibration [7.29, 7.30]. Out-of-plane [7.31, 7.32] and in-plane displacements [7.33, 7.34] have been measured by several investigators using various double exposure and speckle interferometric techniques. Speckle-shearing interferometric methods have been applied to study the directional derivatives of the object displacement [7.35–7.39]. Multiplexing various interferometric fringe patterns onto a single piece of recording medium has been reported by several authors. Recently, Gerhart et al. [7.40] presented a technique for multiplexing double exposure and time-averaged holographic interferometric fringe patterns. They utilized a white-light processing technique for the holographic reconstruction, which permitted pseudocolor encoding of the interferometric fringes. Addams and Maddux [7.41] suggested a technique for multiplexing holographic interferometry and speckle photography; however, no experimental results were reported.

In this section, we propose a technique for multiplexing holographic and speckle interferometry using a narrow angle holographic recording scheme. This process permits reconstruction of the interferometric fringe patterns with an extended white-light source. Thus it allows color encoding of the interferometric fringes. Experimental demonstrations for defocused, double exposure, and time-averaged interferograms are also provided [7.42].

7.4.1. Interferometric Recording and Reconstruction

The defocused speckle interferometric techniques were chosen to best illustrate the advantage of multiplexing with color encoding. On-focus speckle photography and on-focus speckle interferometry can also be multiplexed with holographic interferometry with the following limitation. Holographic interferometry records out-of-plane displacements on the order of several microns while speckle interferometry records in-plane displacements on the order of several tens or hundreds of microns. Certain overlap conditions do arise where both holographic and speckle interferometric fringe patterns may be recorded. With the proper selection of recording parameters, multiplexed interferograms can be constructed.

The multiplexing of defocused speckle and holographic interferograms is better suited to show the advantage of multiplexed interferometric fringe patterns. Here, a holographic interferogram is multiplexed with a speckle interferogram that shows the derivative of the object displacement. In this technique there is also a trade-off between the quality of the separate interfer-

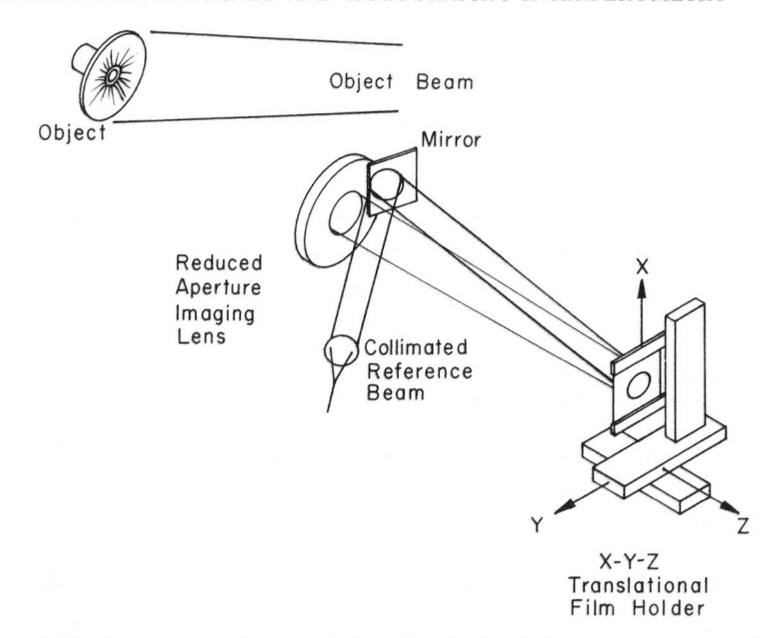

Figure 7.16. Optical setup for construction of multiplexed time-average holographic and speckle interferograms.

ometric images. Proper selection of construction parameters will optimize both images, yielding valuable fringe information.

Constructing a multiplexed interferogram of a vibrating object is schematically shown in Fig. 7.16. Here a vibrating object is illuminated with a coherent source of light. The reflected light field from the vibrational object is imaged using a reduced aperture lens. The purpose of the reduced aperture lens is to permit a reference beam to be brought in at an angle of about 4° with respect to the optical axis. This narrow angle holographic construction process is essential for proper reconstruction of the interferometric images with a white-light optical processor. The image of the vibrating object is also defocused from the image plane by a small distance Δz. The aim of this defocused distance is to generate a defocused speckle interferogram. Here lies the trade-off between the holographic and speckle fringe patterns. The holographic fringes focus at $\Delta z \simeq 0$, while the speckle fringes are optimized at $\Delta z \simeq 3$ cm. A compromised defocus distance $\Delta z = 1$ cm was chosen to optimize the multiplexed image.

Similarly, double exposure multiplexed images can be constructed as shown in Fig. 7.17. However, the double exposure technique has an additional limitation, namely the dual-aperture screen used in the recording process.

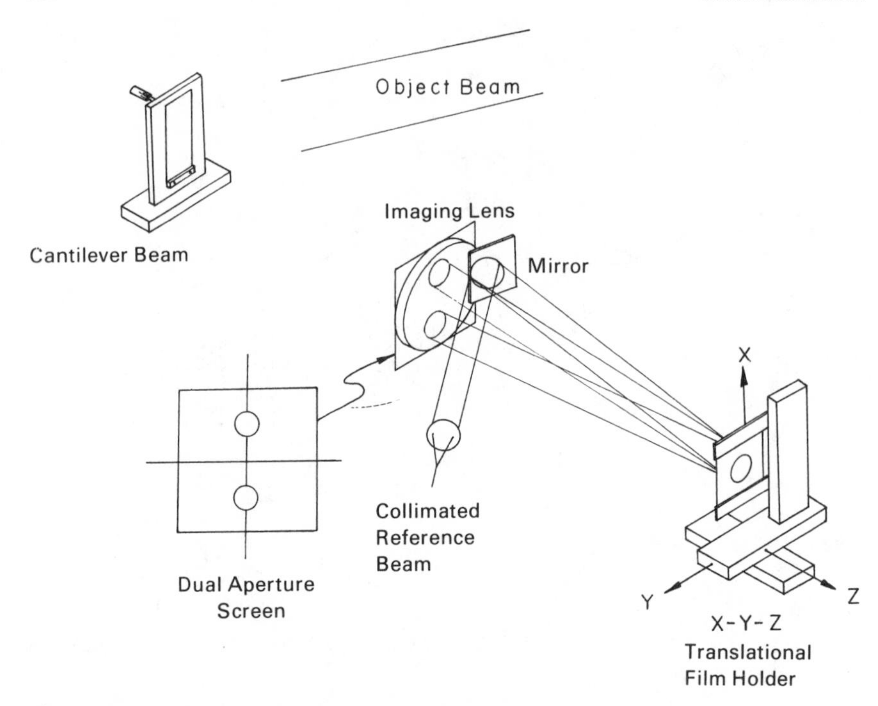

Figure 7.17. Optical setup for construction of multiplexed double exposure holographic and speckle interferograms.

This screen limits the number of directional derivatives of the object motion that can be recorded. We note that Hung et al. [7.38] have used a four-aperture mask to increase the number of directional derivatives that can be measured. In the multiplexing technique this creates numerous cross-modulation terms that can overlap. If a white-light optical processor is not used for the reconstruction process, the reference beam to optical axis angle can be increased. Thus, it reduces the possibility of the overlapping of the cross-modulation terms. Otherwise, sequential exposures should be made for different angular orientations of the encoding dual-apertures.

The techniques of both Figs. 7.16 and 7.17 used an *x-y-z* translational plate holder to hold the recording film. With proper adjustment of the recording system and spatial shielding of the recording medium, up to 12 or more sequential exposures can be spatially encoded on a 4 × 5 in. sheet of photographic film. This represents a large number of interferometric tests that may be completed on one piece of film.

After the photographic procesing, the multiplexed interferogram is placed at the input plane of a white-light optical processor, as shown in Figs. 7.18

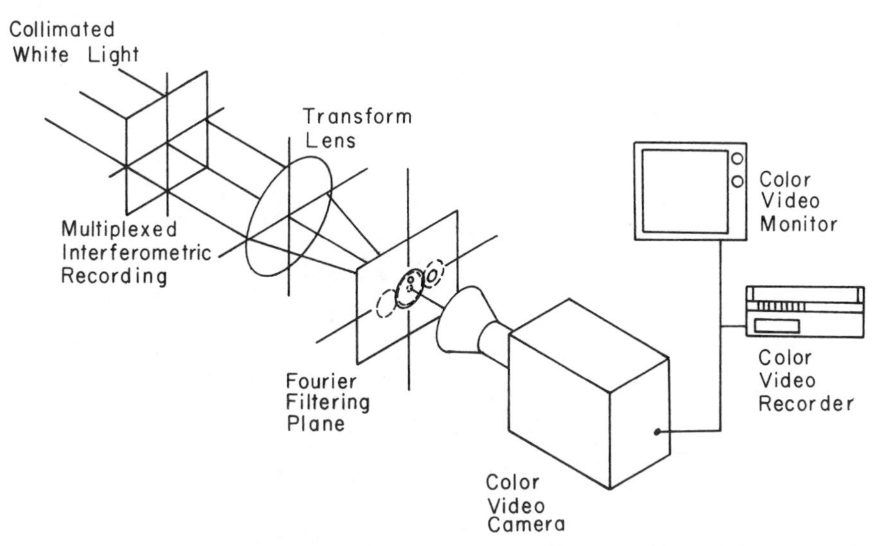

Figure 7.18. White-light optical processor for reconstruction of multiplexed time-averaged interferograms.

and 7.19. A collimated beam of white-light emanating from an extended xenon arc source is used to illuminate the multiplexed interferometric transparency. Spatial filtering and color encoding are then performed in the spatial frequency plane. The result is a set of color coded interferometric fringe data that can be observed at the output image plane.

Figure 7.18 shows the reconstruction scheme for a multiplexed interferogram. We note that the speckle interferometric fringes are available in the halo surrounding the optical axis of the Fourier plane. The directional derivative of the object motion fringe patterns can be reconstructed by placing a small aperture in the halo. This aperture can be rotated to show the various directional derivatives of the object motion. A slowly driven spatial frequency sampling mechanism, Fig. 7.20, in conjunction with a color video system (as shown in Fig. 7.18) yields an excellent method for storage and display of the experimental data. We also note that the holographic interferometric fringe pattern is available in the higher spatial frequency space. Color encoding can be accomplished by placing appropriate color filters over the sampling apertures. The intensity of the fringe pattern can be adjusted by varying the size of the sampling apertures. The fringe resolution and brightness of the reconstructed images is controlled by the aperture size and is optimally adjusted for proper color encoding. With this technique it is possible to view either the speckle or holographic interferometric fringe patterns separately or simultaneously in a multiplexed manner at the output image plane of the white-light

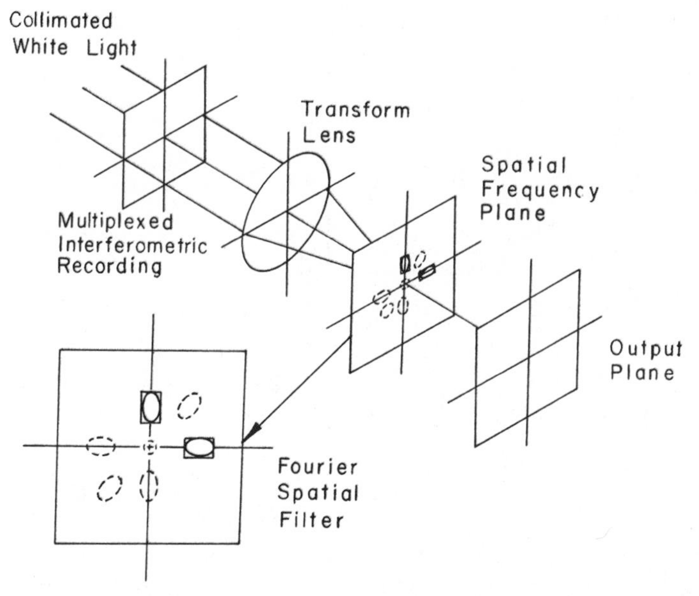

Figure 7.19. White-light optical processor for reconstruction of multiplexed double exposure interferograms.

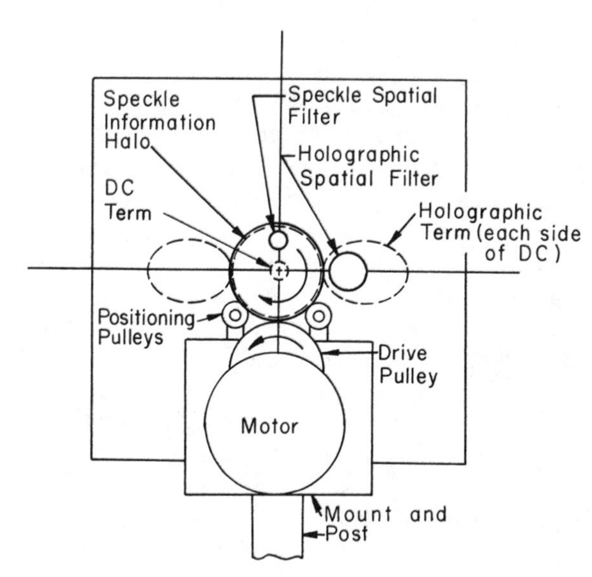

Figure 7.20. Spatial filter with motor drive for reconstruction of multiplexed time-averaged interferograms.

optical processor. These color coded fringe patterns are suitable for photographing or recording with a color video system.

The reconstruction scheme for the double exposure interferogram, Fig. 7.19 is slightly different from the time-averaged case. The double exposure interferogram has four diffracted orders visible in the spatial frequency plane. The holographic interferometric fringe information is available on the horizontal axis, while the speckle interferometric fringe information is available on the vertical axis. The cross-modulation terms are blocked by the stop band spatial filter, so that they can be ignored. The color encoding process is basically identical to that of the time-averaged case.

7.4.2. Experimental Demonstrations

We now present an experimental demonstration in Fig. 7.21, which shows some of the interferometric fringe patterns that can be obtained with a time-averaged interferometric technique. The center photograph, Fig. 7.21*b* shows the holographic interferometric fringes obtained from a 64-mm diameter speaker vibrating at 4 kHz with a 2.2-V $p - p$ sinusoidal input signal. The photographs above and traveling in a clockwise direction from 12 o'clock to 5 o'clock, Fig. 7.21*a*, are the speckle interferometric fringe patterns. From 6 o'clock to 11 o'clock, Fig. 7.21*c*, the photographs show the multiplexed holographic and speckle interferometric patterns. Both types of interferometric fringe patterns were multiplexed onto the recording film during a single construction exposure. All of these fringe patterns can be directly viewed by a color video monitor, photographed, or studied by the human eye. Color encoding of the separate fringe patterns allows easy access for interpretation of the overlapped interferometric fringe data. Although the black-and-white photographs do show the basic advantages involved, in reality the color coded fringe patterns enhance the observability of the interferometric properties. The diagrams beside the photographs show the position and orientation of the sampling apertures as they were placed in the spatial frequency plane to obtain the respective interferometric fringe patterns.

Perhaps the most suitable use for this technique is in finding the location of the J_0-order holographic fringe. Since speckle shearing interferometry measures the derivative of the object motion, the fringe pattern contains no information where the object has been stationary. This dark area corresponds to the J_0-order Bessel function in the holographic interferometric fringe pattern. In the color encoded multiplexed interferogram the areas of pure holographic color are the J_0-order fringes. (Holographic color = the color filter used in encoding the holographic information.) This is very beneficial when the J_0 term is not the brightest fringe.

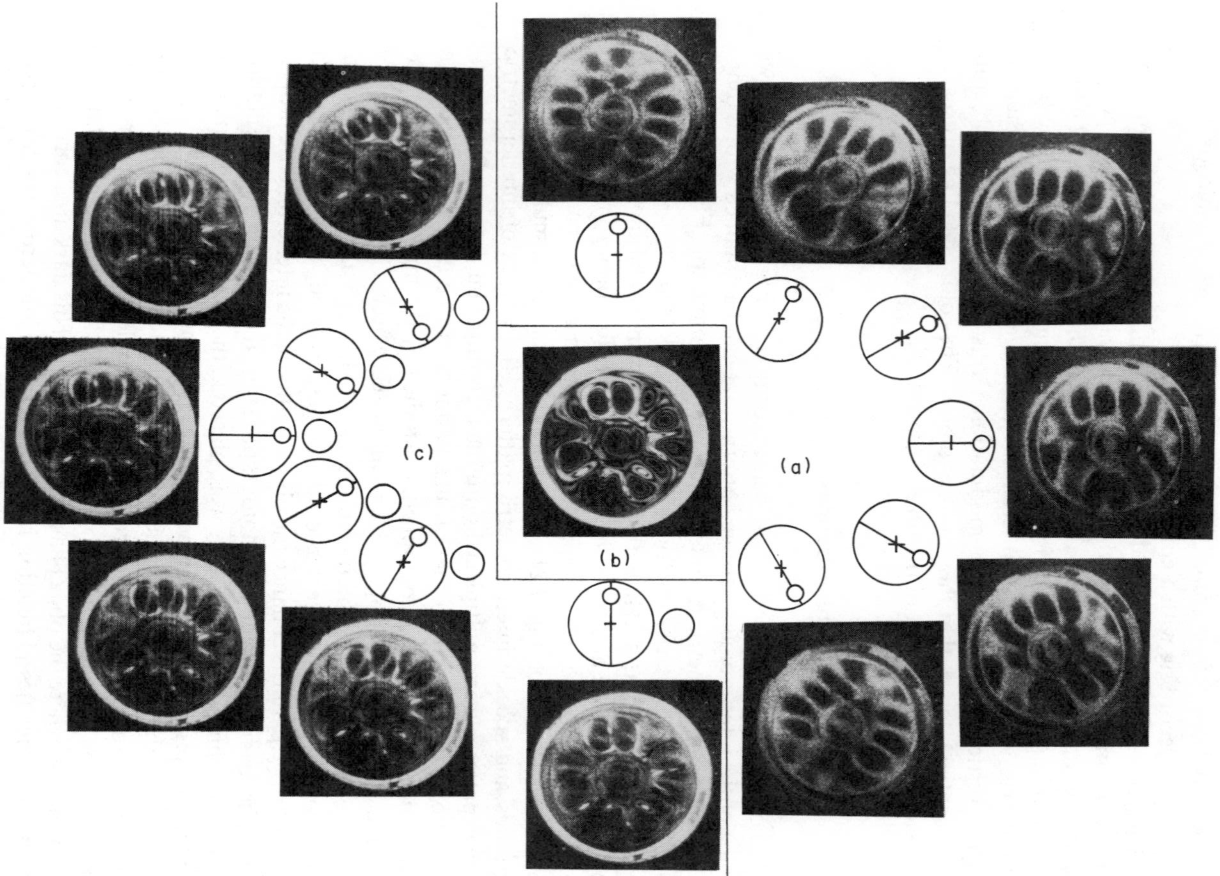

Figure 7.21. Black-and-white pictures of color coded time-averaged interferogram (*a*) 12 o'clock to 5 o'clock showing speckle interferometric fringe patterns. (*b*) Center photograph showing holographic interferometric fringe patterns. (*c*) 6 o'clock to 11 o'clock showing multiplexed interferograms.

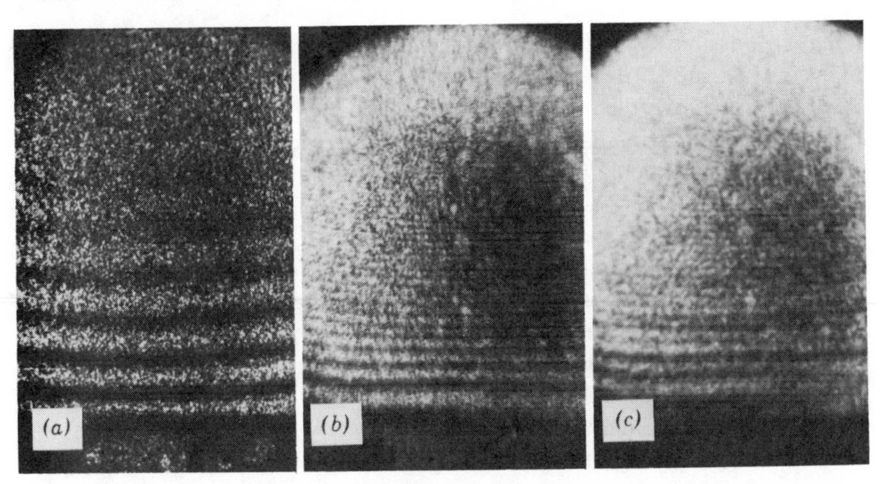

Figure 7.22. Black-and-white pictures of color coded double exposure interferogram. (*a*) Speckle fringe pattern. (*b*) Holographic fringe pattern. (*c*) Multiplexed interferogram.

The double exposure multiplexed interferometric pattern is illustrated in Fig. 7.22. These photographs are the lower portion of a clamped cantilever beam (i.e., the bottom of the photographs are the clamp). These pictures were obtained from a single multiplexed interferogram that was constructed using the narrow angle double exposure holographic interferometric technique. Figure 7.22*a* shows the speckle interferometric fringe pattern that was constructed. Due to the use of the dual-aperture recording technique, we have restricted the directional derivatives of our recording to the direction of our dual-aperture. Therefore, we do not have a sequence of directional derivatives to analyze as in the time-averaged technique. Figure 7.22*b* shows the holographic interferometric fringe patterns that were recorded. Figure 7.22*c* shows a black-and-white picture of the corresponding multiplexed color encoded images. Although it is rather difficult to interpret the results with these black-and-white pictures, it is much easier to study them when the fringe patterns are displayed in color.

In concluding this section, we stress that one of the advantages of this multiplexed interferometric technique is the combination of the holographic and speckle interferometers into one optical system. The ease in operating the system and the time economy achieved are desirable. With the use of an x-y-z translational stage, up to 12 or more experimental tests can be easily accomplished in a single run with a 4×5 in. sheet of photographic film.

The narrow angle holographic recording scheme has additional merits. For example, the angle we used was about 4°, which gives a spatial frequency of approximately 110 lines/mm. With the low spatial frequency requirement it *may* be possible to use a lower resolution and more economical film (e.g., a high contrast, moderate resolution film like Technical Pan). The other advantage is that it allows the use of a white-light optical processor for the interferometric reconstruction. The white-light optical processor offers a significant reduction in the coherent artifact effect and permits color encoding.

The combination of the continuous sampling mechanism and the color video system offers great flexibility for data handling and storage. For example, this system allows a continuous sample of the directional derivatives of the object motion. Also, the use of video cassette recording for data storage allows the removal of the data analysis from the laboratory environment. In short, we feel that this technique offers a convenient method for measuring holographic and speckle interferometric fringe data, particularly suitable for objects in motion. Finally, the ability to accurately locate the J_0-order holographic fringe is an important advantage.

7.5. WHITE-LIGHT PROCESSING WITH MAGNETO-OPTIC DEVICE

The recent development of a magneto-optic spatial light modulator (also called Light-MOD) has stimulated research into potential applications for real-time optical signal processing [7.43, 7.44]. The Light-MOD consists of a layer of magnetic iron-garnet thin film deposited on a transparent nonmagnetic substrate. The magnetic thin film is, however, formed into a two-dimensional array of separated magnetically bistable mesas or pixels. As a plane polarized light transmits through the array of the mesas, the polarization of the mesas can be spatially modulated by magnetically switching through the Faraday effect. Since the Light-MOD is a transparent-type device unlike the other spatial light modulators [7.45, 7.46], the device is very suitable for application to real-time object pattern generation and spatial filter synthesis.

Because of the cross array formation of the mesas, the device has a two-dimensional mesh structure, for which the array of mesas or pixels can be switched on and off with a x-y matrix of currents. Thus each pixel of the Light-MOD would take one of the only two possible states, depending on the direction in which the pixel is magnetized. Hence we see that the Light-MOD is essentially a binary-type spatial light modulator (SLM), in contrast with the other SLMs.

Since the Light-MOD can be electrically switched, an object pattern can be written on the device with a computer; thus the Light-MOD is also a programmable SLM, which is particularly attractive for application to programmable optical processing.

In this section, however, we illustrate some of the applications of this magneto-optics device to a white-light optical signal processor [7.47]. We show that the device is capable of generating various elementary spatial filters and object patterns for optical processings. Since the device would be used in a white-light optical processor, we first demonstrate the effect of the device under the polarized white-light illumination. We show that the device responds to a wide range of polarized light, and that it is suitable for generating color coded spatial filters and color coded object patterns.

7.5.1. Effect Due to White-Light Illumination

We first illustrate the effect of spectra distribution under white-light illumination. Since the device is essentially an $n \times m$ array of transparent pixels (or mesas), the Fourier spectra of the device would have a distribution similar to that of a two-dimensional cross grating. If we insert the Light-MOD in the input plane of a white-light processor, we would expect a set of smeared rainbow color Fourier spectra distributed at the spatial frequency plane. Since the current Light-MOD under test would not respond to wavelengths smaller than the green light, their Fourier spectra would therefore only be smeared from red to green. Nevertheless, this range of spectral lines would provide us with a wide variety of colors for the polychromatic processing.

To simplify our discussion on the effect under the white-light illumination, we provide three simple object patterns of the device, as illustrated in the left-hand column of Fig. 7.23. The center column is a set of equations representing the Fourier transforms of these object patterns, where $1/l$ is the spatial frequency of the inherent grating structure of the device, l is the pixel size, λ is the wavelength of illumination, and f is the focal length of the transform lens. The corresponding smeared Fourier spectra distributions are also shown in the right-hand column of this same figure.

Let us now discuss the effect of the device under the polarized white-light illumination. We assume that an object pattern, used either as a spatial filter or as an input object, is generated by the device with a programmable computer. If the device is illuminated by a polarized white-light, then one would see the color of the object pattern change as the direction of the polarization changes. Since the device responds to a broad spectral bandwidth of light from red to green, a wide variety of color object patterns can be generated, in which the color characterization of the device is as shown in Fig. 7.24. In this figure, the outer region represents the magnetized pixels (i.e., object pattern), and the inner region represents the unmagnetized pixels (i.e., background). For example, if the polarizer is set at about an 88° angle relative to the vertical axis of the device, then the object pattern would be in yellow, which has the greatest transmittance, while the background would be in black, as can be

	Device	Spectra Contents	Spectra Distribution
(1)	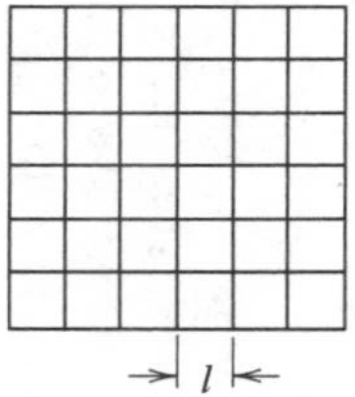 $\rightarrow\| l \|\leftarrow$	$\left(\frac{a}{l}\right)^2 \operatorname{sinc}\left(\frac{a\alpha}{\lambda f}\right)\operatorname{sinc}\left(\frac{a\beta}{\lambda f}\right) \sum_{n,m} \delta\left(\alpha - n\frac{\lambda f}{l}, \beta - m\frac{\lambda f}{l}\right)$	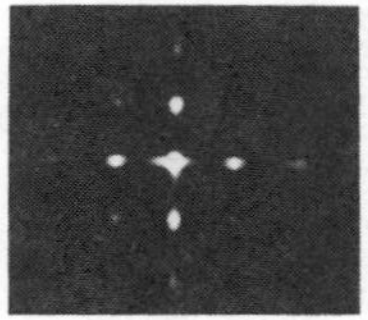
(2)	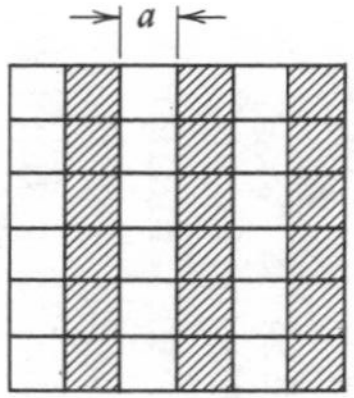 $\rightarrow\| a \|\leftarrow$	$\frac{1}{2}\left(\frac{a}{l}\right)^2 \operatorname{sinc}\left(\frac{a\alpha}{\lambda f}\right)\operatorname{sinc}\left(\frac{a\beta}{\lambda f}\right) \sum_{n,m} \delta\left(\alpha - n\frac{\lambda f}{2l}, \beta - m\frac{\lambda f}{l}\right)$	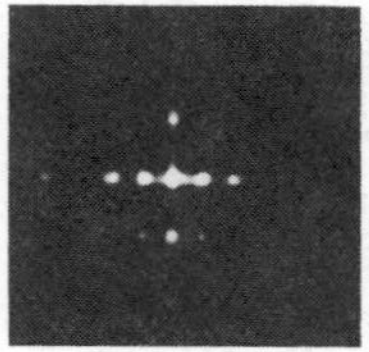
(3)	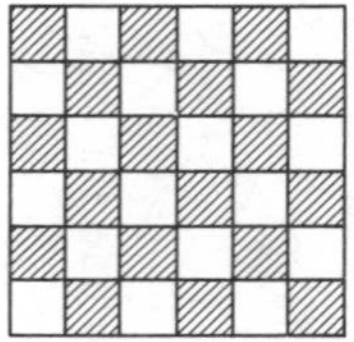	$\left(\frac{a}{2l}\right)^2 \operatorname{sinc}\left(\frac{a\alpha}{\lambda f}\right)\operatorname{sinc}\left(\frac{a\beta}{\lambda f}\right) \sum_{n,m} \delta\left(\alpha - n\frac{\lambda f}{2l}, \beta - m\frac{\lambda f}{2l}\right) \times \left\{1 + \exp\left[-i\frac{2\pi}{\lambda f}(\alpha + \beta)l\right]\right\}$	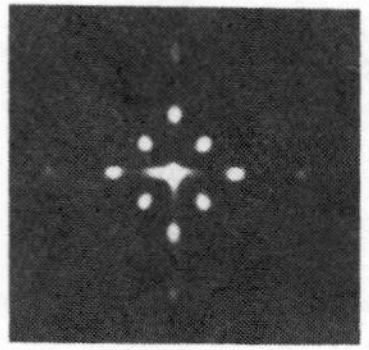

Figure 7.23. Elementary patterns of the device and the corresponding smeared Fourier spectra.

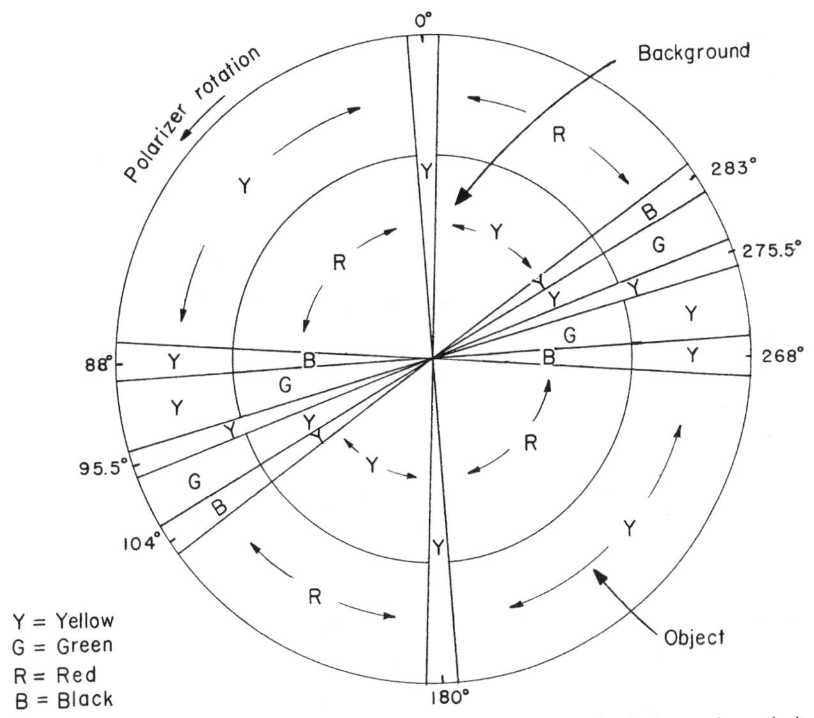

Figure 7.24. Color transmittant characteristic of the magneto-optic device under polarized white-light illumination.

seen in Fig. 7.24. If the polarizer is turned slightly counterclockwise, one would see that the object pattern remains in yellow while the background changes to dark green (or red, if the polarizer is rotated slightly clockwise). If the polarizer is advanced further counterclockwise, the background would become light green while the object pattern still remains in yellow. We further note that, if the polarizer is set at about 104°, a contrast reverse of the color object pattern appears. If the polarizer is set between 180° and 95.5°, the entire device would become yellowish; then a contrast reverse color pattern would occur with further turning of the polarilizer. As can be seen from Fig. 7.24, the color characteristics of the device are repeated for every rotation of the polarizer. Thus a wide range of color object patterns can be generated for polychromatic signal processing.

7.5.2. Elementary Spatial Filtering

We now discuss the utilization of the Light-MOD as a programmable spatial filter for white-light optical processing, as shown in Fig. 7.25. We note that

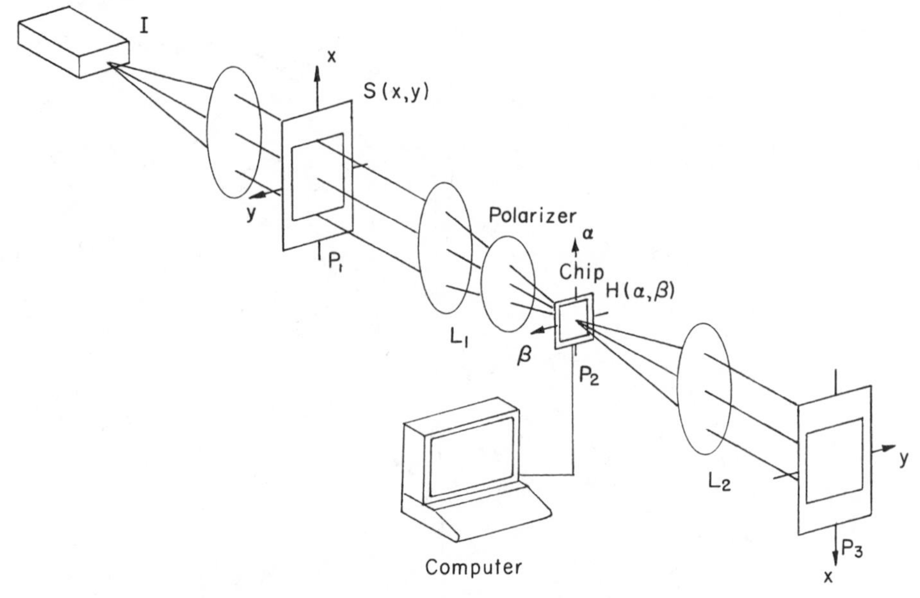

Figure 7.25. A white-light optical signal processor utilizing a programmable spatial filter. I, white-light source; $H(\alpha, \beta)$, spatial filter generated by the device; L; transform lens.

the device is driven by a computer, to generate elementary binary spatial filters (e.g., high pass and wedge filters).

Since the inherent grating structure of the device produces higher order diffractions (as shown in Fig. 7.23), the separation of these higher order signals depends on the sampling frequency (i.e., the grating structure) of the device and the spatial frequency content of the arriving complex wave field. In other words, it depends on the pixel size of the device, on the frequency content of the object spectra, and on the focal length of the transform lens.

We now investigate the effect of the output image irradiance due to the device, used as a programmable spatial filter. With reference to Fig. 7.25, the intensity distribution at the output plane can be written as

$$\begin{aligned}\mathscr{F}^{-1}[S(\alpha, \beta; \lambda)H(\alpha, \beta)] &= s(x, y) * h(x, y; \lambda) \\ &= s(x, y) * \sum_{n,m} \left(\frac{a}{l}\right)^2 \operatorname{sinc}\left(\frac{a}{\lambda f} x\right) \operatorname{sinc}\left(\frac{a}{\lambda f} y\right) \delta\left(x - n\frac{\lambda f}{l'}, y - m\frac{\lambda f}{l}\right) \\ &= \left(\frac{a}{l}\right)^2 \sum_{n,m} \operatorname{sinc}\left(n\frac{a}{l}\right) \operatorname{sinc}\left(m\frac{a}{l}\right) s\left(x - n\frac{\lambda f}{l}, y - m\frac{\lambda f}{l}\right),\end{aligned} \tag{7.34}$$

where $\mathscr{F}^{-1}$ denotes the inverse Fourier transformation, $*$ represents the convolution operation, $S(\alpha, \beta; \lambda)$ is the Fourier spectrum of the input signal $s(x, y)$, $H(\alpha, \beta)$ is the amplitude transmittance of the device used as a spatial filter, $h(x, y; \lambda)$ is the corresponding impulse response, δ is the Dirac delta function, and (x, y) and (α, β) are the spatial coordinate systems. From this equation, we see that the input signal is carried out by an $(n \times m)$ array of smeared delta functions, which are located at $(n(\lambda f/l), m(\lambda f/l))$ in the output plane P_3. We also note that the intensity of the multiple array of output signals (or images) is proportional to the magnitude square of the sinc factors in Eq. (7.34), that is,

$$I_{n,m}(x, y) \propto \left|\operatorname{sinc}\left(n\frac{a}{l}\right)\right|^2 \left|\operatorname{sinc}\left(m\frac{a}{l}\right)\right|^2. \tag{7.35}$$

Thus the intensity of the higher order images decreases rapidly as n and m increase. We would however treat the zero-order (i.e., $m = n = 0$) as the signal and the rest of the diffracted orders as noise. Thus to improve the output signal-to-noise (i.e., the unwanted signals) ratio (SNR), one would make the ratio a/l (where $a \leqq l$) approach unity. However, making the size of the pixel approach the size of the mesh cell of the device is equivalent to making the x-y drive lines very thin, which is very difficult to achieve in practice. Nevertheless, if the mesh cell of the device l is made smaller and the focal length of the transform lens f is made longer, then the separation of the output signal with respect to the unwanted noise (i.e., higher order diffractions) can be obtained. As a numerical example, given $l = 0.127$ mm for a 48×48 array Light-MOD and $f = 1000$ mm, the mean separation of the output smeared signals would be $\bar{\lambda} f/l = 4.3$ mm for $\bar{\lambda} = 5500$ Å. If $l = 0.02$ mm, then the mean separation of smeared signals would be about 27.5 mm for the same focal length. Thus the input signal size could be as large as 14 mm.

7.5.3. Experimental Demonstrations

In our experiments, we show that a piece of 48×48 array magneto-optic Light-MOD is used to generate an elementary spatial filter in the Fourier plane of a white-light optical processor, as depicted in Fig. 7.25. The pixel size of the device is about 0.127 mm and the Light-MOD is about $\frac{1}{4}$ in. square. Needless to say, elementary spatial filters can be generated by the device with a programmable computer. Now if we assume that a high pass filter, as shown in Fig. 7.26a, is being generated in the Fourier plane, then an edge enhanced image can be obtained at the output plane of the processor. Since the magneto-optic device responds to a broad band of light waves (i.e., from

Figure 7.26. Edge enhancement. (*a*) A black-and-white picture of a color coded high pass filter generated by the device. In reality, the low pass region is green and the high pass region is yellow. (*b*) An input object transparency. (*c*) A black-and-white color coded edge enhanced image. In reality, the edges of the building are yellowish, the rest of the picture is greenish.

red to green), it is a simple matter of utilizing the device to generate a color coded high pass spatial filter. For example, if the polarizer is set at an angle between 95.5° and 104° of Fig. 7.24, then the bandpass color coded filter of Fig. 7.26*a* can be generated (i.e., the low pass region is in dark green while the high pass region is in yellow). And it is apparent that a color coded edge enhanced image can be observed at the output plane.

For experimental demonstration, Fig. 7.26*b* shows a gray-scale black-and-white input object. Figure 7.26*c* is a black-and-white picture of the corresponding color coded edge enhanced image obtained with the programmable spatial filter with white-light processing. In the original color image, the edges of the building are mostly coded in yellow, while the rest of picture is greenish. Needless to say, if the polarizer is set at a different angle, then a different shade of color coded edge enhanced image can be obtained.

We now show a second experimental result, obtained with the directional, or wedge, filter [7.48, 7.49] of Fig. 7.27. Figure 7.27*a* shows a two-shade radial spectral filter, generated by the device. Figure 7.27*b* shows a black-and-white input object consisting of two sinusoidal waves taken from an oscilloscope. These two sinusoidal waveforms are of the same frequency; however, the slopes of the two waveforms are very distinctive. Thus these two sinusoids of different amplitudes can be exclusively extracted with a wedge filter. Figure 7.27*c* and *d* show the results obtained by the wedge filter of Fig. 7.27*a*, with two sequential settings of the polarizer: Fig. 7.27*c* is obtained when the vertical region of the wedge filter is coded in black while the horizontal region is coded in yellow; Fig. 7.27*d* is obtained by reversing the contrasts of the filter of Fig. 7.27*a*.

We now demonstrate a theta modulation [7.50] pseudocolor encoding technique with this device, as shown in Fig. 7.28. First the binary black-and-white input object shown in Fig. 7.29*a* is placed at the observation plane *P*. Then a pattern of the input object can be generated by the Light-MOD with a programmable computer, as shown in Fig. 7.28. The object pattern of the device consists of three distinctive spatially modulated regions, as illustrated in Fig. 7.29*b*. The inner region (i.e., the bird) is modulated with a uniform cross grating frequency in both directions, while the intermediate region is not modulated, and the outer region is modulated at a lower grating frequency in the horizontal direction. Thus the corresponding smeared object pattern spectra would be spatially separated along the horizontal axis in the Fourier plane. In pseudocolor encoding, we allow the zero-order (i.e., the intermediate region) and two first-order (i.e., the bird and the outer region) Fourier spectra to pass through three preselected color filters, (e.g., red, green, and yellow), respectively, at the spatial frequency plane of Fig. 7.28. Then a pseudocolor coded image can be obtained at the output plane, as shown in a black-and-white picture of Fig. 7.29*c*. Needless to say, different shades of the pseudocolor encoded image can be obtained with different sets of color filters and different polarizer orientations. Thus a wide range of pseudocolor images can be obtained with the theta modulation technique.

In conclusion, we have incorporated a programmable magneto-optic spatial light modulator with a white-light optical signal processor. We have shown that the device responds to polarized white light, which offers the

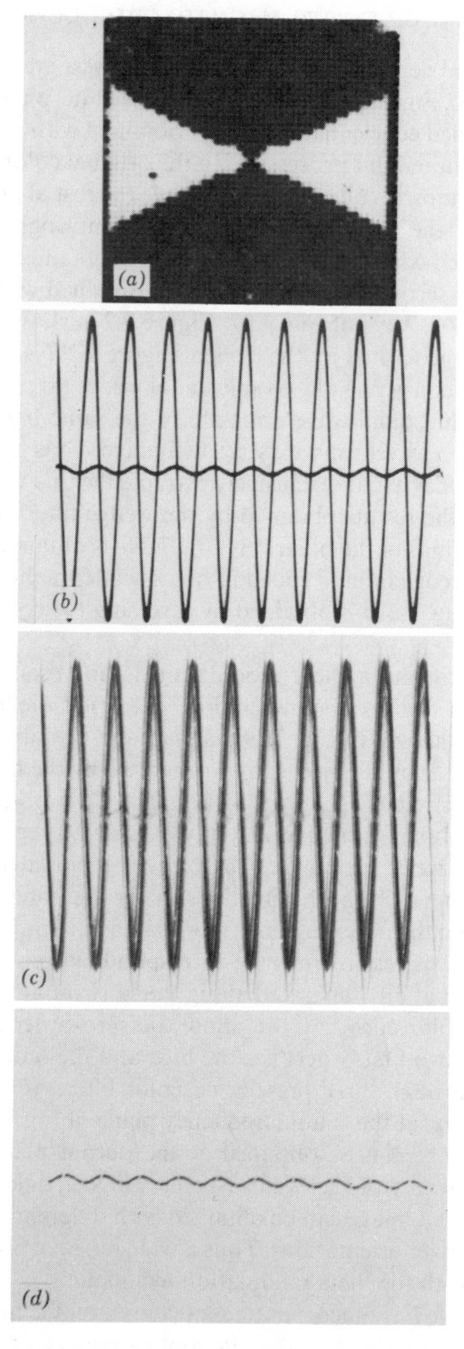

Figure 7.27. Directional filtering. (*a*) A black-and-white picture of a wedge filter. (*b*) Input signal. (*c*) Extraction of the larger signal. (*d*) Extraction of the smaller signal.

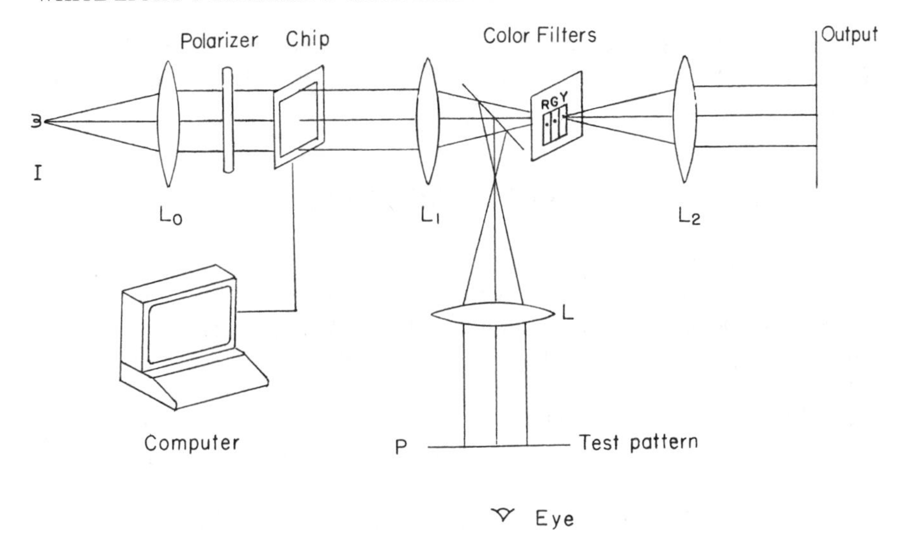

Figure 7.28. A white-light processor with a programmable input object for pseudocolor encoding. *I*, white-light source; *L*, transform lens.

advantages of color coded spatial filter synthesis and pseudocolor object image generation. The device can be used as an input programmable object and also as a programmable spatial filter for real-time optical signal processing. Even though the resolution of this device is rather limited, we have shown some interesting applications with the white-light illumination. Nevertheless, one of the important assets of the device must be the programmability, which, in principle, would offer a wide range of real-time processing capabilities. If

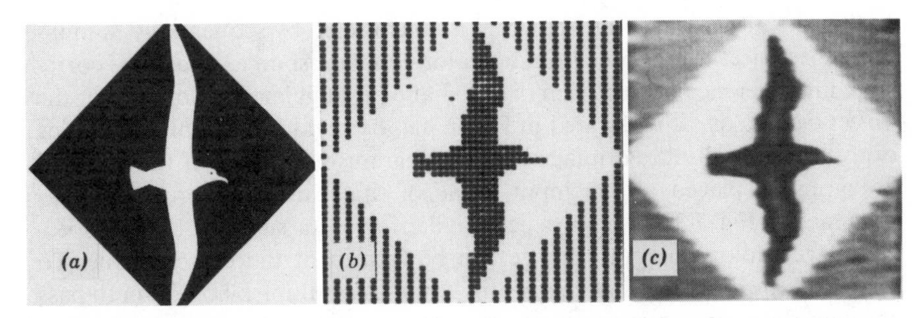

Figure 7.29. Pseudocolor encoding. (*a*) Input object. (*b*) Object pattern generated by the device. (*c*) A black-and-white color coded image. In color, the bird is yellow, the square region is green, and the outer region is reddish.

the resolution of the device can be further improved, then the device could have significant impact on the applications of modern optical signal processing.

7.6. APPLICATION TO COLOR HOLOGRAPHIC IMAGING

The technique of wavefront reconstruction, or rather the holographic process, is now over three decades old, having been discovered by Dennis Gabor in 1948 as a possible means of improving the resolving power of the electron microscope [7.51]. It was not until the early 1960s, when Leith and Upatnieks produced the first high-quality holographic image using a strong coherent light source (i.e., the laser), that wider attention was given to holography [7.52, 7.53]. Since then, many applications have been developed for holography.

Although holography has provided a great step toward practical three-dimensional imagery, its acceptance for commercial and educational uses has been slow for a number of reasons. The high cost of the holographic process, low hologram image luminance, and the necessity for special illuminators for quality images are among the primary causes for its rejection. There are, however, methods of producing color hologram images with simple inexpensive white-light sources [7.54].

One of the major advantages of white-light optical signal processing must be the color imaging. Since the white-light source emanates all the visible wavelengths, the white-light processor is particularly suitable for color holographic imagining. In the following we describe a few techniques for producing true color holographic images utilizing a simple white-light processor.

7.6.1. Spatial Holographic Encoding

In this technique, the color hologram is constructed by sequentially illuminating the object using three primary color coherent sources and three corresponding reference beams, each oriented at equi-angular positions about the object beam axis, as illustrated in Fig. 7.30*a*. By making three primary color exposures, a multiplexed image plane hologram is encoded. If the encoded hologram is placed at the input plane of the white-light processor diagrammed in Fig. 7.30*b*, then by spectral filtering of the smeared Fourier spectra, a true color hologram image can be formed at the output plane. To enable all the diffracted light emanating from the multiplex hologram to pass through the achromatic transform lens during the reconstruction process, a very narrow angle is required between the object beam and the reference beam used in the constructing process. We note that a broader angle may also be used if large aperture and short focal length lens can be found. For experimental demonstration, Fig. 7.30*c* shows a black-and-white photograph of a color holographic image obtained by this white-light processing technique. This technique allows a color holographic image to be reconstructed using a white-light source, while, in principle, eliminating the marginal resolution loss and color blur found in the rainbow holographic system [7.2].

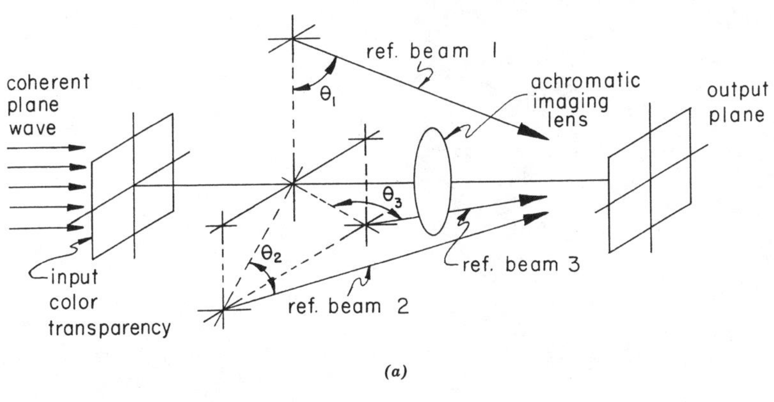

(*a*)

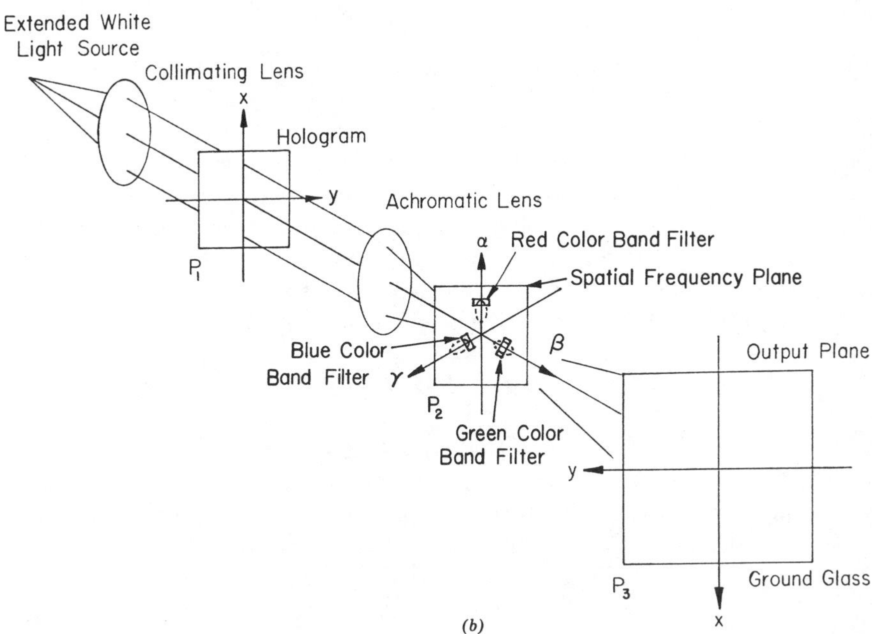

(*b*)

Figure 7.30. Color holographic imaging by holographic encoding. (*a*) Construction of a holographic encoding color hologram. (*b*) Color holographic imaging with a white-light processor.

Figure 7.30. (*c*) Black-and-white photograph of a color holographic image obtained by the holographic encoding technique.

However, some of the previous problems found in processing holographic images still exist. These include the elaborate recording scheme and the undesirable speckle effect that is inherent to coherent construction and that cannot be totally eliminated, although it is somewhat reduced in the white-light reconstruction. This technique also has a limited range of object-to-reference beam angles available for holographic construction. During the recording process the three complex holographic gratings that are formed will also introduce moiré fringe patterns. But, by an ingenious design of the holographic encoding process it may be possible to eliminate these fringe patterns.

7.6.2. Coherent and White-Light Speckle Encoding

We first demonstrate a simple technique that utilizes coherent speckles to encode color images onto a black-and-white film [7.55]. In encoding, a diffuse color object illuminated by coherent light is imaged onto a photographic plate through a narrow slit by an imaging lens as shown in Fig. 7.31*a*. Let us assume that the recording was sequentially performed using only red and green coherent illuminations with the slit first oriented in one direction, and then rotated 90°, thereby giving us red and green encoded color images multiplexed onto one photographic film. Due to the different orientations of the slit aperture, the red encoded specklegram will have speckles elongated in one

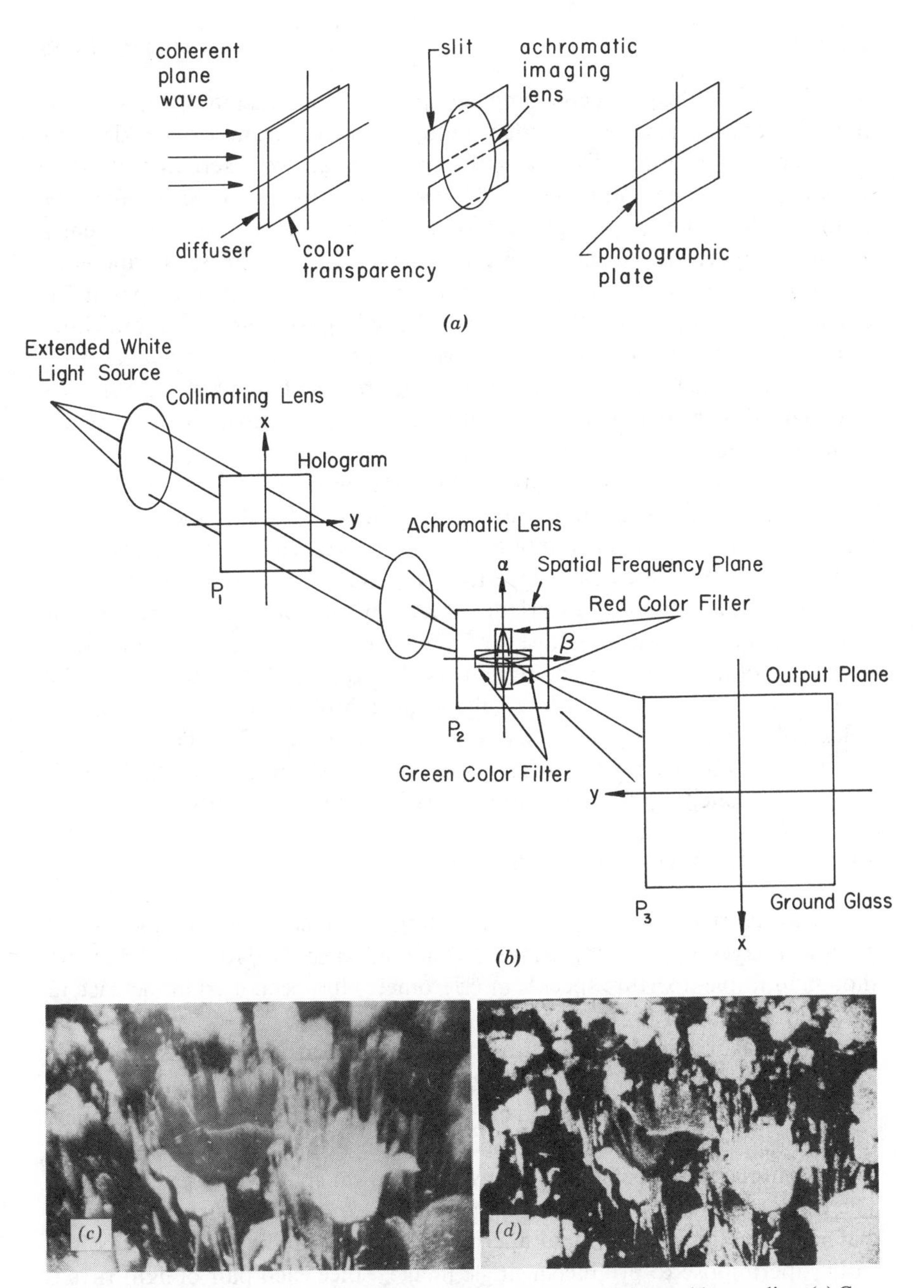

Figure 7.31. Color holographic imaging by coherent and white-light speckle encoding. (*a*) Construction of a speckle encoding color hologram. (*b*) Color holographic imaging with a white-light processor. (*c*) Black-and-white picture of a color holographic image obtained by coherent speckle encoding technique. (*d*) Black-and-white picture of a color holographic image obtained by white-light speckle encoding technique.

direction, while the green encoding has speckles elongated in the other direction (90° apart). Thus an encoded multiplex specklegram, or speckle hologram, can be recorded. Decoding the color image, we insert the multiplex specklegram in the input plane P_1 of a white-light optical processor, depicted in Fig. 7.31*b*. Since the elongated speckles of each specklegram are oriented about 90° apart, the corresponding Fourier spectra would be distributed in confined directions perpendicular to these elongations in the spatial frequency plane P_2. That is, the spectra of the red image is spread in one direction, and the spectra of the green image is spread in the other direction. By color filtering each set of Fourier spectra with respect to the red and the green filters, as shown in Fig. 7.31*b*, a full color hologram image can be reconstructed at the output image plane P_3.

For experimental demonstration, Fig. 7.31*c* shows a black-and-white photograph of a multicolor hologram image obtained by this speckle-graphic technique. Although the resolution of the reconstructed color image suffered a severe drawback, the color reproduction is relatively faithful. With the use of a finer diffuser and appropriate slit size in the construction process, an optimum color hologram image reproduction may be obtained. Finally, we note that color hologram imaging can also be obtained by white-light speckle encoding. With the same white-light optical processing technique, a true color hologram image can be viewed, as shown in Fig. 7.31*d*. This technique has many of the same drawbacks as the coherent speckles. It does, however, have the simplicity and economy of the white-light illumination.

7.6.3. Double-Aperture Encoding

We now describe a technique for recording and reconstructing color holographic images by utilizing fringe modulated speckle patterns [7.56]. Although multiple-aperture speckle interferometry has been used in the past, its application to color holographic imaging is quite new.

A process for constructing a fringe modulated speckle pattern is illustrated in Fig. 7.32*a*. The diffusing element produces a wide-band spatial frequency noise with a sufficient bandwidth for the light from each object point to pass through both apertures. The irradiance function at the recording plane generates a complicated speckle pattern by the random spatial frequency carriers. The two apertures are separated by a distance $2b$, which is greater than the diameter $2a$ of each individual aperture. These two light beams will coherently superimpose to produce a single image. Since each pair of light rays is mutually coherent, it will interfere to produce a random speckle pattern that is modulated by parallel fringes. These interference fringes are perpendicular to a line joining the aperture centers, which can be observed by placing an optical microscope in the image plane. Figure 7.32*b* shows a typical speckle

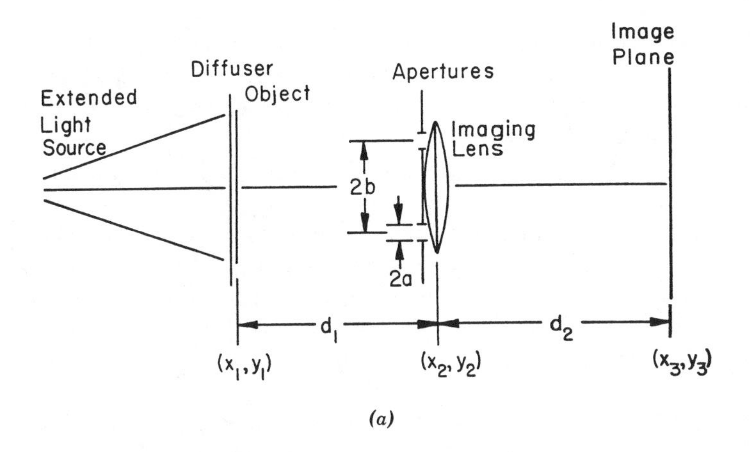

(a)

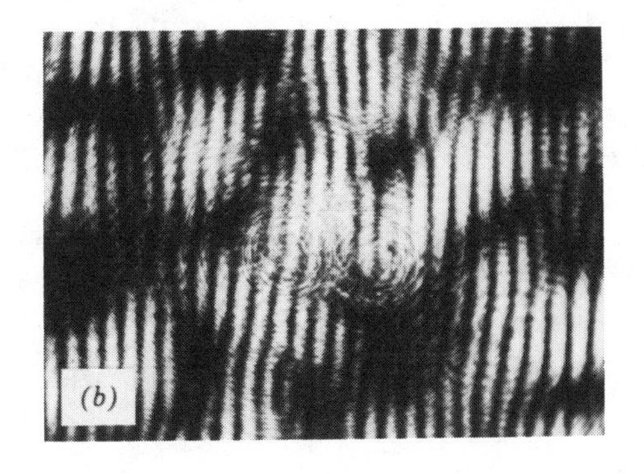

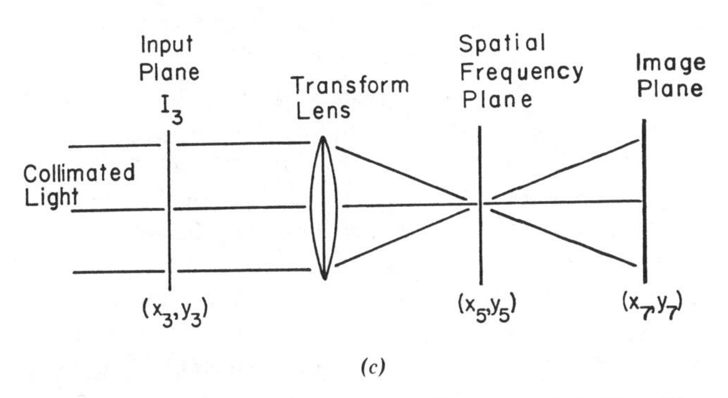

(c)

Figure 7.32. Dual-aperture holographic imaging. (*a*) Dual-aperture holographic construction. (*b*) Typical modulated speckle pattern. (*c*) Hologram image reconstruction by a white-light processor.

fringe pattern with a spatial frequency of 90 cycles/mm and a speckle diameter of 55 μm. The lower spatial frequency content of the input object and the random speckle pattern are modulated onto these holographic fringes, which act as a periodic spatial frequency carrier. Thus a hologram encoding can be recorded at the image plane, and the hologram image can be retrieved by a white-light processor, as shown in Fig. 7.32*c*.

The procedure for a three-dimensional color hologram image encoding is similar to the two-dimensional case where variations in the axial position of the object are encoded by variations in the speckle fringes spacing on the corresponding image points, as shown in Fig. 7.33*a*. A Rubik's cube with six distinct colors is illuminated by red, green, and blue coherent sources. The holographic encoding can take place either sequentially or simultaneously with the coherent illuminations. Three pairs of apertures, each containing a red, blue, or green filter, are positioned in front of the imaging lens. The various diffraction orders are recorded onto a black-and-white holographic film. If the recorded color hologram is placed at the input plane of a white-light processor, similar to that of Fig. 7.30*b*, a multicolor holographic image can be viewed at the output image plane as shown in Fig. 7.33*b*. Although this technique suffers some degree of resolution loss, the holographic encoding offers a much less stringent temporal and spatial coherence requirement, and thus can ultilize a white-light illumination [7.57]. In addition, the simplicity of this technique may offer a new dimension in color holographic imaging applications.

7.6.4. Dual-Beam Encoding

Although color hologram image encoding can be easily obtained by double-aperture speckle modulation, the technique suffers one major drawback, namely the hologram image resolution is limited by the size of the apertures. In this section we describe a technique of dual-beam encoding to alleviate this disadvantage, as illustrated in Fig. 7.34*a*.

With reference to this figure, the color holographic encodings can take place sequentially or simultaneously by dual-beam coherent illuminations, each with a specific spatial sampling direction. In other words, the encoding can be performed by spatial sampling of the color object with the red, green, and blue coherent lights. Since these primary colors of coherent lights are mutually incoherent, a spatially encoded color image can be recorded on a black-and-white photographic plate to form a multiplexed hologram. If this recorded hologram is inserted at the input plane of a white-light processor (see Fig. 7.30*b*), then a true color hologram image with high quality can be observed at the output image plane.

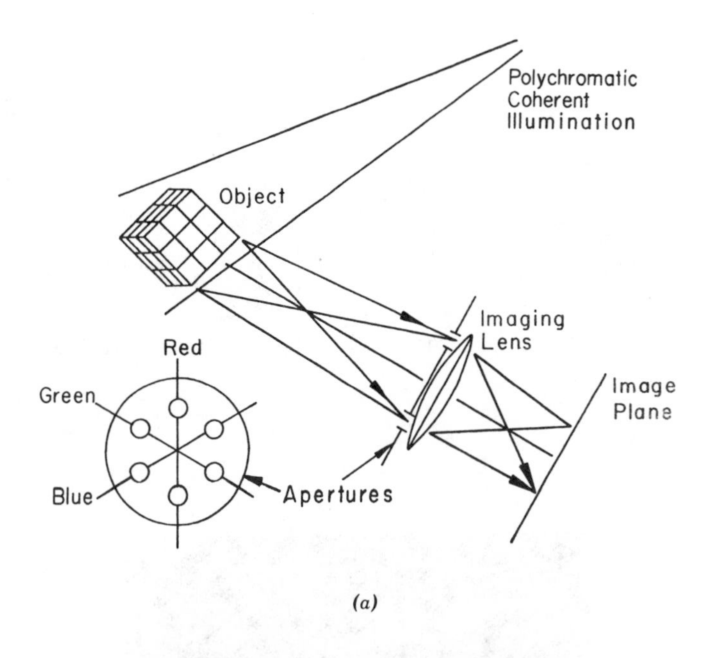

(a)

Figure 7.33. Color hologram generation by multi-aperture encoding. (*a*) Color hologram construction. (*b*) Black-and-white photograph of a color hologram image obtained by the multi-aperture encoding technique.

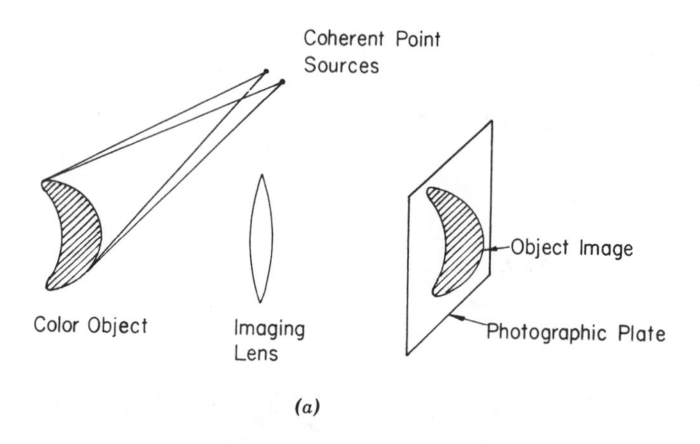

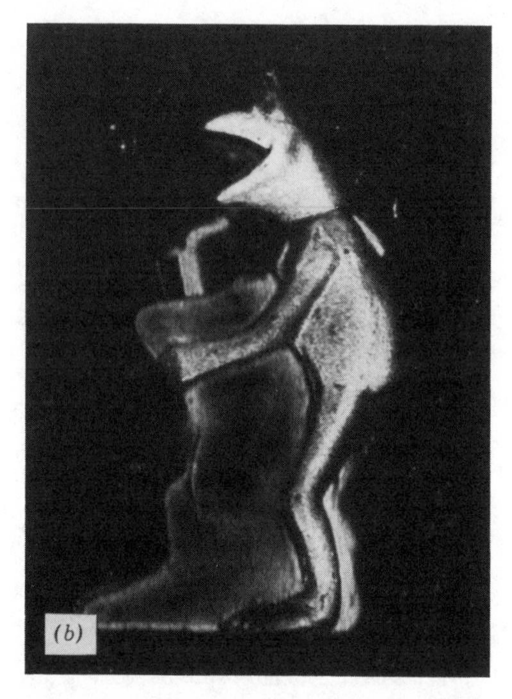

Figure 7.34. Color holographic imaging by dual-beam encoding. (*a*) Color hologram construction. (*b*) Black-and-white photograph of a color hologram image obtained by the dual-beam encoding technique.

For experimental demonstration, a black-and-white photograph of a multicolor hologram image obtained by the dual-beam encoding technique is shown in Fig. 7.34*b*. For simplicity of demonstration, we have used a red and green dual-beam coherent sources for the encodings. Although the quality of this preliminary reconstructed color hologram image appears to be somewhat poor, it may be improved by using a finer encoding frequency and a better optical setup. We feel that an optimum color hologram technique using dual-beam encoding can be found. And, in any cases, the color reproduction obtained in the hologram image is rather faithful.

Finally, we note that the advantages of the dual-beam encoding technique, in principle, can produce a high resolution color hologram image limited only by the imaging lens. This technique is suitable for three-dimensional color hologram image reconstruction with limited depth. In addition this technique eliminates the use of a reference beam, which simplifies the hologram construction. However, with use of coherent illumination for the hologram construction, the inherent coherent noise can not be eliminated.

7.6.5. Incoherent Spatial Encoding

In this section we illustrate an incoherent sampling technique such that a color image can be spatially encoded onto a black-and-white film by a tricolor grid with white-light illumination, as shown in Fig. 7.35*a*. In other words, an image sampling hologram may be obtained by spatially sampling the primary colors of a color image with three spatial directions onto a black-and-white film. For example, the red image of the object is sampled in one direction, while the green and the blue images of the object are sampled in 60° and 120° directions, respectively. Thus the three primary color images of the object are spatially multiplexed onto a black-and-white film to form a multiplexed color hologram. However, this encoded on-focused hologram is a negative image hologram, which is not suitable for color hologram image reconstruction by a white-light processing technique. To alleviate this drawback, either a contact printing process to obtain a positive encoded image or a bleaching process to convert the encoded transparency into a phase object hologram can be used. If the positive encoded transparency or the phase object hologram is inserted in the input plane of the white-light processor shown in Fig. 7.30*b*, then a true color hologram image can be viewed at the output plane of the processor, as shown in Fig. 7.35*b*. For simplicity, this true color hologram image was obtained utilizing only a red and green cross grid. The advantage of this technique is that the color multiplexed hologram can be constructed by simple white-light illumination. However, the hologram image loses its three-dimensionality, due to the incoherent construction.

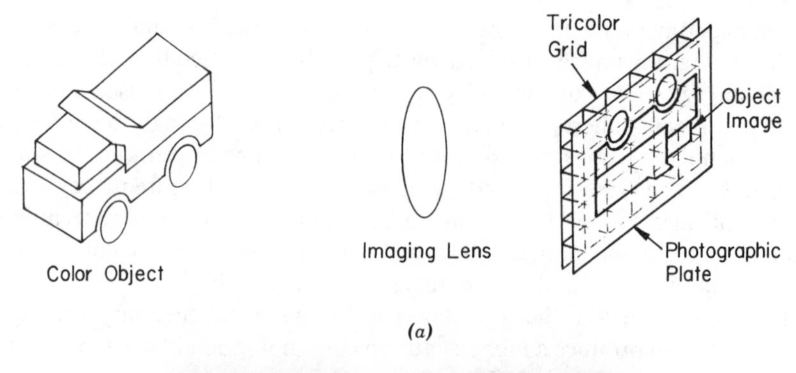

Figure 7.35. Color holographic imaging by incoherent spatial encoding. (*a*) Construction of a color hologram with a tricolor grid. (*b*) Black-and-white picture of a color hologram image obtained by the incoherent spatial encoding technique.

Nevertheless, this technique offers a high quality color hologram image with virtually no coherent artifact noise.

7.6.6. Incoherent Source Encoding

We have, in Section 6.3, described a color image subtraction technique utilizing encoded extended incoherent sources. We have shown that with an appropriate source encoding a point-pair spatial coherence function can be obtained at the input plane of an incoherent optical processor. We utilize this source encoding concept to establish the coherent requirement and to form a color hologram. The true color hologram image can then be reconstructed with a white-light processor.

Again, for simplicity, we use only two primary color light sources for the color hologram encoding. This source encoding system, shown in Fig. 7.36*a*, is similar to the system for color image subtraction [7.58], except that the diffraction gratings are placed at a distance l from the Fourier plane P_2. With

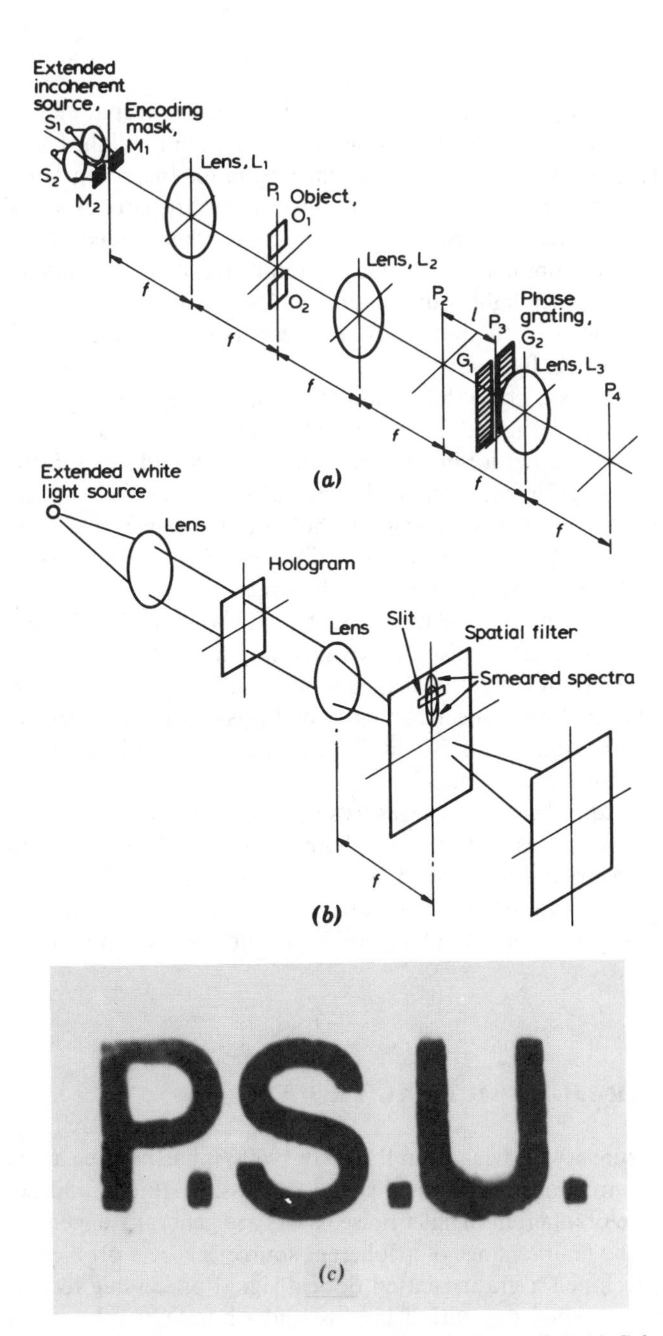

Figure 7.36. Color holographic imaging by incoherent source encoding. (*a*) Color hologram construction with incoherent source encoding. (*b*) Color holographic imaging with a white-light processor. (*c*) Black-and-white picture of a color hologram image obtained by the incoherent source encoding technique, where "P." is green, "S." is red, and "U." is yellow.

reference to this figure, one sees that if a color object transparency is inserted in one of the input open apertures, say $O_1(x, y)$, then the light field from the open aperture O_2 would act as a reference beam for the hologram construction. The two beams of light, one from the object transparency and the other from the open aperture, are mutually coherent when added at the output plane around the optical axis; therefore, a multiplexed color hologram, from the red and the green light sources, is formed at the output plane. The carrier spatial frequencies of the hologram are dependent on the distance l of the wavelength λ of the light sources.

If we insert this encoded hologram at the input plane of a white-light processor, as shown in Fig. 7.36*b*, two sets of Fourier spectra, one for red image encoding and the other for green image encoding, would smear into rainbow color. By pre-implementation of the spatial frequency carriers of red and green image components during the holographic construction, it can be shown that the red and green spectra of the smeared Fourier spectra appear at the same band in the Fourier plane. By using a narrow slit filtering, a true color hologram image can be reconstructed, as shown in Fig. 7.36*c*. Needless to say, this technique can also be extended for three-color hologram encoding. By properly adjusting the distances of the diffraction gratings, it is possible to make the three spectral bands of the smeared Fourier spectra be diffracted in the same band. By slit filtering, a true color hologram image can be viewed at the output plane.

This technique, however, possesses several drawbacks, although the color hologram construction utilizes incoherent sources. The holographic construction is restricted to a two-dimensional object. It can be seen from Fig. 7.36*c* that the hologram image quality is rather poor. This technique does, however, show another way of generating a color hologram with incoherent light.

7.7. SOLAR-LIGHT OPTICAL PROCESSING

Since the discovery of the laser in the early 1960s, it has become an indispensable light source for most optical signal processing. But in addition to the disadvantage of inherent artifact noise, lasers are generally expensive, and in some cases the maintenance of a coherent source can be a problem. Recently we have developed a grating-based optical signal processing technique that can be easily carried out with a broad-band white-light source. The major advantages of the white-light processing, in contrast with its coherent counterpart, are the simplicity, versatility, polychromaticity, cost efficiency, and artifact noise immunity.

In this section we demonstrate that several elementary optical image processings can be easily carried out by solar, or sunlight, illumination [7.59].

7.7.1. As Applied to Color Image Retrieval

Let us briefly describe a color image retrieval system utilizing solar light. Let us assume that a spatially encoded transmittance of a color object transparency is (see Section 6.4)

$$\begin{aligned} t(x, y) \cong \exp[iM\{T_r(x, y)][1 + \operatorname{sgn}(\cos p_r y)] \\ + T_b(x, y)[1 + \operatorname{sgn}(\cos p_b x)] \\ + T_g(x, y)[1 + \operatorname{sgn}(\cos p_g x)]\}], \end{aligned} \tag{7.36}$$

where: M is an arbitrary constant; T_r, T_b, and T_g are the red, blue, and green image irradiances of the color object; p_r, p_b, and p_g are the respective carrier spatial frequencies; (x, y) is the spatial coordinate system of the encoded film; and

$$\operatorname{sgn}(\cos x) \triangleq \begin{cases} 1, & \cos x \geqq 0, \\ -1, & \cos x < 0. \end{cases}$$

If we place the encoded film of Eq. (7.36) at the input plane of a solar optical processor, as shown in Fig. 7.37, then the first-order complex wave field at the Fourier plane, for every λ, would be

$$\begin{aligned} S(\alpha, \beta, \lambda) \cong\ & \hat{T}_r\left(\alpha, \beta \pm \frac{\lambda f}{2\pi} p_r\right) + \hat{T}_b\left(\alpha \pm \frac{\lambda f}{2\pi} p_b, \beta\right) \\ & + \hat{T}_g\left(\alpha \pm \frac{\lambda f}{2\pi} p_g, \beta\right) + \hat{T}_r\left(\alpha, \beta \pm \frac{\lambda f}{2\pi} p_r\right) * \hat{T}_b\left(\alpha \pm \frac{\lambda f}{2\pi} p_b, \beta\right) \\ & + \hat{T}_r\left(\alpha, \beta \pm \frac{\lambda f}{2\pi} p_r\right) * \hat{T}_g\left(\alpha \pm \frac{\lambda f}{2\pi} p_g, \beta\right) \\ & + \hat{T}_b\left(\alpha \pm \frac{\lambda f}{2\pi} p_b, \beta\right) * \hat{T}_g\left(\alpha \pm \frac{\lambda f}{2\pi} p_g, \beta\right), \end{aligned} \tag{7.37}$$

where: $\hat{T}_r$, $\hat{T}_b$, and $\hat{T}_g$ are the Fourier transforms of T_r, T_b, and T_g, respectively; $*$ denotes the convolution operation; and the proportional constants have been neglected for simplicity. We note that the last cross product term of Eq. (7.37) would introduce a moiré fringe pattern, which can be easily masked out at the Fourier plane. Thus by proper color filtering of the smeared Fourier

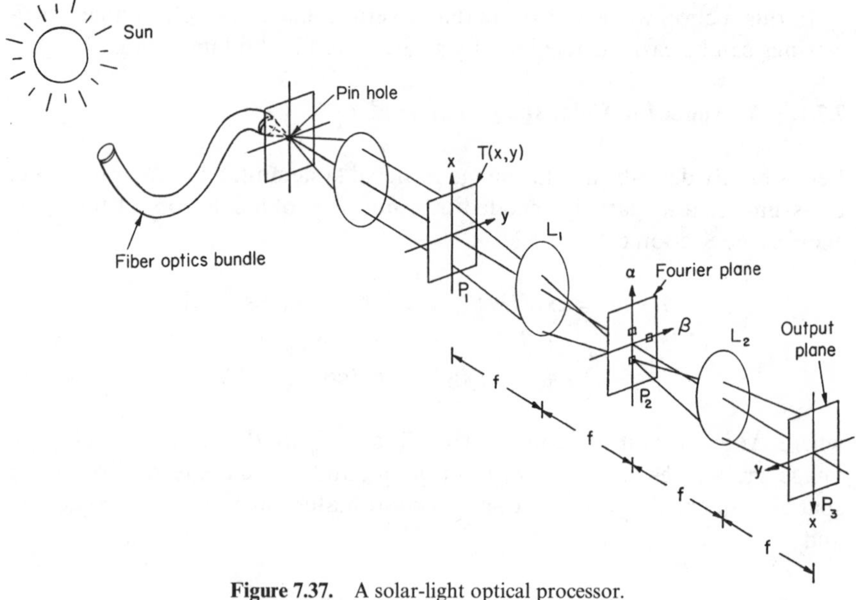

Figure 7.37. A solar-light optical processor.

spectra, as illustrated in Fig. 7.37, a true color image can be retrieved at the output image plane P_3. The corresponding output image irradiance would be

$$I(x, y) = T_r^2(x, y) + T_b^2(x, y) + T_g^2(x, y), \tag{7.38}$$

which is a superposition of three primary encoded color images. Thus we see that a moiré-free color image can be retrieved from natural solar light.

For experimental demonstration, we show, in Figure 7.38, a retrieved color image obtained from the solar-light optical processing system. In view of the retrieved color image, we see that the reproduced color image is spectacularly faithful with respect to the original color object, and the image contains virtually no coherent artifact noise.

7.7.2. As Applied to Pseudocolor Encoding

Let a three-level (i.e., positive, negative, and intermediate level) spatially encoded transparency be (see Section 6.5.1)

$$\begin{aligned} t(x, y) \cong \exp[iM\{&T_1(x, y)[1 + \operatorname{sgn}(\cos p_1 y)] \\ &+ T_2(x, y)[1 + \operatorname{sgn}(\cos p_2 x)] \\ &+ T_3(x, y)[1 + \operatorname{sgn}(\cos p_3 x)\}], \end{aligned} \tag{7.39}$$

Figure 7.38. A black-and-white picture of a retrieved color image.

where: M is an arbitrary constant; T_1, T_2, and T_3 are the positive, the negative and the negative and positive product image exposures; and p_1, p_2, and p_3 are the respective carrier spatial frequencies.

Again by insertion of the encoded transparency of Eq. (7.39) at the input plane P_1 of the solar-light processor of Fig. 7.37, and by various color filterings of the smeared Fourier spectra in the Fourier plane, a density color coded image can be obtained at the output plane, such as

$$I(x, y) = T_{1r}^2(x, y) + T_{2b}^2(x, y) + T_{3g}^2(x, y), \tag{7.40}$$

where T_{1r}^2, T_{2b}^2, and T_{3g}^2 are the red, blue, and green intensity distributions of the three spatially encoded images. Thus we see that a density color coded image can be easily obtained with solar light.

Figure 7.39 shows a density pseudocolor encoded x-ray image of a woman's pelvis, obtained with the solar-light processing. In this color encoded image, the positive image is encoded in red, the negative image is encoded in blue, while the product image is encoded in green. By comparing the pseudocolor coded image with the original x-ray picture, it appears that the soft tissues can be better differentiated by the color coded image.

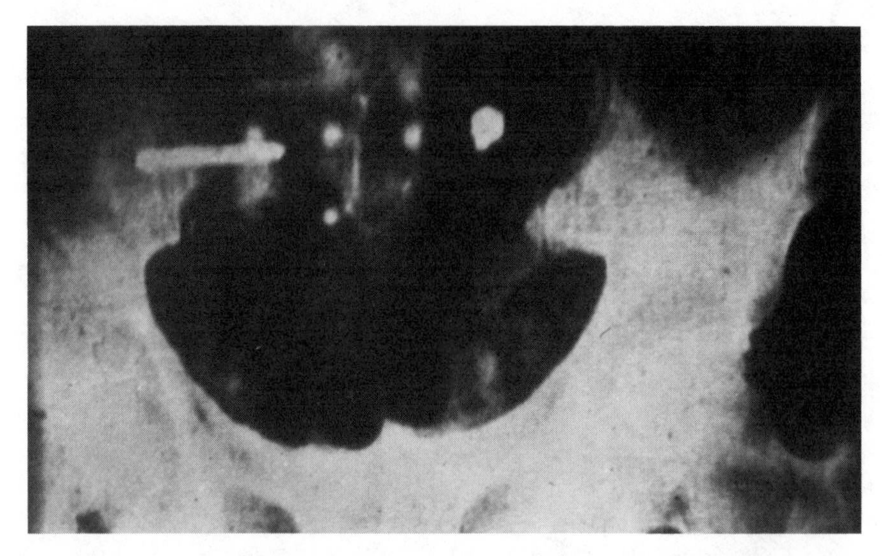

Figure 7.39. A black-and-white picture of a density pseudocolor encoded x-ray image.

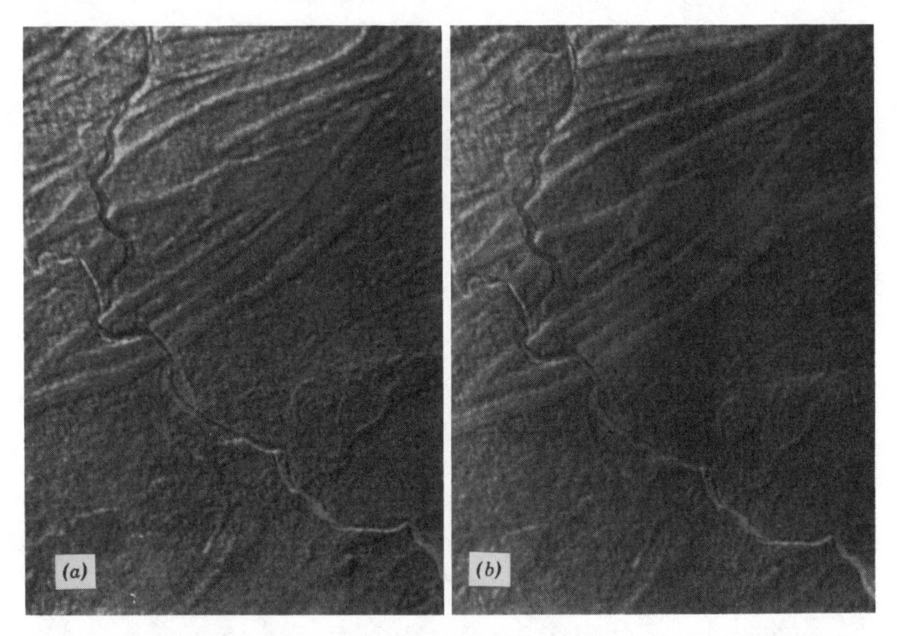

Figure 7.40. Black-and-white pictures of pseudocolor Landsat images. (*a*) Band 4 is encoded in green while band 5 is encoded in red. (*b*) Band 4 is encoded in green, band 5 is encoded in red, and band 7 is encoded in blue.

7.7.3. As Applied to Multispectral Landsat Images

Pseudocolor encoded Landsat images would allow us to discriminate various earth surface features. For example, forests, agricultural lands, water, urban areas, and strip mines can be shown on the color coded images with each feature displayed in a different color. With reference to the spatial encoding method described previously, an N-spectral band encoded transparency may be described as (see Section 6.5.4)

$$t(x, y) \cong \exp\left[iM \left\{ \sum_{n=2k+1}^{N/2} T_n(x, y)[1 + \operatorname{sgn}(\cos p_n y)] + \sum_{n=2k}^{N/2} T_n(x, y)[1 + \operatorname{sgn}(\cos p_n x)]^2 \right\} \right], \qquad k = 1, 2, \ldots, \tag{7.41}$$

where the T_n are the multispectral band image irradiances. We note that the use of the orthogonal samplings (i.e., spatial encoding directions) is to avoid the moiré fringes of the output image. Needless to say, if the encoded film of Eq. (7.41) is inserted at the input plane of the solar-light optical processor, then pseudocolor encoding can take place in the Fourier plane. The output pseudocolor coded image irradiances can be shown to be

$$I(x, y) = \sum_{n=1}^{N} T_n^2(x, y; \lambda_n), \tag{7.42}$$

where T_n^2 represents the irradiance of the nth spectral band image and λ_n denotes the corresponding coded color. Thus we see that a pseudocolor coded multispectral image can be easily obtained with solar-light processing.

For simplicity, three bands of multispectral scanner Landsat data were processed for pseudocoloring. These bands were from the blue-green (band 4: 0.5–0.6 μm), red (band 5: 0.6–0.7 μm), and reflected infrared (band 7: 0.8–1.1 μm) spectral regions. The scene is a 78 × 107 km subsample of a Landsat scene showing a section of the Susquehanna River Valley in Southeastern Pennsylvania. Figure 7.40 shows the results of the pseudocolor coded Landsat data obtained with the solar optical processing technique. In Fig. 7.40*a* where band 4 is encoded green and band 5 is encoded red, the Susquehanna River and small bodies of water are delineated as deep red. The islands in the Susquehanna River are easily distinguished. Strip mines are dark red, urban areas (Harrisburg) are medium red, and agricultural lands are light red, orange, and yellow. Forested areas are green. In Fig. 7.40*b*, where band 4 is encoded in green, band 5 is encoded in red, and band 7 is encoded in blue, the Susquehanna River appears as violet, and the other bodies of water as shades of blue. The surface mines and urban areas are dark red. The agricultural valleys are orange and the forested regions are green. Thus again we see that pseudocolor coded images can be easily obtained with a solar optical processor.

In conclusion we show that the proposed white-light processor can also utilize natural solar light for image processing. One of the obvious advantages of the solar-light processing is that the system does not require its own light source. Thus the proposed solar optical processor is very suitable for spaceborne or satellite optical processing applications. One can imagine that if an orbiting spaceborne satellite optical processor were required to carry its own light source, for example a powerful laser, then, aside the heavier payload, the question would arise of how long the light source would last. If natural solar light is used, we can easily see that the optical system would continue to function for a great number of years, possibly beyond our present civilization. Although the development of solar optical processing is still in its infancy, it is not difficult to predict that it will offer many useful applications, particularly in the areas of space communications and signal processing.

7.8. A LOW-COST WHITE-LIGHT OPTICAL PROCESSOR

Modern optical signal processing systems are generally composed of a complex array of equipment. A simple optical processor offers distinct advantages over complex systems for some processing techniques. Such a simple system is of immense interest to educators if it can be offered at a low cost. The motivation to develop such a low-cost processor (LCP) may be suspect to some people, in that high technology and sophisticated equipment are visible in nearly all of our lives. Yet to the educator who requires optical equipment for an undergraduate-level optics laboratory, a less sophisticated (i.e., less expensive) but adequate optical processing system is attractive. The LCP is thus based on the philosophy of seeing how much can be done with how little [7.60]. There is an elegance in returning to the basic tools that the early investigators in optical processing used and applying more recently developed techniques to them.

7.8.1. Construction

The LCP is built on a base of $\frac{3}{4}$ in. plywood. The lenses are held in wooden mounts that are slotted to allow adjustments along the wooden rails attached to the base. Each rail is faced on one side with a metal strip to allow a screw in each lens to tighten on the rail, fixing the mount in place. Object transparen-

Figure 7.41. The low-cost white-light optical processor (LCP).

cies and filters are held in mounts that are similar to the lens mounts in their operation. Translators were fabricated in a machine shop and run on a simple uncalibrated machine screw. The entire base is mounted on height-adjustable bolts, serving as legs, at each corner. The complete LCP is shown in Fig. 7.41. Normally, for good vibration isolation, the slide projector is placed on an adjacent stand. A complete list of equipment is shown below:

Equipment	Quantity	Model
Lens and mount	2	Wollensak 209-mm *f*/4.5 Raptar copy lens
Slide projector	1	Kodak 600H with 300-W lamp
Transparency mount	3	
LCP base with rails	1	
Translators	2	
Vibrating mirror	1	General Scanning, Inc., Model G330
Photodetector	1	
Signal generator	1	
Oscilloscope	1	

The copy lenses are not achromatic, so some chromatic aberration of color images is inherent, although not limiting. The photodetector is a photodiode powered by self-contained batteries, but any detector available is suitable. The vibrating mirror is mounted on a 60-Hz solenoid, driven by a sinewave generator. The signal generator and oscilloscope used in this system were not low-cost, as is the other equipment, but low-cost versions of these two items do exist. Thus, we see that the LCP is truly a low-cost, simple, and portable white-light optical processor.

Four optical signal processing demonstrations with the LCP are shown in the following: In the first the LCP is used as an optical scanner correlator. Second, the LCP is used as a color schlieren optical system. Third, the LCP is used to process bubble chamber photographs with several different spatial frequency filtering techniques. Finally, the LCP is used as a pseudocolor encoder. These techniques demonstrate a wide range of applications in which the LCP can be used to aid in teaching optical processing.

7.8.2. Scanner Optical Correlator

The LCP is configured as an optical correlator [7.61], as shown in Fig. 7.42. The signal generator drives the vibrating mirror to scan the image of $f(x, y)$ across the transparency $g(x, y)$. By simply reversing $f(x, y)$ to form $f(x, -y)$, the convolution of the two functions can be observed.

The signal transparencies $f(x, y)$ and $g(x, y)$ are shown in Fig. 7.43*a*, along with a calculated result for the correlation and convolution of these transparencies. The output irradiance of the optical correlator is shown in Fig. 7.43*b*. The output shows the result of the correlation of the two transparencies and their convolution. Slight differences occur between the expected results and the scan profiles; these must be attributed to the photodetector. It is clear that

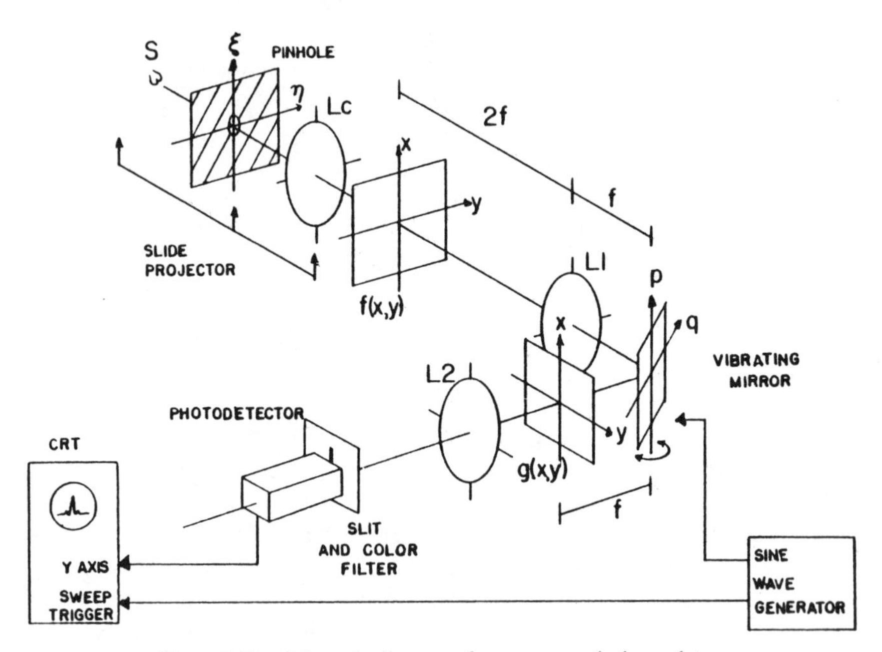

Figure 7.42. Schematic diagram of a scanner optical correlator.

the LCP can effectively demonstrate correlation and convolution in one dimension for a simple slit pattern.

A more complex problem is the correlation of transparencies in both the x- and y-directions. To demonstrate, the second transparency $g(x, y)$ was mounted on a translator to allow adjustment in the x-direction. The maximum output of the photodetector was found by manually adjusting the translator in the x-direction. That position then represented the highest level of correlation in the x- and y-directions of the two transparencies. Fig. 7.44 shows the signal transparencies and the results of this two-dimensional correlation. Note that even though the triangle and star shape can be circumscribed inside the circle, the autocorrelation function is higher in amplitude than any of the cross correlations with the other shapes.

A problem related to correlation is the performances of correlation of color transparencies. In this case, we consider not only a spatial, but a spectral correlation as well.

The response of the photodetector was made nearly panchromatic by placing a blue-green filter at the input of the detector. The transmittance of the first transparency was controlled by matching a neutral density filter with the red and green filters, as shown in Fig. 7.45. The results of the correlation of

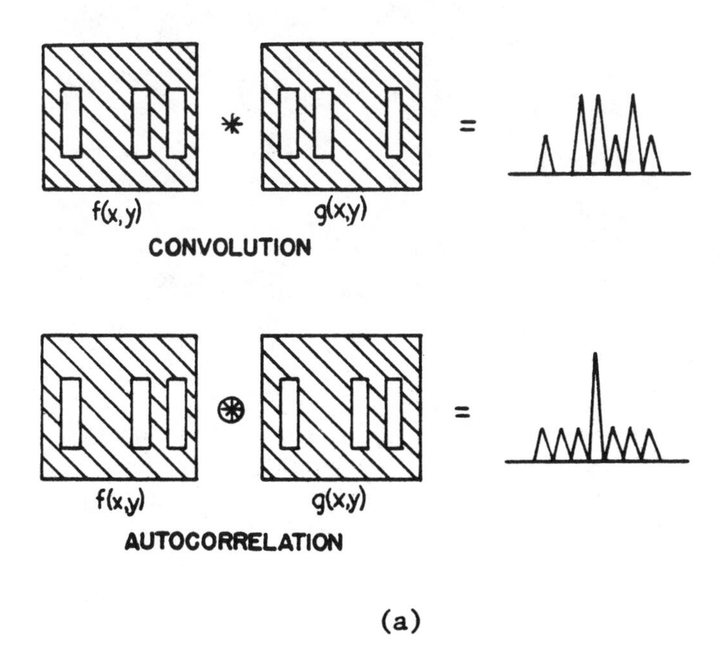

(a)

Figure 7.43. (*a*) Calculated result for convolution and autocorrelation of three-slit transparency.

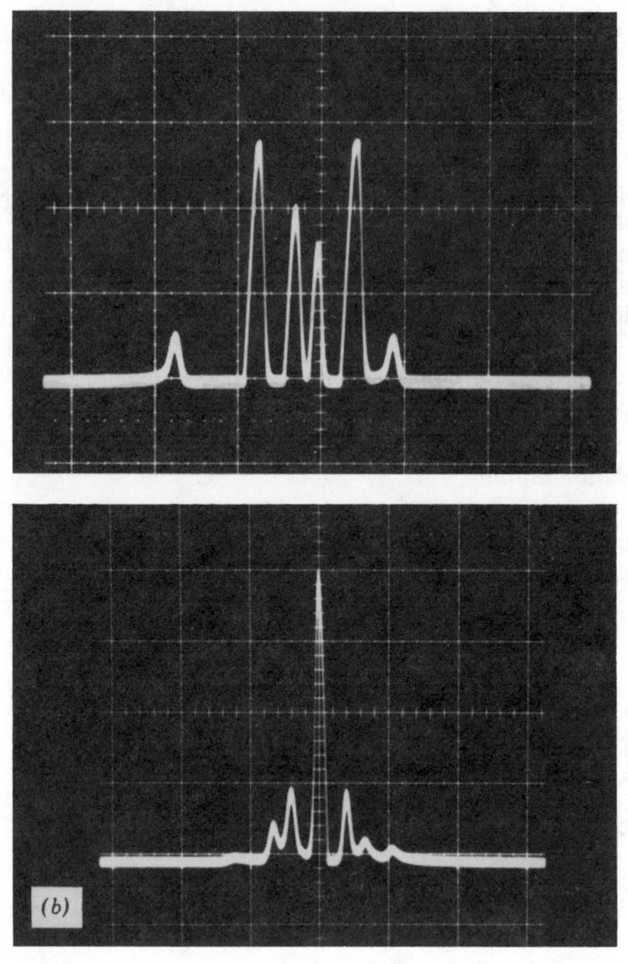

Figure 7.43. (*b*) Output intensity distributions of the convolution and the autocorrelation obtained from the LCP.

these transparencies, seen in Fig. 7.45, are expanded to show a single sweep of the correlator. The relative heights of the outputs show that, even though a high degree of spatial correlation exists in the two transparencies (in fact they are identical), the degree of spectral correlation causes a difference in the correlation of the two transparencies.

It is clear, in each case, that the LCP can be used as a low-cost method of demonstrating and teaching the principles of correlation and convolution. The fact that even more complex problems in correlation of two-dimensional and color transparencies can be shown is a bonus.

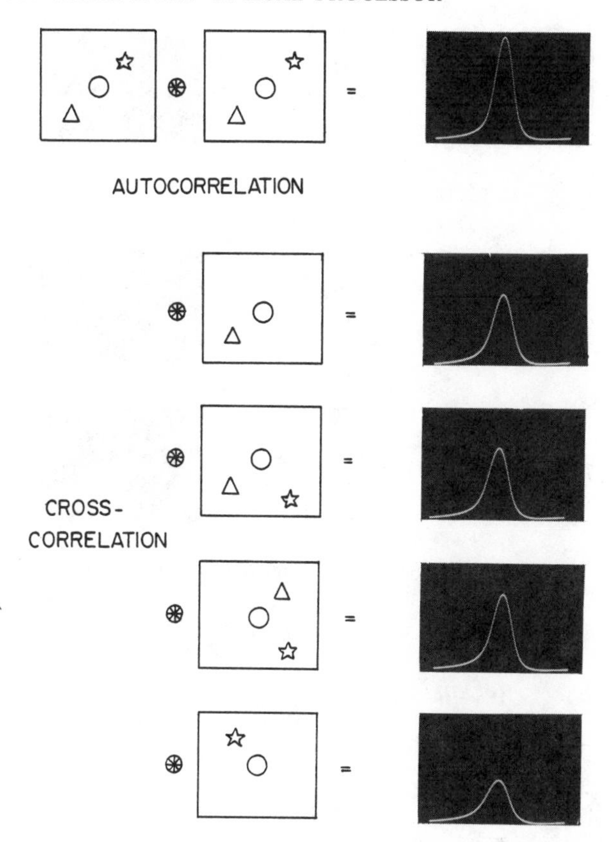

Figure 7.44. Output photometer traces of two-dimensional correlation for various signal transparencies.

7.8.3. Color Schlieren Optical Processing

The German word "schliere" has come to mean, in a transparent medium, a local inhomogeneity that causes a diffraction of light [7.62]. The schlieren method is used extensively in visualization of flow in aerodynamic and thermodynamic research, but the technique can be applied to many flow visualization problems.

The schlieren method is an incoherent processing teechnique, and the processor is therefore linear in irradiance. The diffraction of the light passing through the transparent object is proportional to the derivative of the object phase, which, in a fluid, is dependent on the change in refractive index [7.63].

In color schlieren optical processing [7.62–7.65], the LCP is configured as shown in Fig. 7.46. The source encoding masks used in the system are shown

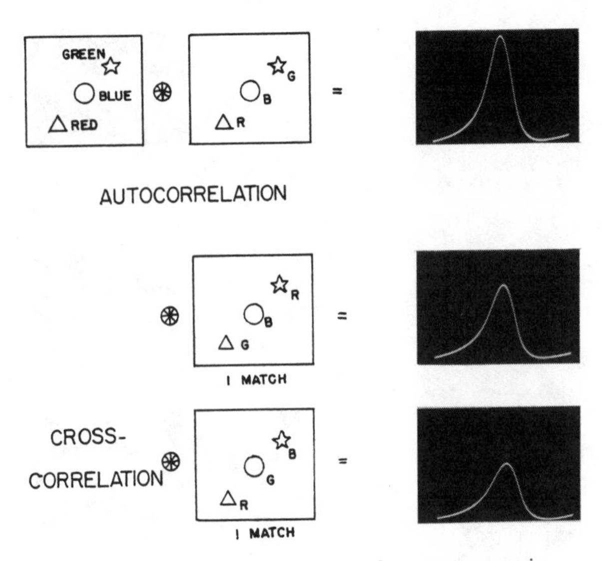

Figure 7.45. Spectral correlation of two transparencies.

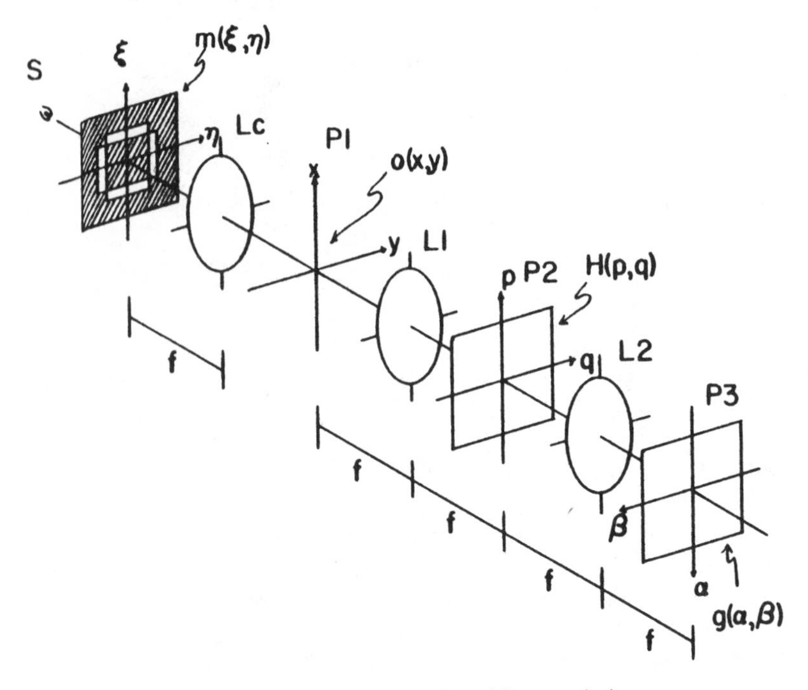

Figure 7.46. Schematic diagram of color schlieren optical processor.

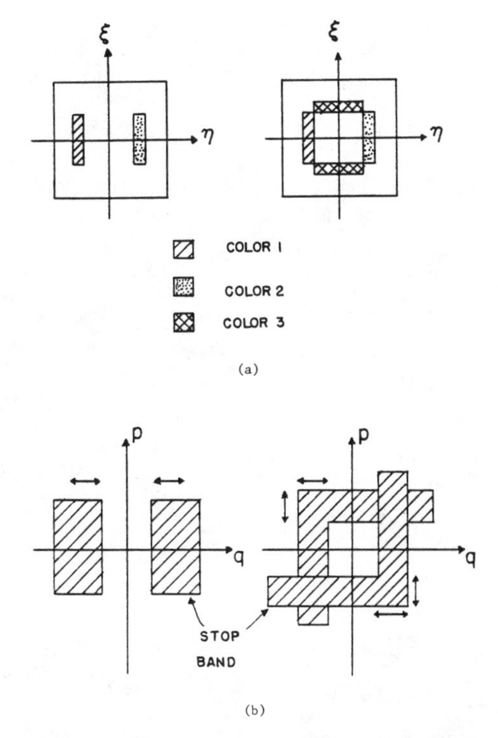

Figure 7.47. (*a*) Source encoding masks for color schlieren optical processing. (*b*) Spatial frequency filters.

in Fig. 7.47*a*, and the corresponding filters used are shown in Fig. 7.47*b*. The filters were placed in uncalibrated translators to allow for adjustment in the p- and q-directions.

To further demonstrate the ability of the color schlieren processing technique to provide information about a transparent object, a stationary three-dimensional object was presented at the input in the form of a waterdrop. The waterdrop can be considered as a transparent object that refracts light due to a gradient in its thickness.

From Fig. 7.48*a* it is clear that some difficulty exists in the interpretation of a schlieren photograph without color. The light and dark areas do impart some information concerning the contours of the waterdrop, but do not clearly indicate the direction of the inhomogeneities. In the black-and-white picture of a color schlieren image, Fig. 7.48*b*, the direction and slope of the contours are much more clearly indicated. The colors are very vivid, and their blending indicates the direction of the contours in two dimensions.

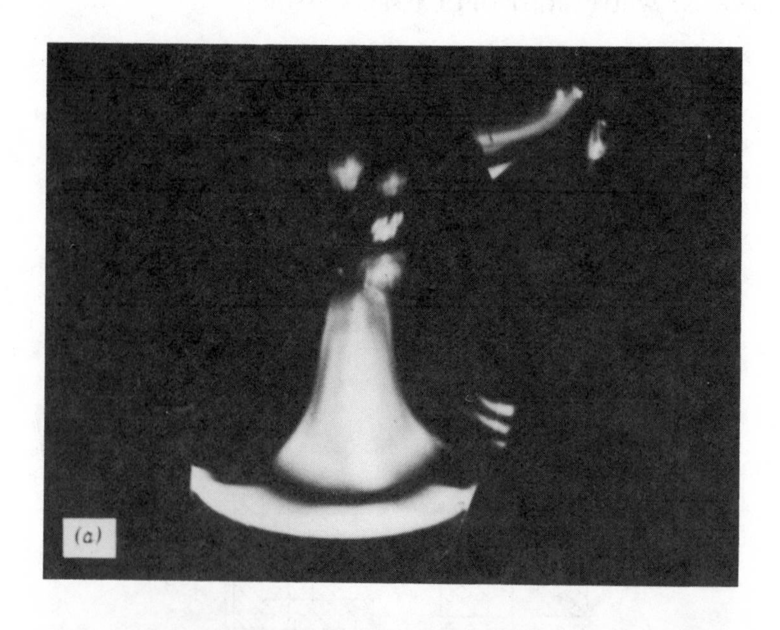

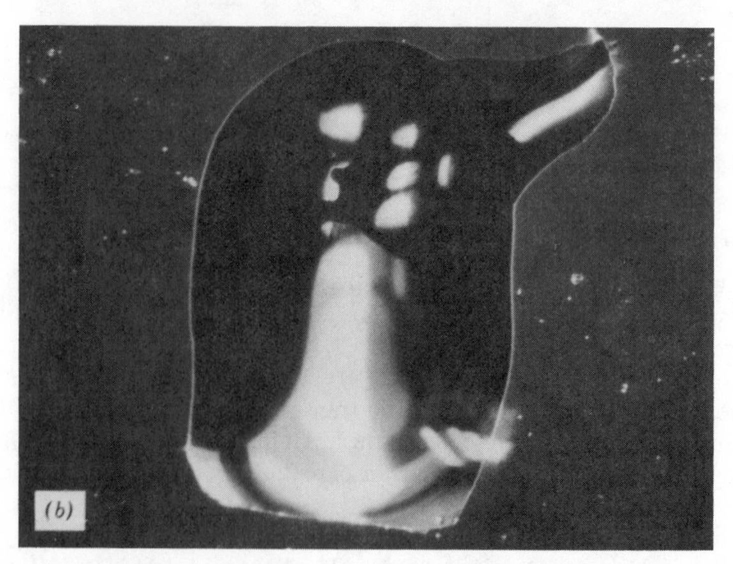

Figure 7.48. (*a*) One-dimensional schlieren optical processing. Source mask had no color. (*b*) Black-and-white picture of a two-dimensional color schlieren optical processing.

Figure 7.49. Black-and-white picture of a color schlieren optical processing showing flow of heated air from the tip of a soldering iron.

Figure 7.49 is a black-and-white picture of a color photo that shows the flow of heated air from a soldering iron placed at the input plane *P*1 of the LCP. The color in the photo changes from blue to green to red, depending on the direction of the heated air flow. There are subtle color changes as the flow changes direction. Although no quantitative data are presented, the value of the visualization of such a flow of gas is obvious, and the applications are numerous. The ability to use the LCP in such a situation makes it a valuable low-cost analytical tool.

7.8.4. Optical Processing of Bubble Chamber Event Photographs

Processing the immense amount of bubble chamber photographs gained from a single high energy experiment has emerged as a major, labor-intensive task [7.66]. An optical technique to obtain the necessary data is attractive in terms of labor saved and in additional features that the optical processing offers over standard hand-processing.

The principles of spatial frequency filtering used in the optical processing of these photographs can be traced to experiments done by Abbe in 1873 [7.67] and Porter in 1906 [7.68]. These principles are the basis of much of optical processing, and this technique is a good demonstration of those principles. In order to process bubble chamber event photographs, the LCP is configured as the standard white-light optical processor as shown in Fig. 7.50. The input to the system is shown in Fig. 7.51 and represents a "draftsman's version" of a bubble chamber event. The horizontal lines are called "beam tracks" and are normally of no interest. The angled "event tracks" are the signals of interest, so their enhancement is the objective of the processing.

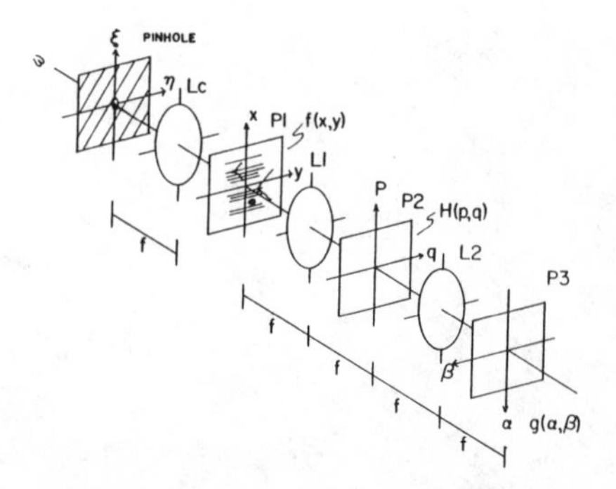

Figure 7.50. Schematic diagram for optical processing of bubble chamber event photographs.

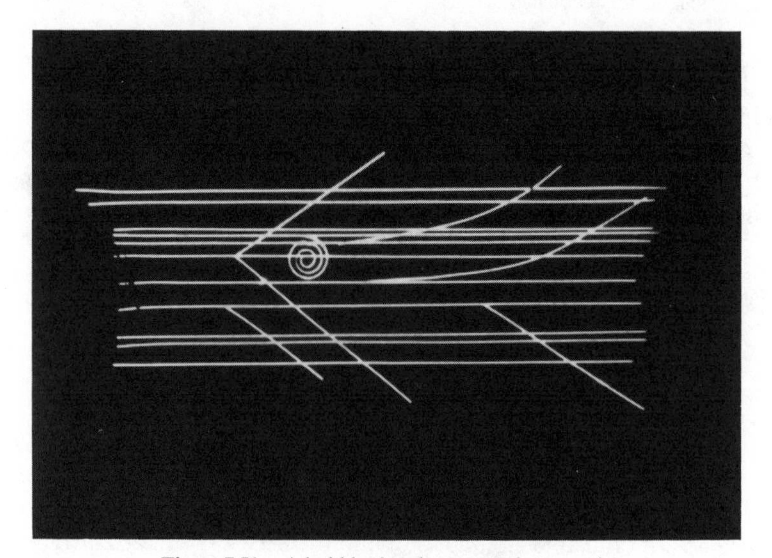

Figure 7.51. A bubble chamber event photograph.

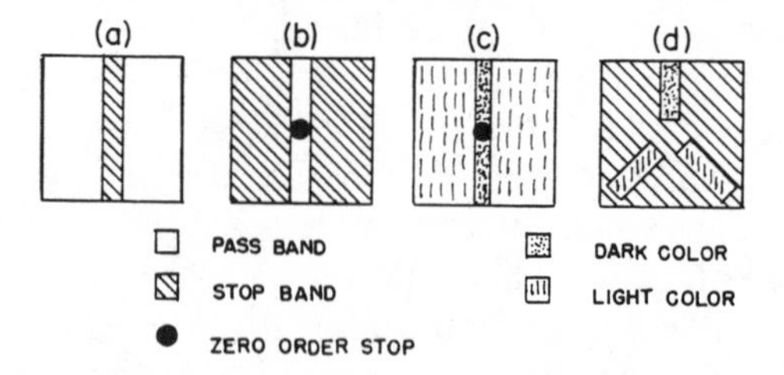

Figure 7.52. Spatial frequency filters used for processing bubble chamber event photographs. (*a*) Stop band filter. (*b*) Directional high pass filter. (*c*) Color filter with zero-order stop. (*d*) Color directional high pass filter.

Four different filtering schemes were used at Fourier plane P_2. The performance of each filter is presented to show the range of processing possible with the LCP.

The first filtering scheme used was a stop band filter, shown in Fig. 7.52*a*. The corresponding output of the LCP is seen in Fig. 7.53*a*. Note that although the beam tracks are not totally suppressed, there is a marked contrast

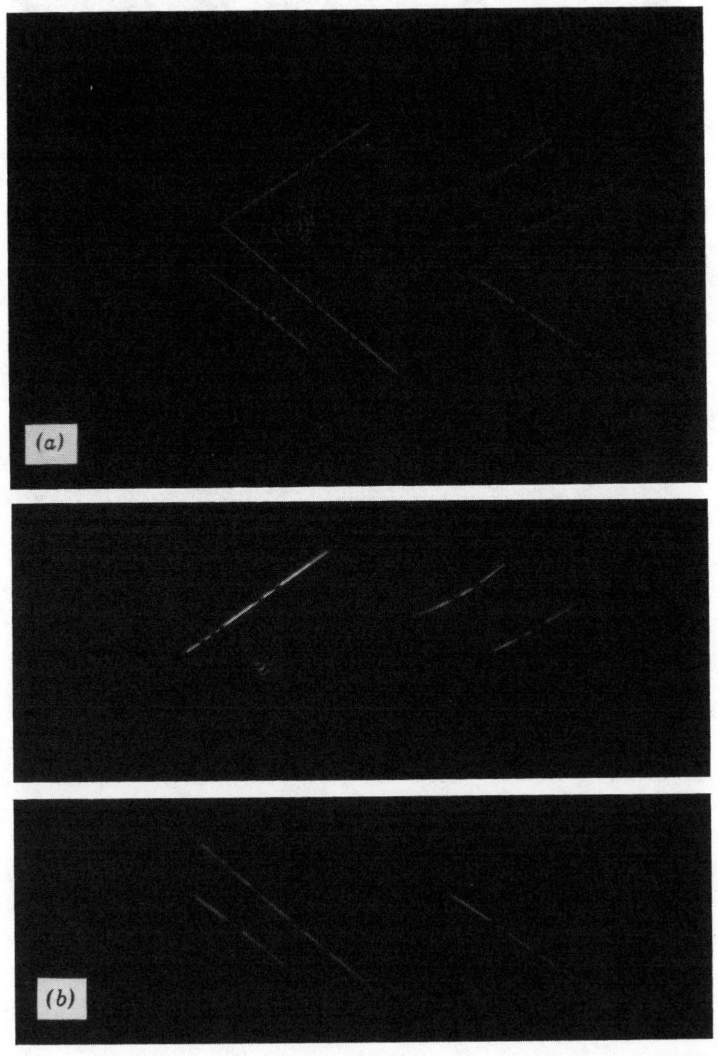

Figure 7.53. (*a*) Output of LCP using stop band filter to suppress beam tracks. (*b*) Output of LCP using directional high pass filter to show only event tracks in a particular direction.

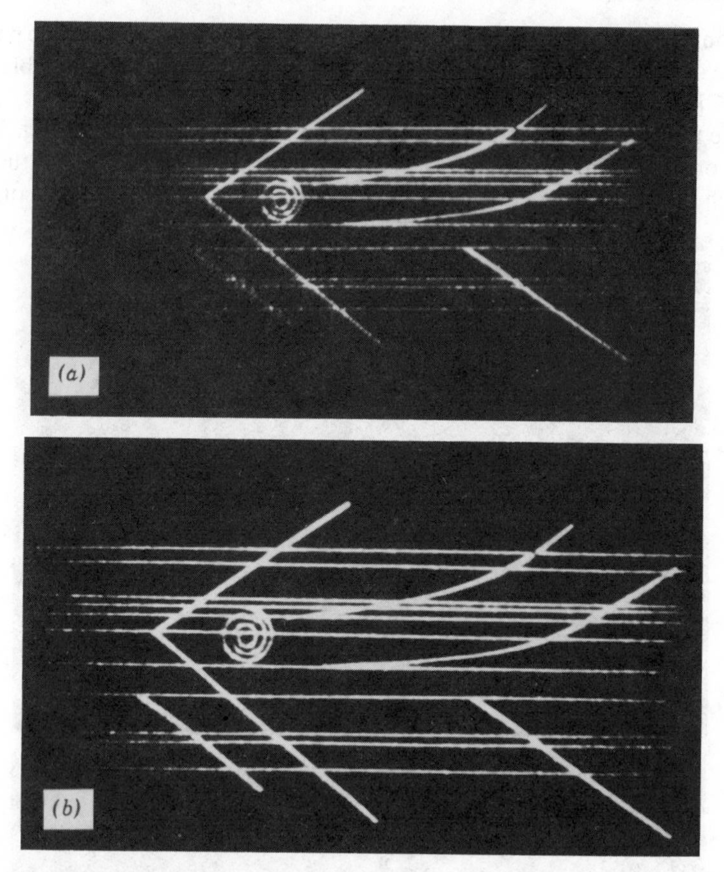

Figure 7.54. (*a*) Black-and-white picture of a color output of LCP using color filter with zero-order stop. Beam tracks are blue and event tracks are red. (*b*) Black-and-white picture of a color output of LCP using color directional high pass filter. Beam tracks are yellow and event tracks are orange.

difference between them and the event tracks. Normally, the enhancement of the event tracks, as shown, is the effect required.

The second filter is a directional high pass filter, as shown in Fig. 7.52*b*, oriented in the direction of the event tracks' spatial frequency diffraction. The corresponding output of the LCP is seen in Fig. 7.53*b*. The beam tracks are almost completely suppressed and only those event tracks at the proper scattering angle are shown. This is most useful in measuring these angles. Note that a two-slit filter could be made that would show both of the event tracks, allowing the angle between them to be found.

The third filter brings in the dimension of color, as shown in Fig. 7.52*c*. Rather than suppress the beam tracks, a color contrast is produced in the tracks, as seen in Fig. 7.54*a*, which is a black-and-white reproduction of a color photo. The beam tracks are blue and the event tracks are red.

The fourth filter is shown in Fig. 7.52*d*. This method is least accurate in portraying the scattering angles, yet provides the clearest color coding and is seen in the black-and-white photo Fig. 7.54*b*. The beam tracks are bright yellow and the event tracks are orange. The addition of color clearly delineates the areas of interest.

As stated before, these filtering methods generalize to any object wherein a regular pattern may obscure a point of interest having a different directional orientation. A scratch on an integrated circuit mask could easily be detected using the LCP in this optical processing mode.

7.8.5. Pseudocolor Encoding

For pseudocolor optical processing through contrast reversal [7.69], the LCP is configured as a standard white-light optical processor. The filter used at the Fourier plane P_2 is seen in Fig. 7.55. The color filters shown in the figure are used during only part of the process.

The object image was contact printed to yield both the positive and negative images of the object. These two images were sequentially recorded on a third sheet of film and encoded by the use of a 40 lines/mm Ronchi grating. Since the sampling theorem states that we must sample at twice the highest frequency that we wish to encode, the highest spatial frequency encoded properly is 20 lines/mm. This is also nearly the limit of accuracy in hand alignment of the positive and negative images during the encoding process. The

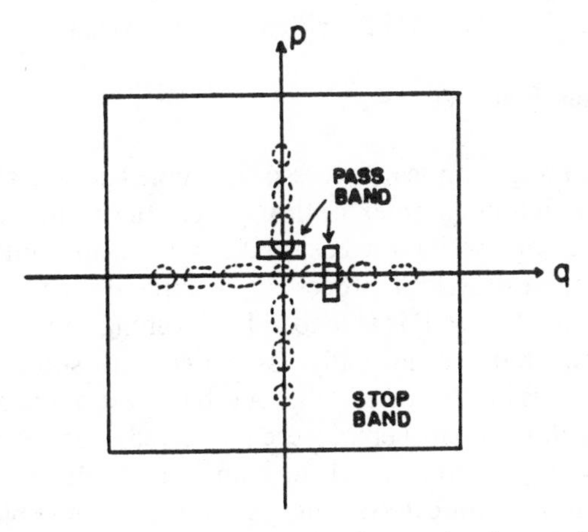

Figure 7.55. Spatial frequency filter for pseudocolor encoding.

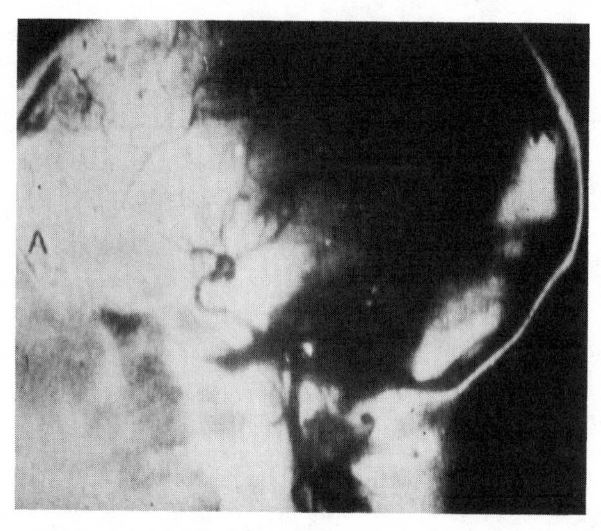

Figure 7.56. Black-and-white picture of a pseudocolor output of LCP. More dense parts of the x-ray picture appear in green, while less dense parts are red.

recorded image was then chemically bleached during the film processing to yield a transparent phase object with a high diffraction efficiency [7.70].

Since, in pseudocolor processing, we are most interested in the quality of the color output of the processor, Fig. 7.56 shows a black-and-white photo of a pseudocolor image of an x-ray. The more dense parts of the head are green, while the less dense parts appear in red. Smaller bones and tissues are more easily seen in color than in a conventional black-and-white x-ray transparency. The colors are very rich and provide very good visual discrimination.

7.8.6. Concluding Remarks

The low-cost white-light optical processor is simple and portable, yet is capable of performing many complex optical processing techniques. The versatility of the processor and its low cost make it especially suitable for educational use. Although only four processing techniques of the LCP are illustrated, it is clear that the LCP is able to aid in teaching complex techniques of optical processing that are normally associated with sophisticated optical components. There is an elegance in the fact that some of the optical processing techniques demonstrated herein were pioneered over a century ago. For the next century they were researched, refined, and applied in increasingly accurate and complex optical systems, yet now are the subject of renewed interest when applied to a system more akin to those early processors.

REFERENCES

7.1. L. J. Cutrona et al., "Optical data processing and filtering systems," *IRE Trans. Inform. Theory* **IT-6**, 386 (1960).

7.2. F. T. S. Yu. *Optical Information Processing*, Wiley-Interscience, New York, 1983, p. 266.

7.3. H. J. Caulfield, *Handbook of Optical Holography*, Academic Press, New York, 1979.

7.4. J. W. Goodman, *Introduction to Fourier Optics*, McGraw-Hill, New York, 1968.

7.5. W. T. Cathey, *Optical Information Processing and Holography*, Wiley-Interscience, New York, 1973.

7.6. A. R. Shulman, *Optical Data Processing*, John Wiley, New York, 1970.

7.7. A. Vander Lugt, "Coherent optical processing," *IEEE Proc.* **62**, 1300 (1974).

7.8. F. T. S. Yu and J. L. Horner, "Optical processing of photographic images," *Opt. Eng.* **20**, 666 (1981).

7.9. M. Born and E. Wolf, *Principles of Optics*, 2nd rev. ed., Pergamon Press, New York, 1964.

7.10. D. Malacara, *Optical Shop Testing*, John Wiley, New York, 1978.

7.11. W. H. Steel, *Interferometry*, Cambridge University Press, London, 1967.

7.12. J. B. DeVelis and G. O. Reynolds, *Theory and Application of Holography*, Addison-Wesley, Reading, Mass., 1967.

7.13. M. Francon, *Progress in Microscopy*, Pergamon Press, New York, 1961.

7.14. L. C. Martin, *Theory of the Microscope*, American Elsevier, New York, 1966.

7.15. A. Bennett, H. Jupink, H. Osterberg, and O. Richards, *Phase Microscopy*, John Wiley, New York, 1951.

7.16. R. A. Sprague and B. J. Thompson, "Quantitative visualization of large variation phase objects," *Appl. Opt.* **11**, 1969, 1972.

7.17. G. Roblin and M. El Sherif, "Restoration of the complex amplitude of a phase object in microscopy by phase modulation interferometry," *Appl. Opt.* **19**, 4247 (1980).

7.18. S. T. Wu and F. T. S. Yu, "Visualization of color coded phase variation with incoherent optical processing technique," *J. Opt.*, in press.

7.19. F. T. S. Yu, S. L. Zhuang, and N. H. Wang, "Integrated circuit (IC) chip inspection with incoherent optical processing," *SPIE* **360**, 310 (1982).

7.20. A. Iwamoto and H. Sekizawa, "Defect-type discriminating optical system," *Appl. Opt.* **20**, 1724 (1981).

7.21. W. Koenig, H. K. Dunn, and L. Y. Lacy, "The sound spectrograph," *J. Acoust. Soc. Am.* **18**, 19 (1946).

7.22. C. E. Thomas, "Optical spectrum analysis of large space-bandwidth signals," *Appl. Opt.* **5**, 1782 (1966).

7.23. F. T. S. Yu, "Synthesis of an optical-sound spectrograph," *J. Acoust. Soc. Am.* **51**, 433 (1972).

7.24. F. T. S. Yu, "Generating speech spectrograms optically," *IEEE Spectrum* **12**, 51 (February 1975).

7.25. F. T. S. Yu, T. W. Lin, and K. B. Xu, "White-light Optical Speech Spectrogram Generation," *Appl. Opt.* **24**, 836 (1985).

7.26. K. R. Hessel, "Some theoretical limitations of the optical power spectrum analyzer," *Appl. Opt.* **13**, 1023 (1974).

7.27. F. T. S. Yu, "Information content of a sound spectrogram," *J. Audio Eng. Soc.* **15**, 407 (1967).

7.28. F. T. S. Yu, *Optics and Information Theory*, Wiley-Interscience, New York, 1976.

7.29. C. M. Vest, *Holographic Interferometry*, Wiley-Interscience, New York, 1979.

7.30. R. K. Erf, *Speckle Metrology*, Academic Press, New York, 1978.

7.31. R. L. Powell and K. A. Stetson, "Interferometric vibration analysis by wavefront reconstruction," *J. Opt. Soc. Am* **55**, 1953 (1965).

7.32. R. E. Brooks, L. O. Heflinger, and R. F. Wueker, "Interferometry with a holographically reconstructed comparison beam," *Appl. Phys. Lett.* **7**, 248 (1965).

7.33. E. Archibald and A. E. Ennos, "Displacement measurement from double-exposure laser photographs," *Opt. Acta* **19**, 253 (1972).

7.34. E. Archibald, J. M. Burch, and A. E. Ennos, "Recording of in-plane surface displacement by double-exposure speckle photography," *Opt. Acta* **17**, 883 (1970).

7.35. D. E. Duffy, "More gauging of in-plane displacement using double aperture imaging," *Appl. Opt.* **11**, 1778 (1972).

7.36. G. Cloud, "Practical speckle interferometry for measuring in-plane deformation," *Appl. Opt.* **14**, 878 (1975).

7.37. F. P. Chiang, K. C. Chin, and W. B. Chang, "Time-average laser specklegram of plate vibration using multiaperture recording," *Appl. Opt.* **20**, 1123 (1981).

7.38. Y. Y. Hung, R. E. Rowlands, and I. M. Daniel, "A new speckle-shearing interferometer; A full-field strain-gage," *Appl. Opt.* **14**, 618 (1975).

7.39. Y. Y. Hung, I. M. Daniel and R. E. Rowlands, "Fill-field optical strain measurement having post-recording sensitivity and directional selectivity," *Exp. Mech.* **18**, 56 (1978).

7.40. G. Gerhart, P. H. Ruterbusch, and F. T. S. Yu, "Color encoding of holographic interferometric fringe patterns with white-light processing," *Appl. Opt.* **20**, 3085 (1981).

7.41. F. D. Addams and G. E. Maddux, "Synthesis of holography and speckle photography to measure 3-D displacement," *Appl. Opt.* **13**, 219 (1974).

7.42. P. H. Ruterbusch, J. A. Tome, and F. T. S. Yu, "Multiplexed speckle and holographic interferometry with color encoding by white-light processing," *Opt. Eng.* **22**, 501 (1983).

7.43. W. E. Ross. D. Psaltis, and R. H. Anderson, "Two-dimensional magneto-optic spatial light modulator for signal processing," *SPIE* **341**, 191 (1982).

7.44. W. E. Ross, D. Psaltis, and R. H. Anderson, "Two-dimensional magneto-optic spatial light modulator for signal processing," *Opt. Eng.* **22**, 485 (1983).

7.45. W. P. Bleha et al., "Application of the liquid crystal light valve to real-time optical data processing," *Opt. Eng.* **17**, 371 (1978).

7.46. C. Warde et al., "Microchannel spatial light modulator," *Opt. Lett.* **3**, 196 (1978).

7.47. F. T. S. Yu, X. J. Lu, and M. F. Cao, "Application of magneto-optic spatial light modulator to white-light optical processing," *Appl. Opt.*, **23**, 4100 (1984).

7.48. F. T. S. Yu, A. Tai, and H. Chen, "Spatial filtered pseudocolor holographic imaging," *J. Opt.* **9**, 269 (1978).

7.49. X. J. Lu, "Pseudocolor encoding with a white-light processing system," *Opt. Comm*, **48**, 13 (1983).

7.50. J. D. Armitage and A. Lohmann, "Theta modulation in optics," *Appl. Opt.* **4**, 399 (1965).

7.51. D. Gabor, "A new microscopic principle," *Nature* **161**, 777 (1948).

7.52. E. N. Leith and J. Upatnieks, "Wavefront reconstruction with continuous-tone objects," *J. Opt. Soc. Am.* **53**, 1377 (1963).

7.53. E. N. Leith and J. Upatnieks, "Wavefront reconstruction with diffused illumination and three-dimensional objects," *J. Opt. Soc. Am.* **54**, 1295 (1964).

7.54. F. T. S. Yu and G. Gerhart, "White-light transmission holographic imaging," *SPIE* **391**, 37 (1983).

7.55. F. T. S. Yu and P. H. Rutherbusch, "Color image retrieval from coherent speckles with white-light processing," Appl. Opt. **21**, 2300 (1982).

7.56. G. Gerhart and P. H. Rutherbusch, "Multiple-aperture three-dimensional image construction utilizing fringe-modulated speckle patterns," *Opt. Lett.* **7**, 599 (1982).

7.57. P. H. Ruterbusch, G. Gerhart, and F. T. S. Yu, "White-light holography with a slide projector," *Opt. Eng.* **22**, 172 (1983).

7.58. F. T. S. Yu and S. T. Wu, "Color image subtraction with encoding extended incoherent source," *J. Opt.* **13**, 183 (1982).

7.59. F. T. S. Yu and X. X. Chen, "Solar optical processing," *Opt. Commun.* **51**, 377 (1984).

7.60. F. T. S. Yu and H. M. Mueller, "A low-cost white-light optical processor for undergraduate optics laboratory," *IEEE Trans. Ed.*, to be published.

7.61. R. E. Haskell, "Fourier analysis using coherent light," *IEEE Trans. Ed.*, **E-14**, 880 (1971).

7.62. W. Merzkirch, *Flow Visualization*, Academic Press, New York, 1974.

7.63. D. W. Holder and R. J. North, *Schlieren Methods*, Her Majesty's Stationary office, London, 1963.

7.64. G. S. Settles, "Color schlieren optics—A review of techniques and applications," in *Flow Visualization II*, W. Merzkirch, Ed., Hemisphere Publications, New York, 1982.

7.65. G. S. Settles, "A direction-indicating color schlieren system," *AAIA J.* **8**, 282 (1970).

7.66. D. G. Falconer, "Optical processing of bubble chamber photographs," *Appl. Opt.* **5**, 9 (1966).

7.67. E. Abbe, *Archiv. Mikrosk. Anat.* **9**, 413 (1873).

7.68. A. B. Porter, "On the diffraction theory of microscope vision," *Phil. Mag.* **11** (6), 154 (1906).

7.69. T. H. Chao, S. L. Zhuang, and F. T. S. Yu, "White-light pseudocolor density encoding through contrast reversal," *Opt. Lett.* **5**, 230 (1980).

7.70. B. J. Chang and K. Winick, "Silver-halide gelatin holograms," *SPIE Proc.* **215**, 172 (1980).

APPENDIX A

In order to show that the following equality holds:

$$\int_{-\infty}^{\infty} \frac{1}{\operatorname{sinc}(W\alpha\pi/\lambda_n f)} \exp\left(-c\frac{2\pi}{f\lambda} x'\alpha\right) d\alpha$$
$$= -\left[\frac{4f\lambda_n}{W} \sum_{n=1}^{\infty} (-1)^n n \sin\left(\frac{2\pi n\lambda_n}{\lambda W} x'\right)\right] \operatorname{sgn}(x'), \tag{A.1}$$

we let $\alpha/\lambda f = v_x$ and $W\lambda/\lambda_n = 1$. The problem is then reduced to evaluating the integration

$$\lambda f l \int_{-\infty}^{\infty} \frac{\pi v_x}{\sin(\pi l v_x)} \exp(i2\pi v_x x') dv_x. \tag{A.2}$$

Since there are an infinite number of poles on the real axis, that is,

$$vx = \frac{n}{l}, \qquad n = 1, 2, \ldots, \tag{A.3}$$

we evaluate (A.1) for two different cases: $x' > 0$ and $x' < 0$. Take the contour integration over the upper half of the complex plane as shown in Fig. A.1; it is given by

$$\left(\int_{-\infty}^{\infty} - \sum_{n=-\infty}^{\infty} \int_{n/l-\varepsilon}^{n/l+\varepsilon} + \int_R\right) \frac{\pi v_x}{\sin(\pi l v_x)} \exp(i2\pi v_x x') dv_x = 0, \tag{A.4}$$

where ε is the radius of the small half circles around the poles, and R is the radius of the larger contour half circle in the upper half plane. We denote that

$$z = R\exp(i\theta), \qquad dz = iR\exp(i\theta)d\theta. \tag{A.5}$$

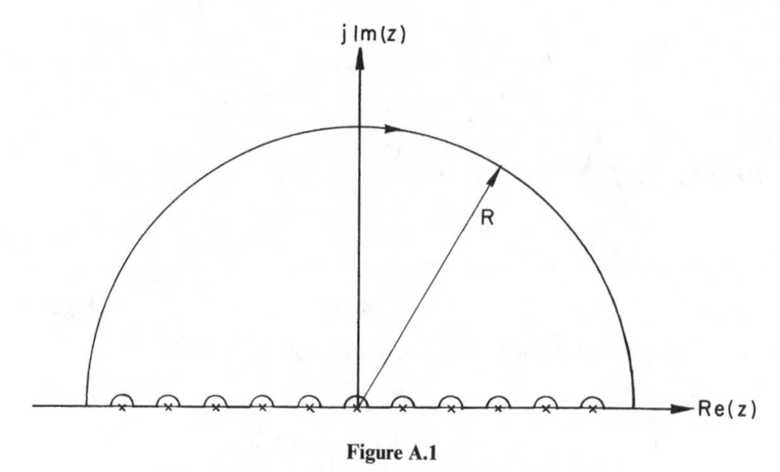

Figure A.1

Then the last term of Eq. (A.4) can be written

$$\int_R \frac{\pi v_x}{\sin(\pi l v_x)} \exp(i2\pi v_x x') dv_x = \int_R \frac{\pi z}{\sin(\pi l z)} \exp(i2\pi z x') dz$$

$$= \int_R \frac{-2\pi R^2 \exp(i2\pi R x' \cos\theta) \exp[-2\pi R \sin(\theta x') \exp(i\theta)]}{\exp[i\pi R l(\cos\theta + i\sin\theta)] - \exp[-i\pi R l(\cos\theta + i\sin\theta)]} d\theta. \tag{A.6}$$

Although

$$\lim_{R\to\infty} \int_R \frac{\pi v_x}{\sin(\pi l v_x)} \exp(i2\pi v_x x') dv_x = 0,$$

the residue's theorem, Eq. (A.4), can be shown as

$$\int_{-\infty}^{\infty} \frac{\pi v_x}{\sin(\pi l v_x)} \exp(i2\pi v_x x') = -2\pi i \sum_{n=-\infty}^{\infty} R_n, \tag{A.7}$$

where R_n is the residue of nth pole, that is,

$$R_n = \lim_{z\to n/l} \left(z - \frac{n}{l}\right) \frac{\pi z \exp(i2\pi z x')}{\sin(\pi l z)} = \frac{(-1)^n n \exp(i2\pi x'/l)}{\pi l^2}. \tag{A.8}$$

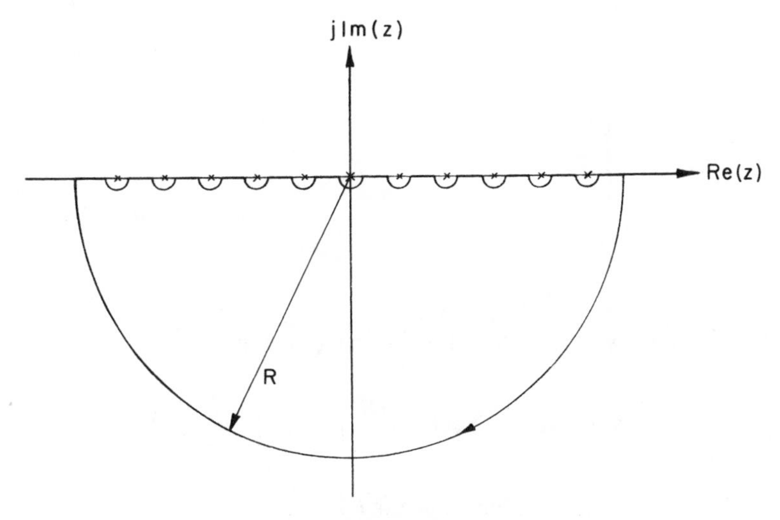

Figure A.2

Therefore

$$\int_{-\infty}^{\infty} \frac{\pi v_x}{\sin(\pi l v_x)} \exp(i2\pi v_x x')dv_x = \frac{4}{l^2} \sum_{n=1}^{\infty} (-1)^n n \sin\left(\frac{2\pi n x'}{l}\right), \qquad \text{for } x' > 0. \tag{A.9}$$

On the other hand, if we take the contour integral over the lower half of the plane, as shown in Fig. A.2, similarly we have

$$\int_{-\infty}^{\infty} \frac{\pi v_x}{\sin(\pi l s_x)} \exp(i2\pi v_x x')dv_x = -\frac{4}{l^2} \sum_{n=1}^{\infty} (-1)^n n \sin\left(\frac{2\pi n x'}{l}\right), \qquad \text{for } x' < 0. \tag{A.10}$$

With reference to these results, we prove that the equality holds for Eq. (A.1).

APPENDIX B

In this appendix we prove the following relationship:

$$f(x) = \text{rect}\left(\frac{x}{a}\right)\text{rect}\left(\frac{x+b}{a}\right) = \text{rect}\left(\frac{x+b/2}{a-|b|}\right). \tag{B.1}$$

1. *For* $b = 0$. It is evident that Eq. (B.1) is valid for $b = 0$.

2. *For* $b > 0$. The left-hand side of Eq. (B.1) can be equivalently expressed within the following set of regions:

$$-\frac{a}{2} < x < \frac{a}{2}, \tag{B.2}$$

$$-\frac{a}{2} - b < x < \frac{a}{2} - b. \tag{B.3}$$

With reference to Fig. B.1, Eq. (B.2) can be reduced to

$$-\frac{a}{2} < x < \frac{a}{2} - b. \tag{B.4}$$

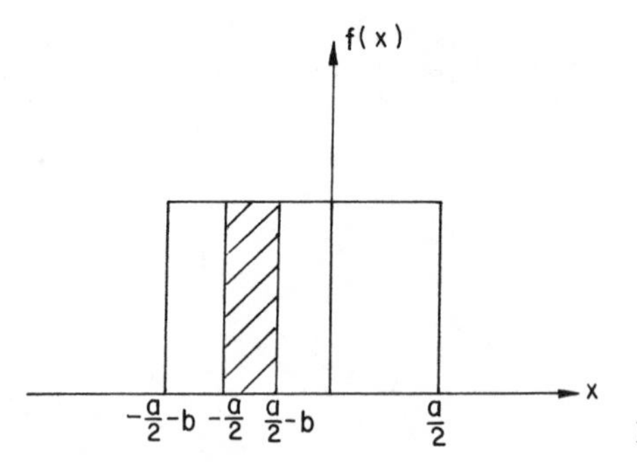

Figure B.1. $f(x)$ for $b > 0$.

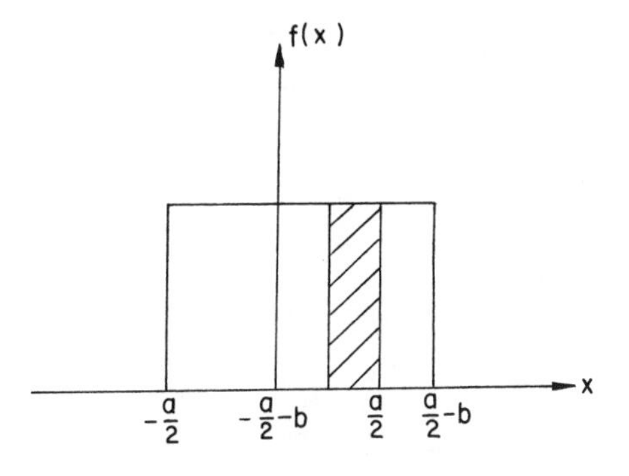

Figure B.2. $f(x)$ for $b < 0$.

Similarly, the right-hand side of Eq. (B.1) is defined within the domain

$$-\frac{1}{2}(a-b) < x + \frac{b}{2} < \frac{1}{2}(a-b). \tag{B.5}$$

Since the inequality of Eq. (B.4) is exactly the same as Eq. (B.3), Eq. (B.1) is true for $b > 0$.

3. *For* $b < 0$. The left-hand side of Eq. (B.1) is plotted in Fig. B.2, which can be written as

$$-\frac{a}{2} - b < x < \frac{a}{2}. \tag{B.6}$$

Since the right-hand side of Eq. (B.1) is equivalent to

$$-\frac{1}{2}(a+b) < x + \frac{b}{2} < \frac{1}{2}(a+b), \tag{B.7}$$

or

$$-\frac{1}{2}a - b < x < \frac{a}{2}. \tag{B.8}$$

Eq. (B.1) is also valid for $b < 0$. Thus we have proved that Eq. (B.1) holds for all b.

Author Index

Subject Index